D0023012

ANALYSIS
AND DESIGN
OF MECHANISMS

PRENTICE-HALL INTERNATIONAL, INC., *London*
PRENTICE-HALL OF AUSTRALIA, PTY. LTD., *Sydney*
PRENTICE-HALL OF CANADA, LTD., *Toronto*
PRENTICE-HALL OF INDIA PRIVATE LTD., *New Delhi*
PRENTICE-HALL OF JAPAN, INC., *Tokyo*

ANALYSIS
AND DESIGN
OF MECHANISMS

SECOND EDITION

DEANE LENT

Professor of Mechanical Engineering
Massachusetts Institute of Technology

PRENTICE-HALL, Inc., Englewood Cliffs, N.J.

© 1970, 1961 by Prentice-Hall, Inc.
Englewood Cliffs, N.J.

All rights reserved. No part of this
book may be reproduced in any form
or by any means without permission
in writing from the publisher.

Current printing (last digit):

10 9

13-032797-2

Library of Congress Catalog Card Number 71-76875

Printed in the United States of America

To
Michael Christopher van Lent

Preface

This book is a teaching text and it is written for students. The presentation is elementary enough to be offered in the first year of any technical school curriculum. It does not presume a knowledge of calculus or physics but exploits graphical techniques to expedite the study of this geometry-based subject. The clarity and simplicity of graphical study extends the student's capacity beyond the usual limits of introductory courses.

The author's objective is to provide a sound basis for the artful application of kinematic principles in analysis and design of mechanisms. This is not a theoretical treatise, exploring the latest research for the accumulation of knowledge, but is intended to provide immediately useful skills to deal with real problems. It is believed that the stimulus of design problems, with their attendant demand for analysis, provide the most effective motivation for engineering study. However sophisticated the techniques that may eventually be employed, the initial graphical approach enables the student to become involved in design at an early stage.

Since so much has already been written on this subject, it would be difficult and doubtless unwise to attempt a completely new treatment. Although considerable material is included which the author believes to be original, many classic techniques are also used with the intention of improving their presentation or broadening their application. It is deemed wiser to endorse and promote successful methods than to seek notoriety through invention. This book is written to instruct the student—not to impress the faculty.

In a number of instances alternate methods of analysis are offered in this edition, allowing the instructor to select that approach which is best suited to his class. Almost none of the first edition has been deleted, so that present users need not necessarily change their instruction or problem assignments. The final chapter on mechanisms is new, offering a collection of useful mechanisms to better familiarize the student with existing designs and stimulate his ingenuity in creative work.

In every writing of this sort there stands behind the author a teammate who types, revises, organizes, and sometimes edits the manuscript. In the preparation of this text, the skill, devotion, and patience of my secretary, Diane R. Mountain, constitute a major contribution which is most gratefully acknowledged.

DEANE LENT

Contents

4   ACCELERATION, 94

5   GRAPHICAL ANALYSIS OF MOTION, 150

6 GEARING, 173

7 GEAR TRAINS, 243

8 LINKAGES, 305

9   CAMS, 346

10   MECHANISMS, 377

APPENDIX, 411

INDEX, 419

ANALYSIS
AND DESIGN
OF MECHANISMS

Introduction

A mechanism is the heart of a machine. It consists of a series of connected moving parts which provide the specific motions and forces to do the work for which the machine is designed. A machine is usually driven by a motor which supplies constant speed and power. It is the *mechanism* which transforms this applied motion into the form demanded to perform the required task.

 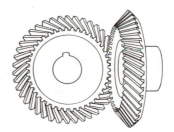

Figure 1.1

The first member of a mechanism, directly attached to the motor, is called the *driver*, and the last member in the unit, which supplies the useful motion or energy, is called the *follower*. Some mechanisms are composed of but two parts, as those in Figure 1.1, while others have many members, as

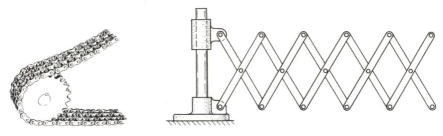

Figure 1.2

shown in Figure 1.2. A complicated machine may employ several mechanisms to perform various functions. The entire machine is designed around the mechanisms which do the work. This active and important role of mechanisms in machine design make this subject interesting and vital.

1.1 Design of Mechanisms

A machine designer's first concern is that of motion. He must select or devise mechanisms which will produce the required displacements and velocities and determine the resulting accelerations. The study of motion of machine members without consideration of the forces and stresses produced is called *kinematics*. It is the primary concern of this text. The existence of forces cannot be ignored. Many mechanisms, such as hoists and presses, are designed for force-producing qualities rather than motion characteristics. All mechanism components must be designed to resist the stresses induced by loads and accelerations.

The study of forces on bodies in motion is called *dynamics*. This is much more involved than kinematics and is the second phase of design. Since many of the dynamic problems are defined by the motion characteristics, it is logical as well as convenient to study kinematics first. As this is intended as a first book in machine design, the treatment of dynamics will be limited.

To design machines one must first have a working "vocabulary" of well-known mechanisms. One must be able to modify and adapt these to the specific requirements at hand. One must then analyze these mechanisms for displacements, velocities, and accelerations in order to predict their behavior and prepare for the dynamic analysis to follow. These are our aims in the study of mechanisms.

1.2 Techniques of Mechanism Analysis

Since much of the work is geometric in nature, drawings are freely used in the study of mechanisms. This graphical technique simplifies the work; first, since problems are more clearly visualized and solutions more easily understood, and second, since difficult or laborious calculations are avoided. The pictorial aspect of the graphical solution explains the process and offers the chance to compare constantly and check results. Vectors and plotted curves afford visual comprehension, which is frequently obscured when numerical methods are used. The substitution of graphical methods eliminates lengthy trigonometric solutions and greatly simplifies the calculus, thereby bringing more problems within the reach of those whose mathematical education is limited.

Calculations should be used wherever they are simple and in all cases where graphical solutions offer no advantage. Thus, all problems will involve some analytical work. One should not seek to avoid calculations, but rather to combine the two methods skillfully in the interests of efficiency and accuracy. The shortest, simplest method is usually the most accurate.

1.3 Drafting Precision Is Demanded

Graphical work cannot be justified unless it is of such quality as to yield accuracy consistent with analytical methods. Results obtained by drawings are unjustly scorned by some as being only approximations. This mistaken attitude is the result of ignorance or abuse of precision drawing methods and clearly indicates the responsibility imposed upon those who employ graphical techniques. It is true that calculations may often be carried to more decimal places than is possible with graphical solutions, but graphical work, executed with skill and judgment, can usually provide the degree of accuracy which the problem requires. Errors of process are more serious than errors inherent in the choice of method, and calculations are quite as prone to this fault as graphical work. The advantages of the faster, simpler graphical technique may only be enjoyed by those who master drawing skills and apply them with intelligence and respect.

A drawing textbook should be consulted for details, but some of the highlights of precision techniques are outlined here. These are indispensable to graphical accuracy.

1.4 Precision Linework

Sharp, fine hairlines are demanded in this work. A line, by definition, has no width or thickness. We cannot achieve this ideal with a visible pencil line, but we can approach it very closely. A long, sharp, tapering point on the pencil lead, as shown in Figure 1.3, is required. It is vitally important that this point be kept sharp as the drawing progresses by frequent honing on a piece of rough paper. Circular arcs must be just as fine as other lines, so the compass lead requires the same constant attention.

Since these drawings are made in order to solve problems and not for

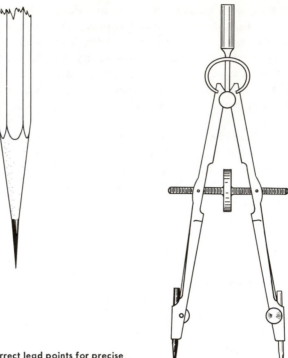

Figure 1.3 Correct lead points for precise drafting

printing purposes, the lines need not be dark in shade. This enables one to use a hard grade of lead which is much easier to keep sharp during the layout. Samples of precision linework are shown in Figure 1.4.

Figure 1.4 Precision linework

1.5 Precision Measurements

Accurate measurement is as important as line quality to the use of graphics in mechanism study. Linear measurements must be accurate within $\frac{1}{100}$ part of an inch. Such precision is possible only through the use of correct methods and instruments.

A scale with inches divided into 50 parts is preferable to one divided into hundredths (see Figure 1.5). It is easier to identify the desired mark, and the fiftieths can be very accurately divided into two equal parts with the naked eye if a needle point is used. A scale with hundredths is difficult to read without a magnifying glass.

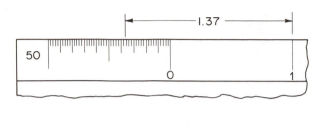

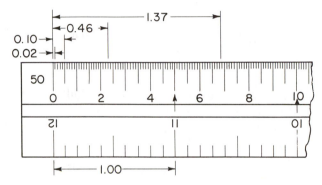

Figure 1.5 Readings on two types of engineers' scales

A needle point is a simple instrument but an absolute necessity for recording measurements precisely (Figure 1.6). The relatively blunt pencil point should never be used for this purpose. With the needle point one can prick a tiny hole in the paper for a permanent and accurate record of the desired dis-

Figure 1.6 Needlepoint

tance. The slender, sharp point helps one to read the scale, especially if he sights down a division line with one eye closed, as shown in Figure 1.7.

The first step in laying out a measurement is to draw a light, fine layout line. This line must be longer than the distance to be measured. The scale is next laid against this line and the ends of the desired distance indicated by pricking small holes exactly on the line with the needle point. Measurements should always be made at that point on the drawing where they are to be

Figure 1.7 "Sighting" along scale to needlepoint

used. Transferring distances from the scale to the layout line with the dividers invites error and wastes effort. Measuring in space without a layout line, or marking points with large holes or pencil marks are very bad practices. Since the tiny holes are not easy to see, they may easily be located if small free-hand circles are drawn around them at the time of layout. Enlarging holes with the pencil sacrifices all accuracy (see Figure 1.8).

Figure 1.8 Locating measurements on layout line

Unless a professional-quality drafting machine with an accurate protractor is being used, angular measurements on precision drawings should be made by the use of a table of chords rather than with the usual small protractors. Arcs of a given radius subtend chordal lengths, which are unique for each subtended acute angle. A table of chord lengths for arcs of 1-in. radius is given in the Appendix for angles up to 45°. In the interest of accuracy, arcs of not less than 5-in. radius should be used, with the tabulated chord lengths multiplied by 5 to correspond with this enlarged scale. Inexpensive protractors are too crude for this quality of work and should not be used. The chord method allows angular measurement to the nearest 10 min and permits the use of the largest practical radius.

Precision measurements can ordinarily be made with the naked eye without strain. However, a small magnifying glass, attached to the needle point, as shown in Figure 1.9, increases accuracy and contributes to comfort during extended periods of this type of drafting.

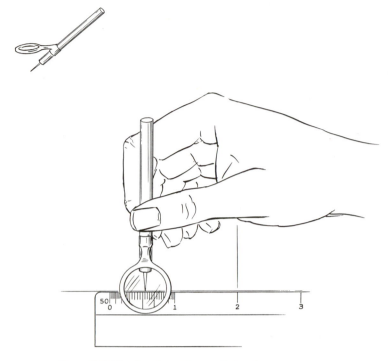

Figure 1.9 Needlepoint with magnifying glass

1.6 Large Scale Drawings Increase Accuracy

A wise choice of scale is a very important factor in the validity of a graphical solution. Within limits, the larger the layout, the more accurate the measured results. A line 10 in. long can be measured to within $\frac{1}{100}$ in. just as easily as a line 1 in. long. The permissible error in the 10-in. line is 1 part in 1000, whereas the error in the 1-in. line is 1 part in 100 (10 times as large!). The slope of a line is much more accurately determined by 2 points 10 in. apart than by points only 2 in. apart. This is another argument for the larger layout.

If the drawing becomes too large, however, special instruments are required and much more time is consumed in making the layout. It is difficult to draw a very long line as straight as a short one or to measure precisely distances which exceed the length of the usual 12-in. scale. Accuracy may actually be in jeopardy if the drawing becomes too large. The quality of the given data, the complexity of the analysis, and the accuracy demanded of the solution, all influence the selection of a proper scale. Care and judgment are demanded in making the choice.

1.7 Divider Technique

There are many instances in kinematic layouts where curved or straight lines must be divided into a given number of equal parts. To do this precisely, the dividers should be used. This instrument also has a valuable function in transferring measurements from one place to another. Two kinds of dividers commonly used are shown in Figure 1.10. The large pair has a friction joint and a thumbscrew for fine adjustment only. The small pair is entirely screw-adjusted.

Figure 1.10 Dividers

The use of this instrument involves trial and error, but this is a sound method because each trial indicates the size of the error and shows how much correction is needed and in which direction the correction should be made. There is no blind guesswork involved.

For example, suppose it is required to divide a line *AB* into 3 equal parts (see Figure 1.11). First set the dividers to what apperars to be about one third of the length *AB*. With this trial setting, start exactly at point *A* and "walk"

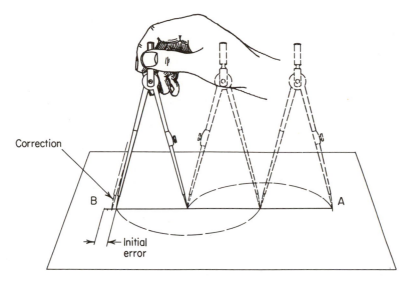

Figure 1.11 Dividing a line into three equal parts

the dividers along the line, swinging the disengaged leg first to one side and then the other (as one would walk a tightrope), until 3 steps have been taken. Two things are important: *Light pressure* must be used to avoid driving the point deeply into the paper, and care must be exercised that the divider points be placed *exactly on the line* at each step (the draftsman fails as miserably as the tightrope walker if he neglects this precaution). The total error is the distance from the third position of the divider point to point *B*. Since there are 3 divisions, the error in the original trial setting equals one third of the total error. This correction is made, again by eye, moving the divider point one third of the distance from the last position toward point *B*. Since this distance is small, it can be judged quite accurately.

Now return to point *A* and step off 3 divisions with the new divider setting. If the third step does not fall exactly on *B*, correct as before (by one third of the total error) and step off from *A* a third time. Usually 3 or 4 trials will yield an exact solution. The entire procedure takes less time to execute than to describe here.

Curved lines can be divided into equal parts by exactly the same process used for straight lines. The length of a step is a straight-line measurement from point to point, so it will be a chordal measurement, not an arc length, since it bridges across the curve. If the curve is circular, as in Figure 1.12, this substitution may be ignored, since equal chords subtend equal arcs.

With irregular curves, the method is still valid but requires adjusting the

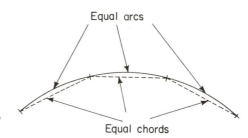

Figure 1.12 Dividing a circular arc into equal parts

length of steps used to the curvature. Suppose one must divide the irregular curve shown in Figure 1.13 into 4 equal parts. If this is done directly, using 4 steps of the dividers, the result would be very inaccurate. The rapid curvature on the left would result in an arc length much greater than the relatively flat right-hand portion if the same chordal setting were used on the dividers. This

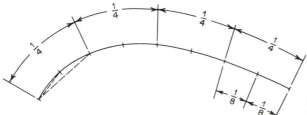

Figure 1.13 Dividing an irregular curve into equal parts

problem can be minimized by selecting more and shorter divider steps in which the lengths of chord and arc are more nearly equal. If a multiple of the required number of parts (such as 8) is used for the division and each consecutive pair of these is designated as one fourth of the curve, the accuracy will be greatly improved. The illustration shows the curve first divided into 8 parts, then each pair marked as one fourth of the length.

The practice of using dividers to transfer measurements is always a time-saver and is sometimes invaluable. Figure 1.14 shows an irregular curved line *CD* whose length is to be obtained. After considering the maximum curvature of the line, select a divider setting (chordal length) such that the difference between the arc and the chord is negligible. Starting at *C*, step off these divisions along the curve and count steps. Assume that there are 12 steps to point *E*, leaving an additional shorter step *ED*. Lay off 12 of these same steps (by no means should the divider setting be altered) on a straight line from point *C'* to *E'*. Now transfer the short segment *DE* on the curve to *D'E'* on the straight line and measure *C'D'* with the scale. This process may be reversed in order to lay out a given measurement on a curve.

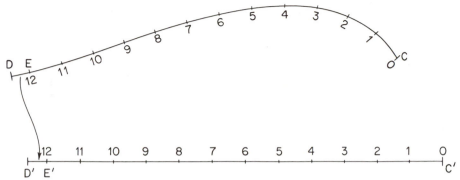

Figure 1.14 Measuring an irregular curve

1.8 Technique of Drawing Irregular Curves

The graphical representation of variable quantities takes the form of an irregular curve defined by a series of plotted points through which a continuous smooth curve must be drawn. In kinematic studies this plot or graph is frequently used to present data or to define solutions. The method of drawing irregular curves differs from other techniques in that it is a trial-and-error process rather than a direct, "one-shot" operation. The curve is built up, piece by piece, with the original trial curve refined and improved until a satisfactory result is obtained.

As with all precision work, an instrument is used to guide the pencil. The most common tool is the "French" curve shown in Figure 1.15. These are

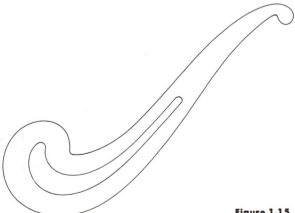

Figure 1.15 An irregular or French curve

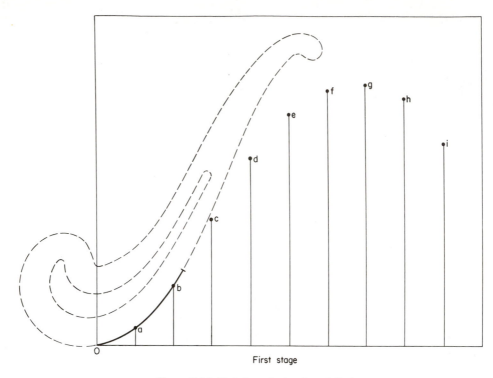

First stage

Figure 1.16 First stage in drawing plotted curve

available in a variety of shapes and sizes, but a draftsman generally uses only one or two, since rarely does any stock instrument really "fit" any considerable portion of the curve to be drawn. Assume that the points *a*, *b*, *c*, *d*, *e*, *f*, *g*, *h*, and *i* have been carefully plotted (needle-pointed) as shown in Figure 1.16 and a smooth curve is to be drawn through them by trial, and select a region of the French curve which will pass exactly through several consecutive points, such as *o*, *a*, and *b* in Figure 1.16. Before drawing a line, note how closely the curve is aimed towards the next point, *c*. If the edge passes close to *c*, one may draw more of a line than if it veers away from *c* markedly. In this case the line can be drawn only a little beyond *b*, since the curvature must be modified beyond this point so as to pass through *c*. Now draw a trial line exactly through *o*, *a*, and *b*. This must be a layout line so that it will be easy to erase if it must be modified. (The draftsman should move about into a convenient position to draw against the curved edge—this cannot be done sitting on the stool).

To continue drawing the curve, the French curve must be moved to a new position so that its edge passes exactly through several more points and also overlaps and matches the end of the curve already drawn. Figure 1.17 shows the instrument in such a position that it matches the first curve from point *b*

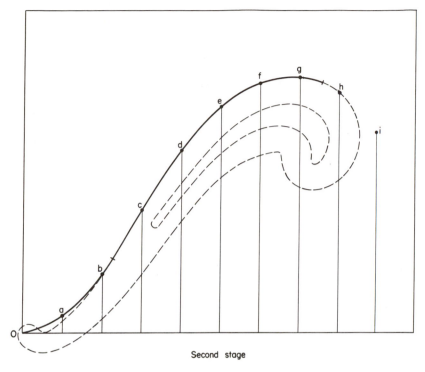

Second stage

Figure 1.17 Second stage in drawing curve

and also passes through *c, d, e, f, g,* and *h*. (It is quite unusual to be able to match the French curve to so many points, but this will serve to shorten the description of the process and avoid needless repetition.) Even though the curve passes through point *h*, note that it is very poorly aimed toward the next point *i*. Consequently, draw the line only as far as halfway between *h* and *i*, so as to provide for a change of curvature through *h* in anticipation of reaching *i* on the next match. The curved line is drawn in layout weight as before to facilitate any possible modification.

The third stage in constructing this curve is shown in Figure 1.18. The French curve has again been repositioned so that it matches the last portion of the curve from a point midway between *g* and *f*, and extends through *h* and *i*. Again, the line is extended to bring the curve to its end at *i*.

This first layout line constitutes a trial only and must be checked and often improved before the finish line is applied. The checking is best done by placing the eye near the level of the paper and sighting along the curve. Humps, cusps, flat spots, and breaks in the curve should be erased and corrected by again positioning the French curve so that it matches exactly existing portions on each side of the erased region, thus assuring a continuous curvature.

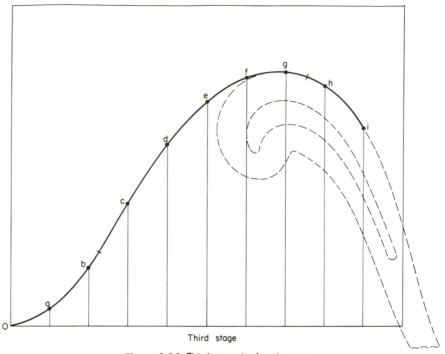

Figure 1.18 Third stage in drawing curve

When the layout curve is satisfactory, a finish line is applied exactly over it. The French curve is repeatedly applied and matched to the layout curve as before, but the positions that were originally used need not be reconstructed in this process. Actually, the end result may be improved if the positions of the instrument are slightly different than the original placements.

1.9 Combine Techniques Intelligently

In mechanism study it is essential to use that method which is best adapted to each part of the problem at hand. One should definitely avoid becoming addicted exclusively to either analytical or graphical methods but should try to select and combine these techniques so as to achieve a simple, efficient and accurate solution.

1.10 Classes of Motion in a Plane

When the paths of all moving points in a body lie in the same or parallel planes, the body is said to have *plane motion.* The analysis of three dimensional motion usually may be resolved into several plane motion studies, so our abilities are not limited if we confine our attention principally to plane motion in this text.

It is helpful to divide plane motion into the following classes:

1. *Translation:*

A body which moves without turning is said to be in translation. All points on a body in translation have identical motion. They may move along straight or curved paths, but at any instant all points move in the same direction and at the same speed. All positions of a given line on a body remain parallel as it moves. The block sliding down the inclined plane in Figure 1.19 is in *rectilinear translation*, since points *A* and *B* travel in straight line paths. The aerial cable car in Figure 1.20 is in *curvilinear translation*, since *A* and *B* travel in curved paths. In each case the line *AB* is parallel to *A'B'* in all positions.

Figure 1.19 Rectilinear translation

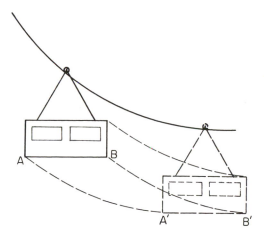

Figure 1.20 Curvilinear translation

2. *Rotation:*

If one point on a body remains stationary while the body turns about that point, the body has a motion of rotation. The paths of all points on the body are circular arcs about the fixed point as a center. An imaginary line through this center, perpendicular to the plane of

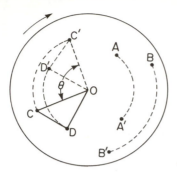

Figure 1.21 Rotation

motion, is called the *axis of rotation*. In this motion, all lines on the body turn at the same speed. The wheel in Figure 1.21 rotates about fixed point O. Points A, B, C, and D describe circular paths about center O. When triangle OCD turns to position $OC'D'$, lines OC, CD, and DO all turn through the same angle θ, since they are all turning at the same speed.

3. Combined Motion (Translation and Rotation): *

When a body moves so that all points change position and all lines turn, that body has combined motion.† In the mechanism shown in Figure 1.22, crank AB rotates about fixed axis A; the block at C translates along the guides; and the connecting rod BC has combined motion. Point C travels to C' while B travels to B' and BC turns

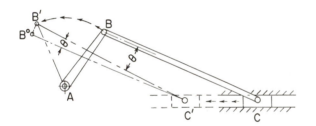

Figure 1.22 Translation, rotation, and combined motion

through angle θ. The rod BC therefore translates and rotates at the same time. It is sometimes convenient to think of combined motion as two separate motions taking place in succession. For example, the rod can first be moved from position BC to a parallel position $B°C'$ by translation; then it can be rotated about C' through angle θ to the final position $B'C'$. The translation and rotation may be considered independently even though they take place simultaneously in reality. Figure 1.22 illustrates all three classes of motion: translation of the block at C, rotation of the crank AB, and combined motion of the connecting rod BC.

*This is also commonly called *plane motion*.

†It is well to note here that the motion of points is limited to translation. Since they have no extent, we cannot observe or measure the rotation of points. Lines can translate, rotate, or do both simultaneously, as in combined motion.

Displacement

Motion is the act of changing position. We must first study the properties of motion so that we may define and measure it before considering the mechanisms by which it is produced. Let us first consider displacement, the result, or end product, of motion. Displacement is a measure of change in position. To describe a displacement completely, we must know the starting position, the direction of motion and the distance, or angle, between the original and final positions.

2.1 Linear Displacement   (Symbol: S)

The distance between two positions of a point as it moves along a line is called *linear displacement*. The path, or locus, of the point may be curved or straight, but the displacement is the straight line distance between two positions, not necessarily the distance actually traveled. Although linear displacement is usually applied to the motion of a point, a body or line in translation may also have linear displacement. Since all points on a body in translation move equal distances in any given interval, the displacement of the entire body is equal to the displacement of any point. Linear displacement is measured in inches or feet.

2.2 Angular Displacement   (Symbol: θ)

The angle between two positions of a rotating line or body is the *angular displacement* for that interval of motion. Since a point has no extent, its angular displacement has no significance. To determine angular displacement of a body we need only measure the angle turned by any line fixed to the body, for when a body rotates, all lines turn through the same angle in a given interval. Triangle OAB (Figure 2.1) rotates about O to position OA_1B_1. In this motion line OA turns through angle AOA_1, equal to θ, and OB turns through angle BOB_1, which can be shown also to equal θ. Angle 1 equals angle 2, since they are the same angle of the triangle in each position. When the triangle is displaced through angle θ, side OA turns through angle 1 plus angle 3. Side OB turns through angle 3 plus angle 2. Since angle 1 plus angle 3 equals angle 3 plus angle 2, lines OA and OB are displaced through the same angle θ.

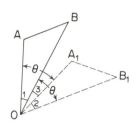

Figure 2.1 Angular displacement

The same triangle is shown in Figure 2.2 and we can prove that sides OA and AB turn through the same angle when the triangle is displaced. Angle 4 equals angle 5, since they are the same exterior angle of the triangle in each position. AB may be displaced to position A_1B_1 in the following three motions:

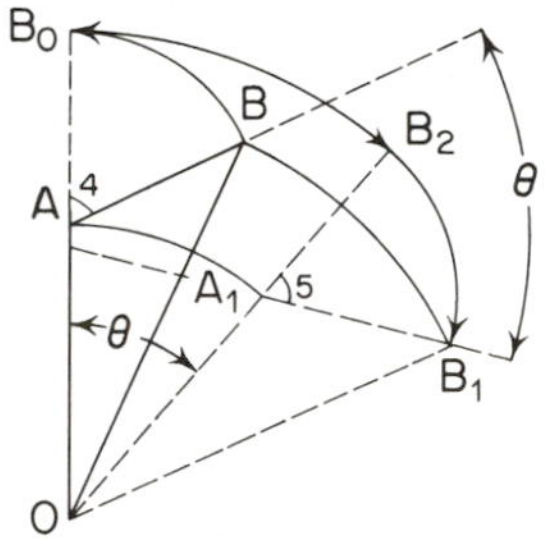

1. Turning counterclockwise about A through angle 4 to AB_0, in line with OA.
2. Turning clockwise about O through angle θ to A_1B_2, in line with OA_1.
3. Turning clockwise about A_1 through angle 5 to A_1B_1.

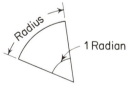

Figure 2.2 All lines on a body turn equal angles

The sum of the counterclockwise motion through angle 4 and the clockwise motion through angle 5 is zero, since angle 4 equals angle 5. This leaves the net displacement of AB from position AB to A_1B_1 equal to angle θ, which is the same displacement as side OA.

The units of angular displacement are degrees or radians. A radian is the angle subtended by an arc of a circle equal in length to the radius of that circle (see Figure 2.3). Since the circumference of a circle equals 2π* times the radius, there will therefore be 2π radians in $360°$. While we usually measure and report angles in degrees, it is often convenient to calculate angles in radians.

Figure 2.3 The radian as an angular unit

*π is taken equal to 3.14 in most calculations.

2.3 Paths of Points on Rotating Bodies

Wheel W in Figure 2.4 turns about a fixed center at O. Any point, such as A, on this wheel remains at a constant distance (r) from O and therefore travels a circular path as the wheel turns. When W makes one complete revolution, the angular displacement of line OA is 360°, or 2π radians. During this motion point A travels a distance equal to the circumference of a circle of radius r, or $2\pi r$ in. When W turns 180°, or π radians, the path of A will be a half circle, or πr in. in length. The path of A will always be directly propor-

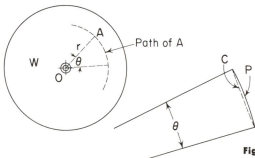

Figure 2.4 Linear and angular displacements

tional to the angular displacement of W. If we express this displacement in radians, we can derive a simple equation for the path of A or any point on the wheel.

In one revolution the path of A equals $2\pi r$ in. in length, and the displacement of W equals 2π radians. Dividing equals by equals:

$$\frac{\text{Path of } A}{\theta_w} = \frac{2\pi r}{2\pi}$$

Canceling 2π:

$$\frac{\text{Path of } A}{\theta_w} = r$$

Multiplying by θ_w:

$$\text{Path of } A = r\theta_w$$

Since A is any point on W, the general equation:

$$P = r\theta$$

gives the distance P traveled by any point at distance r from the axis of rotation of a body turning through an angle of θ radians.

In Figure 2.4 we note that when angle θ is very small, the subtended arc P will be very nearly equal to chord C. Therefore, since arc $P = r\theta$

$$\text{chord } C = r\theta$$

when θ approaches zero and is expressed in radians.

By definition, the displacement of A (S_A) equals the straight-line distance between two positions of A. When A travels along an arc, this displacement would be the chord of that arc. For very small angles (as θ approaches zero), the chord ls approximately equal to the arc, so the displacement becomes:

$$S = r\theta$$

when θ is in radians. This equation gives a relationship between linear and angular displacement on bodies in rotation.

2.4 Displacements and Paths in Combined Motion

A body in combined motion is translating and rotating simultaneously. We are interested in linear displacements of points on the body and the angular displacement of the body itself.

Linear displacements may be obtained graphically, by making an accurate drawing of the body in the initial and final positions and measuring the distance between the two positions of the point in question. Linear displacement may be calculated when both the path of the point and its velocity are known, but this method will be clearer after the study of velocity.

When the path of a point is unknown, it may be determined by drawing the body in a series of positions at short intervals and plotting a curve through successive positions of the point.

In combined motion, rotation of a body is entirely independent of translation. Line AB is fixed on body M shown in Figure 2.5(a). When M moves to a new position M_1, line AB is shown at A_1B_1, having undergone both translation and rotation to gain the new position. The angle θ between AB and A_1B_1 is the angular displacement of AB and of body M. If we consider translation and rotation separately, AB might first translate from position AB to A_1B_0, then rotate about A_1 through angle θ_1 to position A_1B_1. Since AB and A_1B_0 are parallel to one another, angle θ_1 equals angle θ. (When two parallel lines are cut by a transversal, the corresponding angles are equal.) Had line AB first translated to A_0B_1 and then rotated through angle θ_2 to A_1B_1, as in Figure 2.5(b), the linear displacement of AB would have been different from the first case, yet the angular displacement would have been the same, since angle θ_2 equals angle θ. (When two parallel lines are cut by a transversal, the alternate interior angles are equal.)

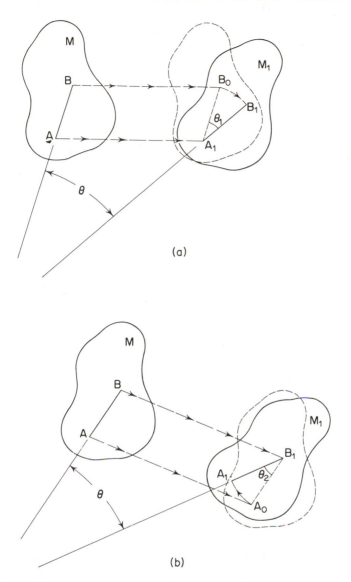

Figure 2.5 Linear and angular displacements in combined motion

Since angular displacement is independent of linear displacement, it is only necessary to measure the angle between two positions of any one line on the body in order to find the angular displacement of the body as a whole. When the motion of a body is dictated by the restraints of adjoining members

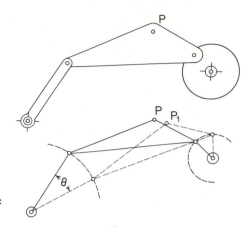

Figure 2.6 A "skeletonized" kinematic layout

of a mechanism, much time may be saved in the layout of several positions if the members are shown in skeletonized from, as illustrated in Figure 2.6.

2.5 Mechanisms Producing Specific Paths

In the design of machines it is frequently required to guide or propel a point along a given path. A number of mechanisms have been devised to produce motion along the common geometric curves, such as a circle, an ellipse, a cycloid, an involute, and, of course, a straight line. The art of designing involves a knowledge of what has been done before as well as the ability to create, so it is important that the student acquaint himself with these known devices and acquire a "vocabulary" of mechanisms. Toward this end some of the more familiar mechanisms will be described here.

2.6 Straight-Line Mechanisms

Guiding a point along a straight line presents few problems today, since it is a simple matter to produce very accurate plane surfaces along which a member may slide. This was not the case before modern machine tools were developed. Before James Watt could build his steam engine, he had to design a linkage to guide a pin along a straight-line path, since there were no machine tools in 1769 capable of producing straight metal slides of sufficient precision.

Straight-line mechanisms are not of historical interest alone, however, since in some machines there are space limitations which prohibit the use of the conventional slides.

1. *Watt's Straight-Line Linkage (Figure 2.7):*

   This is one of the simplest of these devices, consisting of two cranks *AB* and *CD*, equal in length, which swing about fixed pins at *A* and *D*, and a connecting rod *BC*, proportioned as shown. Point *E*, the midpoint of *BC*, follows an approximate straight-line path for a limited distance, as shown by the dotted line segment *GH*.

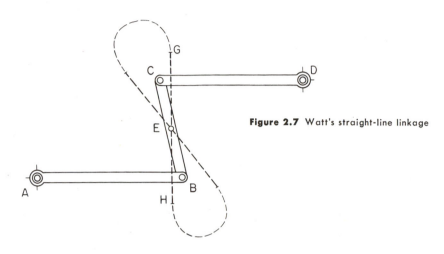

Figure 2.7 Watt's straight-line linkage

2. *Roberts' Straight-Line Linkage (Figure 2.8):*

   In this mechanism cranks *AB* and *CD* are of equal length. The connecting rod *BC* has a projecting strut rigidly attached at 90° at its midpoint to form a T-shaped member. The path of point *P* is approximately a straight line over a portion of its travel.

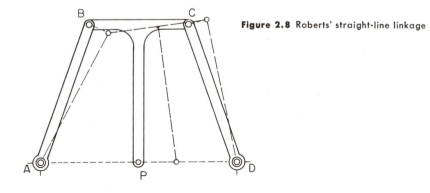

Figure 2.8 Roberts' straight-line linkage

3. *The Isosceles, or Scott-Russell, Straight-Line Linkage* (*Figure 2.9*):
Lengths *AB*, *CB*, and *BF* are equal in this mechanism, forming two isosceles triangles, from which the device receives its name. Point *F* travels in an exact vertical straight line through *A* for the entire range of its motion.

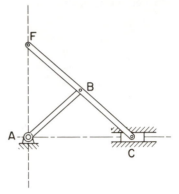

Figure 2.9 Isosceles straight-line linkage

4. *Peaucellier's Straight-Line Linkage* (*Figure 2.10*):
While more complex, this device guides pin *P* in an exact straight line path, perpendicular to *AB*. Fixed axes are at *A* and *B*, and members labeled with like letters are equal in length.

5. *The Epicyclic Straight-Line Mechanism* (*Figure 2.11*):
The diameter of wheel *W* is exactly equal to the radius of the large stationary ring *M*. If *W* rolls around the surface of *M* without slipping, point *P* on *W* follows an exact straight line across the diameter of *M*.

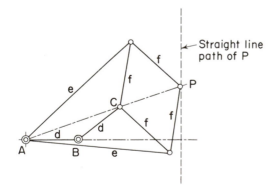

Figure 2.10 Peaucellier's straight-line linkage

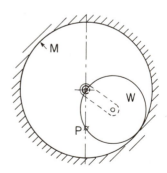

Figure 2.11 Epicyclic straight-line mechanism

2.7 Mechanism to Describe Arcs of Large Radius

Since any point on a body rotating about a fixed axis follows a circular path, devices to describe arcs of small radius present no problems. If an arc of extremely large radius is desired, this simple solution obviously becomes unwieldy.

A modification of Peaucellier's linkage (Figure 2.12) in which AB is not equal to BC causes P to follow an exact arc of a circle instead of a straight line. If BC is made less than AB, the center of the arc is to the right, on line AB extended. If BC is greater than AB, the center lies to the left. Other members labeled with like letters are of equal length.

Figure 2.12 Linkage for arcs of long radius

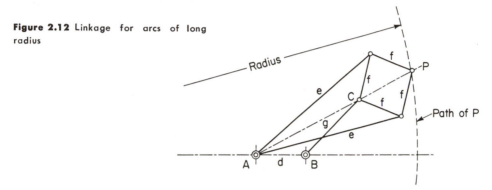

2.8 Mechanism to Describe Elliptical Paths

A modification of the isosceles linkage guides point E along an elliptical path, as shown in Figure 2.13. AB is made equal to BC as before, and E may be located at any point along CF or CF extended (except at B, C, or F). The major axis of the ellipse equals two times the sum of AB and BE. The minor axis equals two times CE. The center of the ellipse is at A, with the major axis along guides.

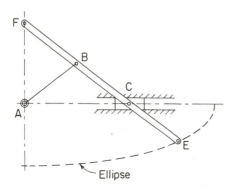

Figure 2.13 Linkage to describe an ellipse

2.9 Mechanism to Describe a Parabola

A parabola is defined as the locus of points which are equidistant from a fixed point, called the *focus*, and a fixed straight line, called the *directrix*. In Figure 2.14 the fixed pin F is the focus, the center line DD of the fixed vertical slot is the directrix, and point P, on rod B, travels along a parabolic path as the slotted arm M is moved up and down. M slides in the vertical slot and is perpendicular to DD at all times. Pin A is on M. $AG = GF = FE = EA$. The sleeve bearings at E and G slide freely on B. Imaginary lines AF and GE are diagonals of the rhombus $AGFE$. GE is the perpendicular bisector of AF. Therefore AP (perpendicular to DD) is equal to PF in all positions.

Figure 2.14 Linkage to describe a parabola

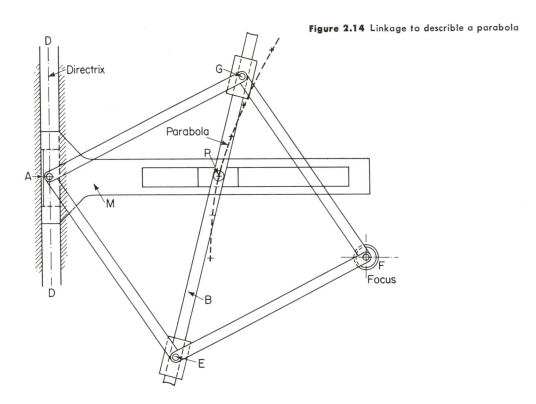

2.10 Mechanism to Generate an Involute

An involute is the curve traced by the free end of a taut, inextensible cord as it is unwrapped from a stationary cylinder. Figure 2.15 shows a chain as it is unwound from a fixed sprocket. The end of the chain P describes an involute. Note that the radius of curvature of the involute at any point equals the length of the unwrapped chain (from P to the point of tangency T, for example).

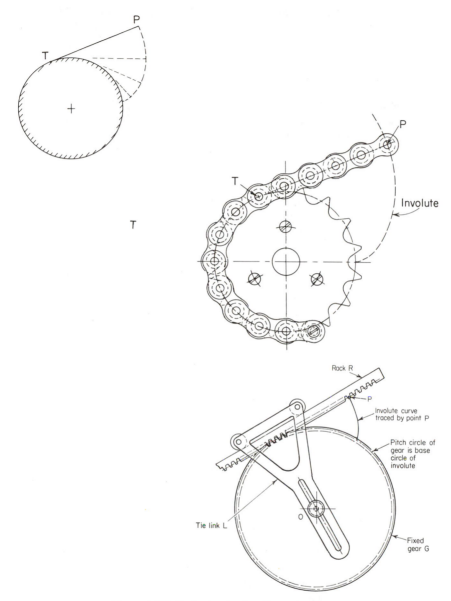

Figure 2.15 Mechanism to describe an involute

A mechanism to describe an involute is also shown in the illustration. Gear G is held stationary while rack R is rolled around it with teeth in mesh to prevent slipping. The tie link L with rollers is rotated about the center of the gear O causing the rack to turn and keeping the teeth engaged. A scriber at point P on any tooth traces an involute. The pitch circle of the gear must have the same diameter as the base circle from which the involute is to be generated. The slot in *L* permits adjusting *L* to different sizes of gears.

2.11 Enlarging or Reducing Mechanism

There are a number of parallel linkages which may be used to change the scale of the drawing of a pattern or profile without altering its proportions. A common example is the *pantograph* shown in Figure 2.16. Point *A* is fixed and pins *A*, *E*, and *P* are in a straight line. The linkage formed by *CE*, *ED*, *DB*, and *BC* is a parallelogram. To enlarge a drawing, the given contour is traced by a stylus at *E* and the enlarged figure will be reproduced by a pencil at *P*. To reduce the size, we trace the given drawing at *P* and the reduced figure will be described by a pencil at *E*. The ratio of sizes of the original and traced drawings is the same as the ratio of distance *AE* to *AP*. In this example that ratio is 2 to 5.

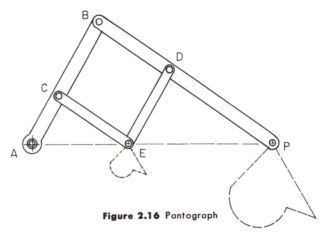

Figure 2.16 Pantograph

2.12 Designing Mechanisms for Given Displacements

We are now ready to consider some elementary design problems involving displacements. In the examples offered below, only one of many possible solutions is suggested and only the problem of producing the required displacement is considered. In an actual design situation, speeds and accelerations might well be vital factors in the choice of a mechanism, but these must be deferred until later, when our study is further advanced.

Example 1

A typical problem demands that a point be propelled back and forth along a straight path (reciprocate) between two points, *M* and *N*, located 4 in. apart.

The mechanism is to be driven by a revolving shaft, located in fixed bearings at a point A on line MN extended, 6 in. from M [Figure 2.17(a)].

A simple mechanism for converting rotation to translation is the *crank and slider linkage*, ABC, shown in Figure 2.17(b). Considering N, the extreme position attainable by C toward the right (C_R), it is evident that crank AB and connecting rod BC must be extended in one line in order to bring C to this position. We cannot predict the dimensions of either AB or BC individually now, but it is apparent that:

$$BC + AB = AM + MN = 6 + 4 \quad \text{or} \quad BC + AB = 10$$

When C is at M, the extreme position to the left (C_L), AB and BC will be folded over one another so that the distance from B to M equals BC. Then

$$BC - AB = AM \quad \text{or} \quad BC - AB = 6$$

We now have two equations involving AB and BC and can solve for both of these distances:

$$
\begin{aligned}
BC + AB &= 10 \\
\underline{BC - AB} &= \underline{\ 6} \\
2BC &= 16 \text{ (Adding equals to equals)} \\
BC &= 8
\end{aligned}
$$

Substituting: $8 + AB = 10$ or $AB = 2$

So a crank and slider mechanism with $AB = 2$ in. and $BC = 8$ in., as shown in Figure 2.17(c), gives the required displacement.

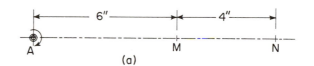

(a)

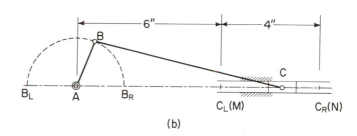

(b)

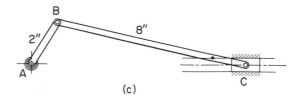

(c)

Figure 2.17 Design of crank and slider linkage for given displacements

Example 2

With a drive shaft at A, turning in fixed bearings, suppose a mechanism is required to swing a crank CD back and forth (oscillate) through angle θ (93.5°), between positions $C_L D$ and $C_R D$, located as shown in Figure 2.18(a).

A four-bar linkage will provide the motion desired. Crank AB, turning about A, is connected to CD by the link BC [Figure 2.18(b)]. AB is to make complete revolutions while CD oscillates. As in the previous example, AB and BC must be extended in line when CD is in the extreme position to the right ($C_R D$). From this we deduce that $BC + AB = AC_R$. If we scale or calculate the distance AC_R and find it equal to 8 in., it then follows that $BC + AB = 8$. When CD is in the extreme left-hand position ($C_L D$), BC must be folded over AB, in which case $BC - AB = AC_L$. If AC_L measures 4 in., we can state that $BC - AB = 4$ in.

We now have two equations involving AB and BC and can solve for both of these lengths:

$$
\begin{aligned}
BC + AB &= 8 \\
\underline{BC - AB} &= \underline{4} \\
2BC &= 12 \ (\text{Adding}) \\
BC &= 6
\end{aligned}
$$

Substituting: $6 + AB = 8$ or $AB = 2$. The required mechanism will then have a driving crank AB, 2 in. long, and a connecting rod BC, 6 in. long, as shown in Figure 2.18(c).

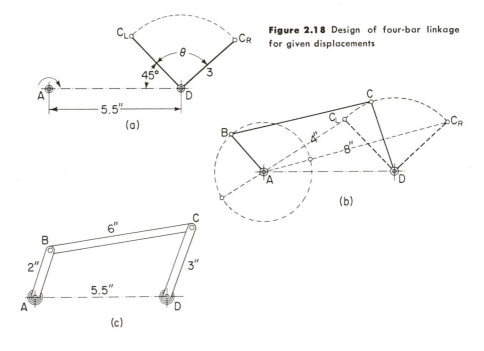

Figure 2.18 Design of four-bar linkage for given displacements

2.13 Designing Mechanisms for Given Paths

If it is required that a point be guided along an irregular curved path, several methods may be employed, the selection depending upon the available space, the degree of accuracy demanded, and cost limitations.

The simplest approach from the point of view of the designer would be to use a stationary template, cut to the desired curve, which would guide the moving member.

A second method involves two revolving plates of specially designed contour, called *cams*. These, through contact with a freely moving follower, provide the proper horizontal and vertical displacements needed to follow the required path.

A third method is to design a linkage which would guide and propel the point along the curve. This method is perhaps the most difficult, but the resulting mechanism would very likely be the most satisfactory.

2.14 The Fixed Template Method

The crank and slider mechanism *ABC* is employed in Figure 2.19 to propel pin *C*. A roller, mounted on pin *C*, is constrained to follow along a groove cut in template *D*. The center line of this groove is the curved path specified, which *C* will follow. The mechanism is simple, but, if a high degree of accuracy is demanded, the precise machine work required to prevent backlash might well be prohibitive in cost. The template must be rigidly mounted on the machine and occupies considerable space in the immediate area of the output motion which might not be available.

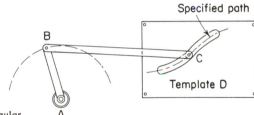

Figure 2.19 Fixed template for irregular curved paths

2.15 The Method of Combined Cams

The cam and follower is a very simple yet versatile mechanism frequently used to obtain an irregular program of displacements. Cams are made in many different forms with various types of followers, so we will study them in more detail later (see Chapter 9).

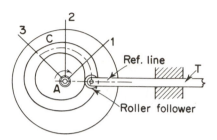

Figure 2.20 Positive-motion cam with roller follower

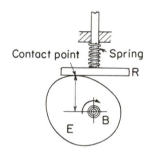

Figure 2.21 Plate cam with flat follower

Figure 2.20 shows a *radial, positive-motion cam* with a roller follower. A groove is milled in the surface of the cam to receive the roller. As cam *C* turns clockwise about fixed shaft *A*, rod *T* is driven from left to right, since the radius to the center of the groove increases as the radial lines 1, 2, 3, etc., successively turn to the reference line position. Since the roller is at all times contained within the groove, this is called a positive-motion cam.

Figure 2.21 shows a *plate cam with a flat follower*, which is held in constant contact with the cam by means of a compression spring. As cam *E* turns clockwise about fixed shaft *B*, follower *R* is caused to rise and fall as the vertical distance between the cam axis *B* and the contact point on the cam surface increases or decreases.

Suppose we use cam *C* to drive a horizontal rod *T* and cam *E* to drive a vertical rod *R* which is mounted in *T* so that it is free to rise and fall. It is possible, if the cams are properly designed, to guide point *P* on rod *R* along almost any curved path, as shown in Figure 2.22. If the cams are synchronized, cam *C* will produce the necessary horizontal displacement, while cam *E* simultaneously provides the corresponding vertical displacement required to locate *P* at any point along the curve. The method of designing these cams is described in detail in Chapter 9.

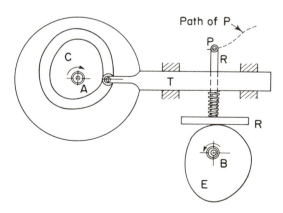

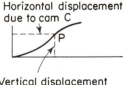

Figure 2.22 Combined cams for irregular paths

2.16 The Design of Linkages to Describe a Given Path

The *four-bar linkage* is the most basic of all link mechanisms. All other linkages can be shown to be modifications or combinations of several four-bar linkages. Figure 2.23 shows a typical example. One member (*AD*) is stationary and usually is not in the form of a bar or link, since it is defined by two fixed points on the frame of the machine. There are two cranks, a driver crank *AB* and a follower *CD*. As *A* and *D* are fixed axes, these cranks are in pure rotation. The moving ends of the cranks are connected by the fourth bar, a connecting rod *BC* (called a *coupler*), which is in combined rotation and translation. This simple mechanism is capable of a tremendous variety of motions which may be obtained by adjusting the relative size of the four links. Different points on the coupler, or on the extensions of the coupler, will trace innumerable irregular curves as the dimensions of the links are altered.

It is conceivable that if we could determine the proper link proportions and select the correct tracing point on the coupler, we could obtain, with this simple mechanism, any path of motion which might be required. This solution would be accurate and inexpensive and would involve a minimum of space. While the resulting mechanism is simple, the process of design of such a device is unfortunately difficult.

Figure 2.23 Paths of points on a four-bar linkage

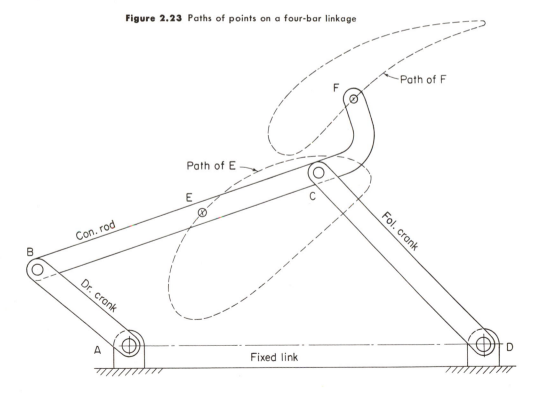

Trial and error methods, using pasteboard models, are very fascinating and instructive but are slow and irrational and offer no assurance of success. In general, solution by trial should not be too hastily discarded if preliminary trials point the way to success, but this approach is discouraging if it degenerates to pure gambling, without suggesting ways to improve.

Analysis of the Four Bar Mechanism, by Hrones and Nelson (Wiley), is a catalogue which shows the paths of points on the coupler of over 700 four-bar linkages and gives the relative dimensions of the links in each case. This book offers a direct solution to this difficult problem. Using this catalogue we find a path curve which matches that curve which we wish to produce. The location of the point on the coupler and the link dimensions producing this path are readily available for the mechanism chosen, so it remains only to adjust the scale for the size of the curve desired and the solution is complete.

Figure 2.24 shows a typical catalogue page, and Figure 2.25 shows a linkage selected for a given curve. An auxiliary linkage, shown dotted, may be added if desired to limit the motion of the tracer point to that path which is required.

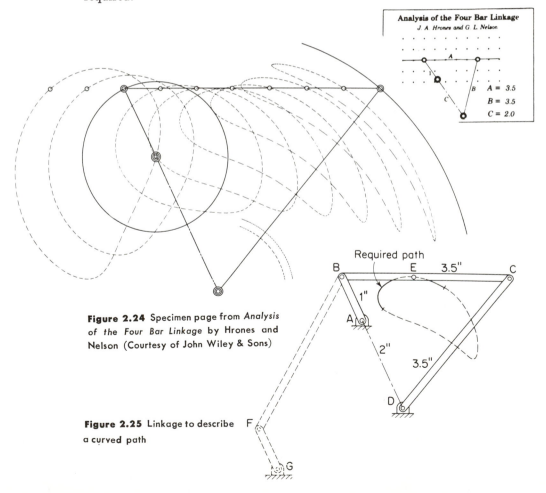

Figure 2.24 Specimen page from *Analysis of the Four Bar Linkage* by Hrones and Nelson (Courtesy of John Wiley & Sons)

Figure 2.25 Linkage to describe a curved path

PROBLEMS

2.1. Tchebicheff's approximate straight-line mechanism is shown in Figure P2.1. *A* and *D* are fixed axes, and *P* is the midpoint of *CB*. *AD* = 6 in., *AB* = *CD* = 7.5 in. (crossed as shown) and *CB* = 5 in. Draw the mechanism full size and plot the path of *P* to determine the length of its straight line motion.

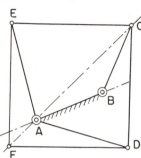

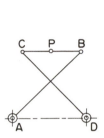

Figure P2.1   **Figure P2.2**

2.2. In the linkage shown in Figure P2.2, *A* and *B* are fixed axes. *AB* = *BC* = 3 in., *AE* = *AD* = 4 in. and *DC* = *CE* = *EF* = *FD* = 5 in. Draw the mechanism full size and plot the path of point *F*. What linkage does this resemble?

2.3. In Figure P2.3, bar *LM* = 6 in., and point *P* is located 2 in. from *L*. The center lines of the fixed guides are at 90° with one another. Plot the path of *P* and draw a smooth curve. Of what linkage is this a modification?

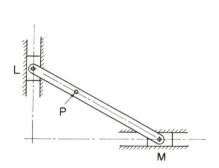

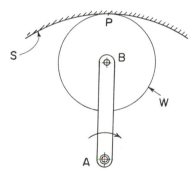

Figure P2.3   **Figure P2.4**

2.4. In Figure P2.4, *A* is the center of the fixed circular surface *S*. Wheel *W* is held in contact with *S* by arm *AB*. These is no slipping between *W* and *S* as *AB* turns about *A*. *P* is a point on the rim of *W* which is in contact with surface *S* in position shown. The radius of *S* = 6 in.; *AB* = 4 in., making the radius of *W* = 2 in. Plot the path of *P* on *W* as *AB* turns through an angle of 30° clockwise. Draw a smooth curve for the path of *P*. What is the correct name for this curve?

2.5. Design a mechanism to drive a pin R back and forth along the straight line path AB, 3.2 in. long. The mechanism is to be driven by a shaft located at O (Figure P2.5) which is continuously revolving in the same direction. Sketch the mechanism, giving all dimensions.

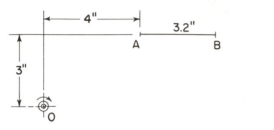

Figure P2.5

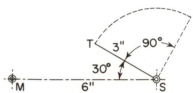

Figure P2.6

2.6. Design a mechanism to cause crank ST, 3 in. long, to oscillate through an angle of 90° positioned as shown in Figure P2.6. The mechanism is to be driven by a continuously revolving shaft located at M. Sketch the required mechanism, giving all dimensions.

2.7. Buchanan's paddle wheel, invented in the days of side-wheel steamboats, was very effective because the blades were held in a vertical plane throughout their motion and thus could enter and leave the water without useless splashing and exert a maximum force against the water while immersed. Successive positions of the blades are shown in Figure P2.7. Design a linkage which will hold a blade in this vertical position as the wheel makes complete revolutions. Specify relative dimensions of all members.

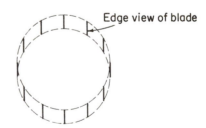

Edge view of blade

Figure P2.7

2.8. Design a mechanism which will locate the perpendicular bisector of lines (up to 6 in. long). This device must align a straightedge along the required perpendicular bisector.

2.9. Design a mechanism which will guide a point in a true elliptical path. The major axis of the ellipse is to be 12 in. and the minor axis 4 in. Sketch the mechanism and specify all dimensions.

2.10. The flat follower *F* is to be raised and lowered through contact with a cam which turns about shaft *A*, located as shown in Figure P2.10. Starting from the position shown, *F* is to rise $1\frac{1}{2}$ in., then return to the original position. This motion is to take place during each revolution of the cam. Design a circular cam to produce the required motion of *F*.

2.11. Design a linkage which will guide a point *N* back and forth along path *ABC*, shown in Figure P2.11. This mechanism is to be driven by a continuously revolving shaft. The path is to be approximately straight from *A* to *B* and a circular arc of $\frac{5}{8}$-in. radius from *B* to *C*.

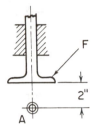

Figure P2.10

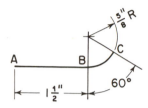

Figure P2.11

2.12. Design a mechanism to guide a point *S* along the path *DEF*, as shown in Figure P2.12. The path from *D* to *E* is a circular arc of 2-in. radius with center at *O*. From *E* to *F* the path is an arc of 1-in. radius with center at *P*.

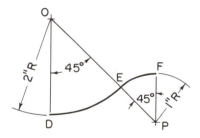

Figure P2.12

2.13. Design a linkage which will guide a point along a true straight line at least 6 in. long. The linkage is to be composed of bars which are pinned together; no slides or rollers are to be employed. It is to be driven by a rotating shaft which turns one of the bars. Draw the required linkage to scale in a typical position. Indicate all fixed points and give true dimensions of the members. Label the axis of the drive shaft and the scribing point. Make a cardboard scale model of the linkage so that you can demonstrate the validity of your design.

2.14. Design a linkage which will guide a point back and forth along an elliptical path. The major axis of the ellipse is to be 6 in. and the minor axis is to be 4 in. Only one quarter of the ellipse is to be traced. The linkage is to be driven by a continuously revolving shaft. Draw the required linkage to scale, indicating all fixed points or surfaces and giving true dimensions of the members. Label the axis of the drive shaft and the scribing point. Make a cardboard scale model of the linkage so as to demonstrate the validity of your design.

2.15. Design a pin jointed linkage which will drive an oscillating arm 6 in. long back and forth through an angle of 75°. The overall time ratio of the forward to the return stroke is to be 2 to 1. The linkage is to be driven by a revolving shaft which turns at constant speed. (The angular displacement of the drive shaft is directly proportional to time.) Draw the required linkage to scale, indicating all fixed axes and giving true dimensions of all members and relative positions of fixed axes. Label the driving crank and the oscillating arm. Make a cardboard model of the linkage to scale so that you can demonstrate its validity.

2.16. Design a linkage which will guide a point along a parabolic path. The focus (point) should be 8 in. from the directrix (line). The linkage must be capable of describing a parabola approximately 6 in. long at least. The linkage may be driven by a reciprocating member traversing a straight path or by a rotating shaft attached to one of the links. Sliding members may be employed as well as pinned joints. Draw the required linkage to scale, indicating all fixed points and surfaces and giving true dimensions of the members. Label the driving member and the point which scribes the parabola.

2.17. Design a linkage which will guide a point along any path which is two and one half times the length and parallel at all times to any given path in slope (pantograph). The driving point of this linkage is to be traced along the given path by hand. Use any irregular triangle as an example of the path to be reproduced at enlarged scale. Draw the required linkage to scale, indicating all fixed points and giving true dimensions of the members. Label the hand driven point D and the tracing point T. Make a cardboard scale model of the linkage so that you can demonstrate the validity of your design.

3

Velocity

The study of velocity is perhaps the most important phase of a course in mechanisms. Displacement and acceleration are so closely related to velocity that they are often more easily determined through velocity study than by a direct approach. For this reason we shall concentrate on developing a firm foundation of velocity analysis as a basic core around which to build.

Velocity is the rate of change of position with respect to time or *displacement per unit of time*. Common units of velocity are feet per second, radians per second, revolutions per minute, and miles per hour. It is not an expression of magnitude alone, however, but denotes direction of motion as well. The term *speed* is often used to express the magnitude of velocity, but this term does not include a designation of direction. The velocity of a point or of a member is an instantaneous value, which may remain unchanged or may vary over a period of time.

3.1 Linear Velocity   (Symbol: V)

A point moving along a line has a linear velocity equal to its linear displacement per unit of time. If the point travels equal distances in successive equal intervals of time, its velocity is described as *uniform* or *constant* and may be determined by measuring its displacement S in a given time T and dividing:

$$V = \frac{S}{T}$$

For example, in Figure 3.1, if a point P travels 4 in. from a to b in 2 sec, and 4 in. from b to c in the next 2 sec, etc., its velocity is constant and equals:

$$V_P = \frac{S}{T} = \frac{4}{2} = 2 \text{ in./sec}$$

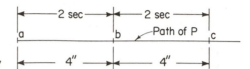

Figure 3.1 Constant linear velocity

The direction of this velocity is along ac toward the right at all times.

If a point travels unequal distances in successive equal time intervals, its velocity is said to be *variable* and has a different value each instant. In Figure 3.2, a point moves 4 in. from a to b in 2 sec and 6 in. from b to c in the next 2 sec. The average velocity between a and b equals:

$$V_{av} = \frac{S}{T} = \frac{4}{2} = 2 \text{ in./sec}$$

This average velocity is not the velocity for all positions between a and b, however, since the velocity is changing continuously as the point moves. It is only a theoretical number.

To determine the instantaneous velocity at any given point e, we consider a very small displacement ΔS* and the corresponding small time interval ΔT which is required to travel this distance. As these small increments ΔS and ΔT approach zero, their ratio will equal the instantaneous value of the velocity of the point at e:

$$V_e = \frac{\Delta S}{\Delta T} \quad \text{(as } \Delta T \text{ approaches zero)}$$

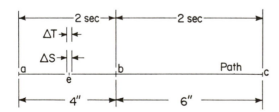

Figure 3.2 Variable linear velocity

In terms of calculus this ratio is expressed as ds/dt (the first derivative of linear displacement with respect to time).

*The Greek letter Δ (delta) is commonly used to denote a small increment of, or a small change in, any quantity.

If a point moves along a curved path the direction of its velocity at any given position is the direction of the curved path at that point, which is best defined by the tangent drawn to the curve. In Figure 3.3, a point traveling clockwise along the arc *abc* has different inclinations of velocity at each position, as shown by the tangents at *a*, *b*, and *c*. If the point travels equal lengths of arc in equal time intervals, the magnitude of the velocity at any point equals:

$$V = \frac{\text{arc } ab}{T} \quad \text{(where } T \text{ equals the time required to travel from } a \text{ to } b\text{)}$$

Since the direction of velocity is different for each location, the displacement *S* between *a* and *b* (the chord *ab*) is not equal to the arc *ab*. Therefore the former equation

$$V = \frac{\Delta S}{\Delta T}$$

can only be applied when ΔS is so small as to be essentially equal to the arc subtended. When this is the case, the general equation applies:

$$V = \frac{\Delta S}{\Delta T} \quad \text{(as } \Delta T \text{ approaches zero).}$$

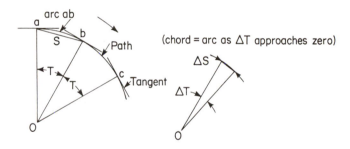

Figure 3.3 Linear velocity on a curved path

A body in translation may be said to have linear velocity. Since the body does not rotate and all points upon it have equal displacements in any given time interval, the ratio $\Delta S/\Delta T$ will be the same for all points at any instant. The velocity of the entire body will therefore equal the velocity of any point upon it.

3.2 Vector Representation of Linear Velocities

Since velocity has the property of direction as well as magnitude, it is called a *vector quantity* and can be graphically represented as the length of a straight line called a *vector*. Suppose a point A on a body has a velocity of 3 in./sec in a direction upward to the right at 45° with an axis of reference XX. Figure 3.4 shows the vector representing this velocity. The magnitude is depicted by drawing the vector to a scale of 1 in. equals 1 in./sec, which results in a line 3 in. long, originating from point A. There are two aspects of direction: first, *inclination*, which is the line along which motion takes place, and second, *sense*, which denotes which way motion is directed along this line. The vector is drawn at the required inclination, and an arrowhead is added to show the sense. The label V_A is an identifying symbol meaning "velocity of point A." Whether drawn to scale or sketched with magnitude noted, vectors are extensively used to afford a clear, graphical representation of linear velocities.

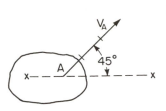

Figure 3.4 Velocity vector

3.3 Angular Velocity (Symbol: ω)*

A line or body in angular motion (rotation) has *angular velocity equal to its angular displacement per unit of time*. If it turns through the same angle in equal intervals of time, its angular velocity is *uniform* or *constant*. The magnitude of this constant angular velocity is determined by measuring the angle turned through in a given time interval and dividing by that time interval:

$$\omega = \frac{\theta}{T}$$

For example, Figure 3.5 shows line AB fixed on body W, which is turning clockwise through an angle of 45° every 2 sec. The angular velocity of the body W is:

$$\omega = \frac{\theta}{T} = \frac{45}{2} = 22.5°/\text{sec, clockwise}$$

Angular velocity is usually expressed in radians per second or in revolutions per minute, rather than degrees per unit of time. If ω is to be in radians per second, θ must be in radians, so the equation above becomes:

$$\omega = \frac{\theta^{\text{rad}}}{T} = \frac{\pi/4}{2} = \frac{\pi}{8} = 0.3925 \text{ radian/sec, clockwise†}$$

*The greek letter ω (omega, pronounced oh-may-ga) is commonly used for angular velocity.

†Unit conversions (use $\pi = 3.14$ in calculations):

$$2\pi \text{ radians} = 1 \text{ revolution} = 360°$$
$$\text{number of radians} = \text{number of revolutions} \times 2\pi$$
$$\frac{\text{angle in radians}}{\text{angle in degrees}} = \frac{2\pi}{360} = \frac{\pi}{180}$$
$$\text{angle in radians} = \text{angle in degrees} \times \frac{\pi}{180}$$

Only a line or a body can have angular velocity. Since a point has no extent, angular motion of a point has no significance.

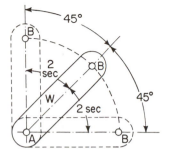

    If a line or a body turns through unequal angles in equal intervals of time, its angular velocity is described as *variable*. For any given period of time the average angular velocity may be found by dividing the angular displacement during that interval (θ) by the time interval (T):

$$\omega^{\mathrm{av}} = \frac{\theta}{T}$$

Figure 3.5 Constant angular velocity

Again, this average velocity is only a theoretical value, and not the angular velocity for all positions of the line during the time interval T. To determine the angular velocity at any specific instant, a very small $\Delta\theta$ and the corresponding ΔT must be used. As these small increments approach zero their quotient becomes the instantaneous angular velocity:

$$\omega = \frac{\Delta\theta}{\Delta T} \quad \text{(as } \Delta T \text{ approaches zero)}$$

In terms of calculus this ratio is expressed as $d\theta/dt$ (the first derivative of angular displacement with respect to time).

3.4 Relationship between Linear and Angular Velocity

    In Figure 3.6, wheel W turns clockwise about fixed axis O with an angular velocity equal to ω. Line OA will have the same angular velocity as W, and point A will travel a circular path of radius OA. As it moves, point A has no motion toward, or away from, the fixed axis O and therefore has no velocity along the radial line OA. The velocity of A is then tangential to its circular path (perpendicular to OA) at all times.

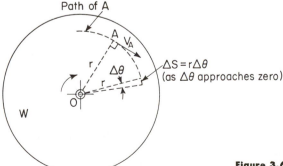

Figure 3.6 Angular and linear velocities

It is clear that, if W turns fast, point A will have a high linear velocity and, if W turns slowly, the velocity of A will be low. From this observation, we may expect that there exists some definite relationship between angular velocity of a body and linear velocity of points on that body.

At any given instant,

$$V_A = \frac{\Delta S}{\Delta T} \quad \text{(Article 3.1)}$$

and

$$\omega_W = \frac{\Delta \theta}{\Delta T} \quad \text{(Article 3.3)}$$

To apply in all cases, these equations specify that ΔT approaches zero, which makes ΔS and $\Delta \theta$ very small. In Article 2.3 we observed that during an angular displacement θ (in radians) a point on a rotating body will traverse a circular arc equal to $r\theta$. For a very small angle like $\Delta \theta$ in Figure 3.6, the circular path of A is essentially equal to the chord ΔS, which is the displacement of A during the angular displacement $\Delta \theta$ of W. As $\Delta \theta$ and ΔS approach zero, we are then justified in stating

$$\Delta S = r \, \Delta \theta$$

In the equation

$$V = \frac{\Delta S}{\Delta T}$$

we may substitute $r \, \Delta \theta$ for ΔS, so that

$$V = \frac{r \, \Delta \theta}{\Delta T}$$

But

$$\frac{\Delta \theta}{\Delta T} = \omega$$

So

$$V = r\omega \quad \text{(where } \omega \text{ is in radians per unit of time)}$$

In the example of Figure 3.6, r equals OA and ω_W is the angular velocity of W, so

$$V_A = OA \times \omega_W$$

This formula not only simplifies the calculation of linear velocities on a rotating body, but it affords a means of determining the angular velocity of a body when the linear velocity of a point and its distance from the axis of rotation are known:

If $V = \omega r$, then $\omega = \dfrac{V}{r}$ (dividing both sides of the equation by r)

In all applications of this formula, ω must be in radians per unit of time.

We have observed (in Article 2.2) that all lines on a body have the same angular displacement in any given time interval. Since all lines will turn through the same angle in a given time, the ratio of $\Delta\theta$ (the angle turned)/ΔT (the time consumed in turning) will be the same for all lines on the body for any instant. Since $\Delta\theta/\Delta T = \omega$, the angular velocity will likewise be the same for all lines. This justifies the use of the linear velocity of any point on a body to find its angular velocity.

3.5 Velocities of Points on Rotating Bodies

In Figure 3.7, a body M rotates clockwise about fixed axis P. When in the position shown, the angular velocity of M equals ω_M. All points on rotating bodies have linear velocities which are perpendicular to lines from these points to the axis of rotation. Therefore, the velocity of L at this instant is perpendicular to PL and (since $V = \omega r$)

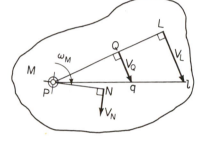

$$V_L = \omega_M \times PL$$

Similarly, point Q has a tangential linear velocity:

$$V_Q = \omega_M \times PQ$$

Figure 3.7 Relationship of linear velocities on rotating bodies

If we wish to note the relationship between V_L and V_Q, we may write the ratio:

$$\frac{V_L}{V_Q} = \frac{\omega_M \times PL}{\omega_M \times PQ} = \frac{PL}{PQ} \quad \text{(canceling } \omega_M \text{)}$$

Similarly,

$$\frac{V_L}{V_N} = \frac{\omega_M \times PL}{\omega_M \times PN} = \frac{PL}{PN}$$

Thus we find that points on a rotating body have linear velocities whose magnitudes are in the same ratio as (are proportional to) their distances from the axis of rotation.

For example, if $V_L = 20$ in./sec, $PL = 5$ in. and $PQ = 3$ in., we can calculate ω_M and V_P, or the linear velocity of any other point on the body, if its location is known.

$$\omega_M = \frac{V_L}{PL} = \frac{20}{5} = 4 \text{ radians/sec}$$

$$\frac{V_Q}{V_L} = \frac{PQ}{PL} \quad \therefore \quad V_Q = \frac{PQ \times V_L}{PL} = \frac{3 \times 20}{5} = 12 \text{ in./sec}$$

Since L and Q lie on the same radial line, it is easy to obtain V_Q graphically. If V_L is represented by a vector Ll, drawn to scale, we can predict that Qq, the vector of V_Q, will terminate on a line lP. This vector Qq may be measured at the same scale to determine the value of V_Q. This is valid, since the triangles PLl and PQq are similar (corresponding angles are equal) and their corresponding sides proportional:

$$\frac{Qq}{Ll} = \frac{V_Q}{V_L} = \frac{PQ}{PL} \quad \text{(as above)}$$

The formula $V = \omega r$ enables us to determine the linear velocity of any point on a body in rotation if we know the angular velocity of the body. The same formula in the form $\omega = V/r$ will give the angular velocity of the body if we know the linear velocity of one point. In each case, of course, we must know the location of the axis of rotation and the direction of motion. This is all we need to make a complete velocity analysis on a rotating body.

3.6 Velocities of Points on Bodies in Combined Motion

There is no simple formula for calculating linear velocities of points on a body moving with combined translation and rotation. For this reason we find that a graphical approach is simple and efficient. This method avoids lengthy and laborious trigonometric calculations and, with a proper choice of scale and precise drafting technique, suffers no detrimental loss of accuracy. Graphical methods should not be used exclusively. Where calculations are fast and simple they should be used to supplement the graphics. A wise combination of these two methods is the key to efficiency.

3.7 Velocity Vectors

In graphical analyses, vectors will be used to represent linear velocities. A vector is a straight line which may express the magnitude of any quantity by means of its length, measured at some specified scale. It also denotes inclination, in which the quantity is directed, by the slope at which it is drawn, and sense, by the arrowhead at one end.

Vectors are not limited to velocity symbols alone, but may be used to depict any quantity which has the properties of magnitude and direction. They may be added, subtracted, and divided into component parts, yet these operations are performed in the same manner regardless of the quantity which is represented. In a sense they are the graphical equivalents of the numbers used in analytical work.

3.8 Vector Addition: Components and Resultants

Figure 3.8 shows two vectors, *A* and *B*, which are drawn with origin at point *O*. Let us add these two vectors graphically. As is the case with numbers, the sum must be exactly equivalent to the original parts.

The sum of *A* and *B* is found graphically by drawing the parallelogram of which *A* and *B* are adjacent sides. The diagonal *R*, drawn from the origin *O*, is the vector sum of *A* and *B* and is usually called their *resultant*. The magnitude and inclination of *R* is correct by this construction, and the sense must be made consistent with the sense of *A* and *B*.

Figure 3.8 Vector addition

In graphical solutions it is wise to use the simplest construction possible in the interest of accuracy as well as of speed. Accordingly, it may be noted that, if a triangle is formed by drawing *A* and *B* in succession (as adjacent sides), as shown in Figure 3.8, and making *R* the closing side, this construction will produce the same resultant *R* as the parallelogram method and will involve less drafting.

In general, to add vectors we start from an origin *O* and lay out the vectors in succession, putting the tail of each succeeding vector at the head end of the previous one. The resultant is a single vector drawn straight from *O* to the head of the final vector in the layout. This resultant always points away from the origin.

The vectors *A* and *B*, which are added to produce *R*, are called *components* of *R*. These are not the only pair of components which will yield the resultant *R*. There are innumerable combinations, some of which are shown in Figure 3.9. Among these are pairs of components at right angles to one another. These rectangular components are of special interest.

Figure 3.9 Components of a given vector

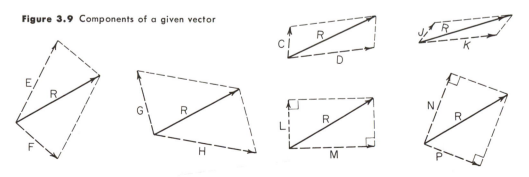

3.9 Effective Components   (Symbol: *ec*)

It is often necessary to determine the effect of a vector along some line which is not in the direction of the vector itself. Figure 3.10 shows a vector *R* with origin at *O* and of length *Oa*. If it is desired to measure the effect of this vector along any line, such as *SS*, we replace *R* by a pair of rectangular components, one of which, *E*, lies along *SS* and the other, *F*, along *TT*, perpendicular to *SS*. In order that these components may together be equivalent to *R*, their length is determined by considering *R* to be the diagonal of a parallelogram of which *E* and *F* are adjacent sides. Accordingly, we draw a line *ab* parallel to *SS* to meet *TT* and *ac* parallel to *TT* to meet *SS*. The parallelogram in this case is a rectangle, since the angles at *b* and *c* are right angles. The senses of *E* and *F* are shown by arrowheads at *b* and *c*, so as to be consistent with the sense of *R*.

Taken together, *E* and *F* are in all respects equivalent to *R*. Thus, the total effect of *E* and *F* along *SS* will equal the effect of *R* along *SS*. Since *F* is perpendicular to *SS*, it has no effect at all in that direction. Therefore, the component *E* will equal the total effect of *R* along *SS*. Components like *E* may logically be called *effective components*, since they measure the total effect of a vector in a given direction.

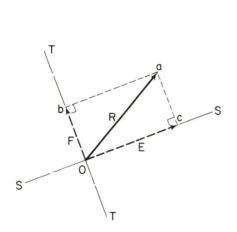

Figure 3.10 A mated pair of effective components

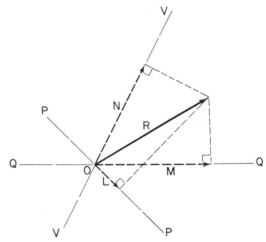

Figure 3.11 Effective components in different directions

To determine an effective component we need only draw a line from the arrowhead end of the given vector *perpendicular to the line along which the effect is to be measured*. The sense of the effective component is determined from the sense of the given vector, by inspection.

It is not necessary to draw the mated component at all, since it has no effect in the desired direction. Figure 3.11 shows several effective components of *R*. *L* is the *ec* along *PP*, *M* is the *ec* along *QQ*, and *N* is the *ec* along *VV*.

3.10 The Rigid-Body Concept

A rigid body is one which does not stretch, contract, bend, or deform in any way. If the body in Figure 3.12 is truly rigid, all points on it (such as *A*, *B*, and *C*) will at all times remain at fixed distances from one another and all straight lines (such as *AB* and *BC*) remain straight no matter what loads are applied or what motion the body may undergo.

This, of course, is a theoretical condition which in actuality does not exist in the full sense of the definition. All bodies suffer some minute deformation when resisting motion or applied forces. In most cases these deformations are small as compared to the dimensions of the body as a whole and therefore may be neglected in kinematic studies, where internal stresses are not under scrutiny, without sacrificing accuracy. Neglecting these small deformations greatly simplifies our work. Accordingly, with the exception of springs, rubber pads, and members which are intentionally designed for flexibility, we shall consider all machine members to be rigid bodies as we study their motion.

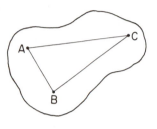

Figure 3.12 Rigid body

The properties of a rigid body impose very definite constraints on the motion of points and lines on that body, so this concept becomes one of the most useful tools in velocity analysis.

3.11 Velocities on a Rigid Body

The bar *AB* in Figure 3.13 is pinned at *A* and *B* to blocks which slide freely in parallel guides. *AB* is therefore in translation, as all positions are parallel. The velocity of *A* is given by the vector V_A. Let us find the velocity of *B*.

If the bar is a rigid body, the dimension *AB* remains the same whatever motion takes place. If we consider motion along the line *AB*, we know that, since this distance remains constant, *A* will never approach any closer to *B*, or

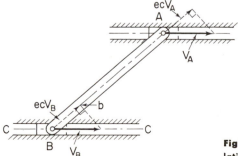

Figure 3.13 Velocities on a body in translation

move any farther away from *B*, regardless of what motion the bar undergoes. If *A* is displaced 2 in. along line *AB* in 1 sec, the displacement of *B* along line *AB* must likewise be 2 in. and in the same sense during the same second. Since $V = \Delta S / \Delta T$, we can state that, *along the line joining them, any two points on a rigid body must have identical velocities at all times.*

The velocity of *A* in the direction *AB* is the effective component of V_A along line *AB*. This is found by drawing a perpendicular to *AB* extended from the head end of the V_A vector (see ecV_A in Figure 3.13). Point *B* on *AB* must have the same effective component in the same sense along *AB* (see ecV_B). The pin *B* is a point on the sliding block as well as on bar *AB*, so the direction of motion of *B* is defined as along the center line (*CC*) of the guides. Since we know that ecV_B is the projection of V_B along *AB* and that the resultant V_B lies along *CC*, we have only to construct a perpendicular to *AB* at *b* to meet line *CC*, and V_B will be determined. The origin of a velocity vector is always at the point whose velocity it defines, so the sense of V_B is towards the right, as shown.

This method applies equally well to bodies in *rotation*, although calculation, using $V = \omega r$, is recommended for finding velocities in such cases. In Figure 3.14, member *EFG* turns about a fixed pin at *E*, and the velocity of *G* is given. It can be proved that V_G has to be perpendicular to line *EG* by means of the rigid body principle. Since *E* is fixed,

$$V_E = 0$$

so ecV_E along *EG* must be equal to zero. Since *E* and *G* are on the same rigid body, ecV_G along *EG* must equal ecV_E along *EG*, so ecV_G equals zero. If this is true, V_G must be perpendicular to *EG*, since it can have an effective component of zero along *EG* only when in this direction.

To find V_F, we observe that *F* and *G* are on the same rigid body and therefore have equal effective components along line *FG*. The ecV_G along *FG* is shown in Figure 3.14, and ecV_F is laid out equal to ecV_G. From previous investigation we know that the direction of the resultant V_F is along line *DD*,

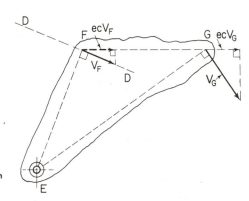

Figure 3.14 Velocities on a body in rotation

perpendicular to line *EF*. The magnitude of V_F is determined by the intersection of a perpendicular to *FG* drawn from the head end of ecV_F and the line *DD*. The sense of V_F is downward to the right, by inspection.

The rigid-body principle is most effective when applied to a body in *combined motion*, such as bar *LM* in Figure 3.15, where *O* and *P* are fixed axes. If V_L is given, let us determine V_M. Since *OL* is in pure rotation, we know that V_L is perpendicular to *OL* as shown. Since pin *L* is common to *OL* and *LM*, this V_L applies to pin *L* on both of these bodies. Applying the rigid body principle to *LM*, we lay out the ecV_L along *LM*. Since *L* and *M* have the same velocity along line *LM*, we lay out ecV_M along *LM* equal to this ecV_L. Bar *PM* is a body in pure rotation about fixed axis *P*, so it is known that the resultant V_M is perpendicular to line *PM*. The length of this resultant is determined by a perpendicular to *LM* from the head end of ecV_M. The arrowhead denoting sense of V_M is applied by inspection, consistent with ecV_M.

We have shown that the rigid-body principle for determining velocities may be applied to any rigid body in any kind of motion. It is indeed a versatile tool.

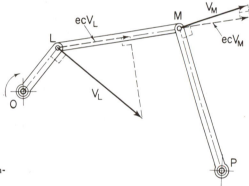

Figure 3.15 Velocities on a body in combined motion

3.12 Determining Velocities When Direction Is Unknown

In the previous cases it was possible to predict the direction of the velocity which was to be determined, since the path of the point in question was known. Let us consider the velocities of points whose path is not obvious.

The velocities of pins *R* and *T* are given in Figure 3.16, and it is desired to find the velocity of pin *S*. On bar *RS* the ecV_R along *RS* can be found. Since this is a rigid body, ecV_S along *RS* may be laid out equal to ecV_R in that direction (see vector *SA*). We know that the head end of V_S must lie somewhere on a line *AA*, a perpendicular to *RS*, drawn through the head end of this ecV_S, but since we cannot predict the direction of the resultant V_S, this construction does not completely define the vector V_S. We now examine the rigid bar *ST*, since the V_T is known and *S* is also a point on this body. Along

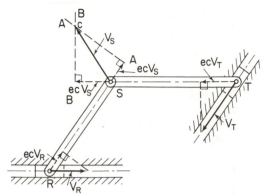

Figure 3.16 Two effective components determine a velocity

the line ST, points S and T have equal velocities, so we can predict that, along ST, ecV_S equals ecV_T. We lay this out as shown in Figure 3.16.

As in the former case, the resultant V_S will terminate somewhere on line BB, a perpendicular to ST drawn through the head end of this ecV_S. Since we know that the resultant V_S must terminate on AA and also on BB, it can only satisfy both of these requirements by terminating at c, the intersection of AA and BB. The line Sc is therefore the vector of V_S and will have the correct magnitude and inclination. The sense of this V_S is upward to the left so as to be consistent with both of its effective components.

The student should note that the effective components of V_S along RS and ST are not a mated pair. The resultant V_S was not found by completing a parallelogram, but by the intersection of two perpendiculars. The perpendicular to RS represents the mate to ecV_S along RS, and the perpendicular to ST represents the mate to ecV_S along ST. We could not determine the length of either of these mating vectors until we found their intersection. The V_S thus determined may be considered the resultant of either *pair* of *mated* effective components, but not the resultant of one member from each pair. This explains why the parallelogram construction was not used.

3.13 Expansion of the Rigid Body

This concept of the absolutely rigid body is an assumption and not a fact. It is an assumption that all points contained within the physical boundaries of a body remain at the same distances from one another when the body is subjected to forces or motion. Similarly, it is permissible to assume that certain points outside of the physical body may also remain at fixed distances from points within the body or, in other words, to imagine that the body is expanded beyond its original size. This theoretical expansion of a rigid body is often a very useful hypothesis.

In the mechanism shown in Figure 3.17, wheel W turns clockwise about fixed axis A with a given angular velocity (in radians/sec). The linear velocity of the pin F is to be determined.

We start this analysis with wheel W, determining first its class of motion. A body with one fixed point can only have a motion of rotation, so the motion of W is rotation. Next we find the linear velocities of B and G. Using $V = \omega r$,

$$V_B = \omega_W \times AB$$

and it is directed perpendicular to AB toward the right, as shown. By the same formula,

$$V_G = \omega_W \times AG$$

and it is perpendicular to AG and points toward the left. On the bar BC we can now determine the ecV_B along BC and lay out ecV_C equal to it in the same direction, employing the rigid-body principle. Since we cannot predict the direction of the absolute velocity of C, we cannot obtain V_C directly. Next we consider bar GE. Along line GE on this bar $ecV_G = ecV_E$, and since E can only move along the fixed guides, we know that the resultant V_E is in that direction. A perpendicular from ecV_E thus determines V_E. Now on bar ED we know that, in the direction ED, $ecV_E = ecV_D$, but since we cannot determine the direction of the motion of D, we cannot find the resultant V_D from this effective component. On the bar CDF we now have one effective component at C and one at D. In the previous example we found that two effective

Figure 3.17 Velocities on an expanded rigid body

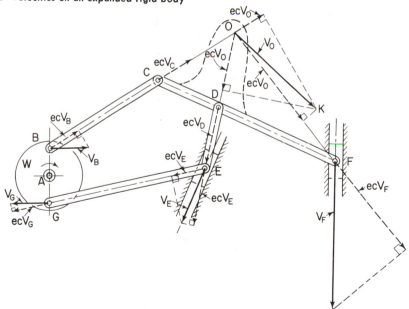

components of the velocity of a single point are needed in order to determine the resultant velocity of that point, when the direction of motion is unknown. We must then find a point on CDF at which we can establish two effective components.

If we enlarge body CDF, as shown (dotted) in Figure 3.17, so as to include a point O, at the intersection of lines BC and ED extended, we have such a point. Since points C and O are on the same rigid body, $ecV_C = ecV_O$ in the direction CO. Similarly, D and O are on the same body, so ecV_O along DO is made equal to ecV_D. We now have two effective components of the velocity of O and can determine the resultant velocity of O by erecting two perpendiculars to these components intersecting at K, as shown in Figure 3.17. Points O and F both lie on this same expanded rigid body, so $ecV_O = ecV_F$ along OF. Since the direction of V_F is determined by the fixed guides, a perpendicular to the ecV_F will define the desired resultant V_F.

3.14 Relationship of Velocities on a Rigid Body

The power of the rigid-body principle in velocity analysis has been illustrated in the preceding studies. Let us summarize these findings in a general example.

Figure 3.18 shows any typical rigid body in motion. The velocity of one point A on this body is known, and the body may be in translation, rotation, or combined motion. Let us explore the velocities of other points on this body. We know that *any two points on the same body have the the same amount of*

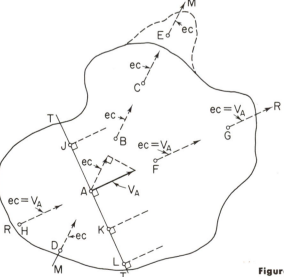

Figure 3.18 Velocity relationships on a rigid body

velocity along the line connecting them. If we consider any line through point *A*, such as *MM*, we can predict that points *B*, *C*, *D*, and even *E* (if the body is expanded) will all have effective components along *MM* equal to ecV_A along *MM*. In fact, this is true for all points on line *MM*. By the same principle, if we draw line *RR* through point *A* in the direction of V_A, points *F*, *G*, *H*, and all other points on this line will have effective components along *RR* equal to V_A.

If we consider a third line *TT* drawn through point *A*, perpendicular to V_A, we discover that points *J*, *K*, *L*, and all other points on this line will have zero velocity in the direction *TT*, since V_A has zero effective component along *TT*. Furthermore, we can state that the resultant velocities of *J*, *K*, *L*, and all other points along *TT* must, like V_A, be perpendicular to line *TT*. This is true since a vector can only have a zero effective component in a direction perpendicular to itself.

3.15 Instant Center of Rotation   (Symbol: *I*)

A body in pure rotation is turning about a fixed axis, or permanent center of rotation, which has zero velocity at all times.

A body in combined motion is rotating and translating at the same time. In Article 2.4 we noted that, as a body moves from one position to another, it may be assumed to turn about any convenient point on the body. Its angular displacement remains the same, but its linear displacement varies with each choice of rotation center. It would appear that, if we selected just the right point as a center of rotation, the attendant translation might be reduced to zero so that the change of position could be brought about by rotation alone.

For example, the bar *AB* in Figure 3.19(a) may be moved to position A_1B_1 by a rotation about point *C* to A_0B_0 and then translation to A_1B_1. The same change of position may be accomplished by a single motion of rotation about point *D*, as shown in Figure 3.19(b). The bar *AB* is assumed to be en-

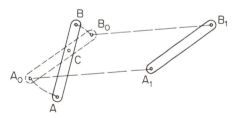

Figure 3.19(a) Successive rotation and translation

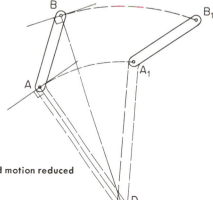

Figure 3.19(b) Combined motion reduced to rotation

larged to include point D which lies at the intersection of the perpendicular bisectors of straight lines between AA_1 and BB_1.

This concept reduces the combined motion of AB to simple rotation in which both A and B travel circular arcs about the stationary center D. The direction of the motion of A is along a straight line tangent to the arc AA_1, and the direction of the motion of B is along a tangent to the arc BB_1. We note that point D lies at the intersection of the two perpendiculars to these tangents from A and B.

At this stage our velocity studies are concerned with one specific position of the mechanism. The velocities which we determine are instantaneous velocities good only for this one position in which the mechanism is being examined. It would greatly simplify our velocity analysis if we could locate the center about which a body in combined motion could be assumed to rotate at a given instant. Point D, described above, is such a center and may be called the *instant center of rotation*. Like all axes of rotation, it is that point on a moving body which has zero velocity.

Figure 3.20 shows a body W upon which the velocities of two points, A and B, are known. If we draw line PP through A, perpendicular to V_A, the rigid-body principle tells us that all points on this line on body W have zero velocity in the direction PP. If we draw another line NN through B, perpendicular to V_B, we likewise know that all points on this line have zero velocity in the direction NN. Let us consider the point I, where PP and NN intersect. I is a point on body W (expanded) which has zero velocity along PP and also

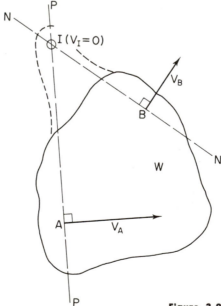

Figure 3.20 Instant center of rotation

zero velocity along *NN*. The resultant velocity of point *I* is therefore zero, since there is no vector which can be drawn from *I* in any direction which will have zero effective component in two different directions. *If a body is in motion and one point on it is stationary, the only possible motion of the body is rotation about the fixed point.* At this instant point *I* has no velocity and other points on *W* do have velocity, so the only possible motion of the body is rotation about *I*. Accordingly we can call *I* the instant center of rotation of body *W*. Since the entire body *W* is turning at this instant about *I*, the velocities of *A*, *B*, and all other points on *W* will be directed perpendicular to lines joining them with point *I*, and their magnitudes will be directly proportional to their distances from *I*, as is the case with all bodies in rotation. Thus, by locating the instant center *I*, we can determine the linear velocities of all points on that body without the use of effective components. This method is simple and fast and eliminates considerable drafting.

3.16 Locating Instant Centers of Rotation

In general, to locate the instant center of a body we need only to know the directions of the velocities of two points on that body. From each point we draw a line perpendicular to the velocity vector of that point. The intersection of these perpendiculars is the instant center of rotation of the body.

For example, in Figure 3.21, if the angular velocity of crank *AB* is given, let it be required to find the linear velocity of *E* on bar *BCE*. We know that, since *AB* is in pure rotation, V_B must be perpendicular to *AB*. The velocity of *C* is directed along the center line *SS* of the fixed guides. The instant center of body *BCE* will be located at the intersection of a line through *B* perpendicular to V_B (line *AB* extended in this case) and a line through *C* perpendicu-

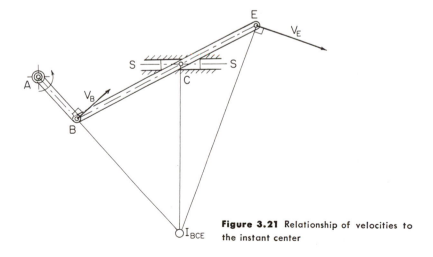

Figure 3.21 Relationship of velocities to the instant center

lar to *SS*. This intersection is shown labeled I_{BCE}. (The subscript $_{BCE}$ is used to denote clearly the body under consideration, as I does not fall within the boundaries of the bar *BCE* in this case.)

It is important to use the subscripts, as each instant center applies to only one body and the centers of several bodies may be required in one analysis. The direction of V_E is now determined, perpendicular to line *IE*, and its magnitude is proportional to V_B directly as their distances from I and may be calculated:

$$\frac{V_E}{V_B} = \frac{IE}{IB}$$

and (multiplying both sides by V_B)

$$V_E = V_B \times \frac{IE}{IB}$$

But $V_B = \omega_{AB} \times AB$, so

$$V_E = \omega_{AB} \times AB \times \frac{IE}{IB}$$

The sense of V_B denotes clockwise rotation of *BCE* around I, so the sense of V_E must likewise denote clockwise rotation about I.

In some cases, where the two known velocities on a body are parallel to one another, we must know the magnitude of these velocities as well as their direction in order to find the instant center.

In Figure 3.22 the velocities of points F and G on body M are known, and it is desired to find the velocity of point J. As shown, points F, G, and J are on the same line *TT* and V_F and V_G are both perpendicular to this line.

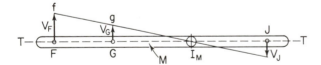

Figure 3.22 Relationship of parallel velocities to instant center

The use of effective components would not give a solution in this case, so we will locate the instant center of the bar. Perpendiculars drawn to V_F and V_G will be the same line *TT*, so there is no intersection to locate the instant center. We only know that it lies somewhere on *TT*. However, we also know that V_F and V_G are proportional in magnitude to the distances of F and G from the instant center, or:

$$\frac{V_F}{V_G} = \frac{IF}{IG}$$

This relationship suggests a simple graphical method of locating I. If we draw a straight line through the head ends of V_F and V_G (f and g) it will intersect TT at I. This line forms two similar triangles FfI and GgI in which corresponding sides are proportional so that:

$$\frac{Ff}{Gg} = \frac{V_F}{V_G} = \frac{IF}{IG}$$

as stated above. Since M is at this instant rotating about I, we can predict that V_J is perpendicular to IJ and can calculate the magnitude of V_J as follows:

$$\frac{V_J}{V_G} = \frac{IJ}{IG} \quad \text{so} \quad V_J = V_G \times \frac{IJ}{IG}$$

(The lengths IJ and IG must be scaled from the layout.) Graphically we could define V_J by extending the line fg, through I, to meet a perpendicular to TT from J. In any case, the sense of V_J must be such as to produce clockwise rotation about I.

It must be clearly understood that an instant center of rotation of a body is a point on that body, whether it falls within the physical boundaries of the body or outside them. It is usually cumbersome to show the body expanded to include its instant center, but the subscripts added to the label I are adequate for identification.

Since the locations of points on a body change position as the body moves and the velocity vectors change inclination, it follows that the position of the instant center is different for each successive position that the body assumes. As the name implies, an instant center is only valid for one position of a body in combined motion.

It may have occurred to the reader that the location of an instant center may be very remote when a body assumes certain positions. If the center were 50 ft away from the mechanism, it would be impossible to show it on a drawing of convenient size. This would limit the use of this valuable tool quite seriously if there were no remedy. However, as the study of velocities progresses, we will show several methods of overcoming this difficulty and alternate solutions for special cases. No one tool is effective for all applications. Resourcefulness and ingenuity are constantly demanded in all engineering analyses and designs.

3.17 Angular Velocity of Bodies in Combined Motion

A body in combined motion has both rotation and translation, but we may evaluate the angular motion of the body apart from its translation. All lines on a rigid body have equal angular displacements in a given time interval and therefore have equal angular velocities at all times, whatever class of motion they undergo.

In many problems it is important to ascertain the angular velocity of bodies in combined motion. If we can locate the instant center of rotation of a body, it is very easy to calculate its angular velocity.

In the four-bar linkage in Figure 3.23 the driving crank OP turns about fixed center O with a given angular velocity. R is also a fixed axis. It is required to find the angular velocity of the connecting rod PT.

Since OP is in pure rotation, V_P is perpendicular to OP. Likewise, RT rotates about fixed center R, so V_T is perpendicular to RT. The instant center I of PT is at the intersection of perpendiculars to V_P and V_T, drawn from P and T as shown. Bar PT may be considered to rotate about I at this instant, and on all rotating bodies $V = \omega r$, so

$$V_P = \omega_{IP} \times IP,$$

or solving for the angular velocity:

$$\omega_{IP} = \frac{V_P}{IP}$$

I is a point of zero linear velocity on rod PT expanded and $\omega_{IP} = \omega_{PT}$ (all lines on a rigid body have equal angular velocities). Therefore, substituting:

$$\omega_{PT} = \frac{V_P}{IP}$$

V_P may be evaluated, since $V_P = \omega_{OP} \times OP$. Therefore,

$$\omega_{PT} = \frac{\omega_{OP} \times OP}{IP}$$

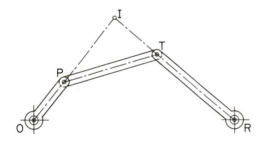

Figure 3.23 Rotation about the instant center

3.18 Absolute and Relative Linear Velocities

Two cars, A and B, traveling in the same direction, are shown in Figure 3.24(a). V_A (the velocity of A) is 40 mph, and V_B is 60 mph. Since these velocities are measured with reference to a fixed point on the road, they are called *absolute* velocities. It is desired to find the velocity of B relative to A (Symbol: $V_{B/A}$).

If we imagine ourselves riding in car A we can visualize the velocity of B relative to A. We will see B passing us in the same direction at 20 mph. $V_{B/A}$ is therefore the difference between the absolute velocities V_B and V_A, or

$$V_{B/A} = V_B - V_A.$$

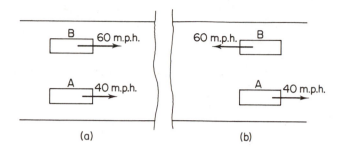

Figure 3.24 Relative velocities on parallel paths

If car B were traveling 60 mph in the opposite direction [Figure 3.24(b)], we would see it passing us at 100 mph in the opposite direction. This can still be considered the difference between the absolute velocities of B and A if we denote sense by $+$ or $-$ signs. If velocities toward the right are denoted as positive and toward the left as negative, in this second case

$$V_{B/A} = (-V_B) - (+V_A) \quad \text{or} \quad V_{B/A} = -V_B - V_A = -100 \text{ mph}$$

We conclude that when both bodies are traveling *parallel* paths their relative velocity equals the *algebraic difference* of their absolute velocities.

It is also useful to determine relative velocities when bodies are not moving in parallel paths. Figure 3.25 shows cars A and B traveling along diverging roads. If we were in car A we would see car B departing in a general direction sloping upward to our left. The $V_{B/A}$ is still the difference of absolute velocities, but since their paths are not parallel, it is a *vector difference* and not algebraic.

To add two vectors, we lay them out from an origin O as adjacent sides of a parallelogram, complete the parallelogram, and draw its diagonal from O, which equals the vector sum. A simpler method of vector addition, recommended for graphical work in Article 3.8, required the layout of the two vectors in succession, starting from an origin O. The resultant vector in this case originated at O and terminated at the head of the second vector, geometrically equivalent to the parallelogram construction (see Figure 3.8).

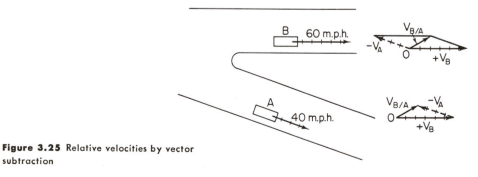

Figure 3.25 Relative velocities by vector subtraction

Vector subtraction is performed in the same manner as addition except that the vector to be subtracted is laid out pointing opposite to its actual direction, or with negative sense. In other words, to subtract, we add a negative vector. Both the parallelogram and the triangle method for subtracting V_A from V_B are shown in Figure 3.25. When the velocities are not parallel, the relative velocity equals the vector difference:

$$V_{B/A} = V_B - \rightarrow V_A \quad \text{(the symbol } - \rightarrow \text{ means "minus vectorially")}$$

3.19 Relative Velocities on a Rigid Body

Velocities of points A and B on the same rigid body are shown in Figure 3.26. Let us investigate the velocity of B relative to A. If we imagine ourselves to be standing on A, since the distance between A and B does not change, B will appear to move in a circular path (with radius AB) about us. This is the only motion B can have about A. Therefore $V_{B/A}$ must be perpendicular to AB, in the direction of the motion of B relative to A. The magnitude of $V_{B/A}$ is found by vector subtraction as with all relative velocities

$$V_{B/A} = V_B - \rightarrow V_A$$

If we add $-V_A$ to $+V_B$, the resultant will be $V_{B/A}$, which is shown in Figure 3.26 to be perpendicular to AB, as predicted.

As A and B are on the same rigid body, the $ecV_A = ecV_B$ along line AB, so it follows that *two points on a rigid body have no relative velocity along the line connecting them.*

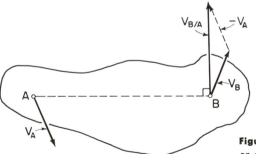

Figure 3.26 Relative velocities of points on a rigid body

3.20 Use of Relative Velocities to Find Angular Velocity

We have shown that angular velocity of a body in combined motion may be obtained by use of the instant center of rotation. This method would, of course, fail if the instant center proved to be inaccessible, as is sometimes the case.

An alternative method is made available by use of the relative velocity concept. Suppose it is desired to find the angular velocity of the body in Figure 3.26. We have observed that, relative to A, B travels a circular arc.

If B travels in a circular arc relative to A, then line AB will rotate about A with an angular velocity:

$$\omega_{AB} = \frac{V_{B/A}}{AB} \quad \left(\text{since } \omega = \frac{V}{r} \right)$$

Since line AB is fixed on the body, ω_{AB} equals the angular velocity of the body. Since the angular motion of a body is independent of any motion of translation, ω_{AB} relative to A is also the absolute angular velocity of the body. By means of the relative velocity concept, angular velocity may be measured with any point of reference as an axis.

3.21 Relative-Velocity Concept: A Tool of Analysis

To this point we have used effective-component relationships based upon the rigid-body principle as a method of determining velocities. This technique has been used to introduce velocity study because it is believed to be the clearest and most illustrative approach, particularly when graphical solutions employing vectors are used. There are, however, disadvantages in the effective-component method in that there is a great deal of drawing involved, some of which is repetitive. The scale of the vectors is limited by the size of the mechanism layout when vectors are drawn in place on the different members. The notation which should accompany these studies also becomes quite intricate.

Once the student has an initial understanding of the vector relationships, a little more sophisticated method may be used in which there is less to draw, there is an abbreviated notation, and greater graphical accuracy results from the use of larger vector scales. This new method still rests upon the rigid-body motion constraints, but we employ relative-velocity vectors instead of effective components. We also combine the vector system in a single polygon, separated from the mechanism. In summary, there are two facts which are useful in the application of the relative-velocity method:

1. If two points lie on the same rigid body, their relative velocity equals the vector difference of their absolute velocities, as shown in Fig. 3.26. Considering points A and B,

$$V_{B/A} = V_B - \rightarrow V_A$$

   When the velocity of A is known, it then follows that:

$$V_B = V_{B/A} + \rightarrow V_A \quad \text{(adding } +V_A \text{ to both sides of the equation)}$$

   In other words, the unknown V_B vector is the sum of the V_A vector and the $V_{B/A}$ vector.

2. *If two points lie on the same rigid body, their relative velocity vector is perpendicular to the line connecting the points.* This results from the fact that these points can have no relative velocity in the direction connecting them. (Their effective components in that direction must be equal.)

3.22 Constructing a Velocity-Vector Polygon

Let us consider the example of the four-bar linkage shown in Figure 3.27 in which crank KL turns clockwise about fixed axis K with a given angular velocity. It is required to determine the velocity of pin M which connects the coupler LM to the other crank MP.

Since KL is in pure rotation, the velocity of L will be perpendicular to KL and equal to $\omega_{KL} \times KL$ ($V = \omega r$). This vector is shown on the mechanism drawing (V_L).

Let us consider relative velocities on the rigid coupler LM. If $V_{M/L} = V_M - \to V_L$, then $V_M = V_L + \to V_{M/L}$. That is, if we add $V_{M/L}$ to V_L vectorially, we will obtain the desired V_M.

Starting at a convenient origin O, we lay out the V_L in a direction perpendicular to crank KL. To this we must add $V_{M/L}$. We do not know the magnitude of $V_{M/L}$, but since L and M are on the same rigid body, we know that the velocity of M relative to L must lie on a line perpendicular to LM, such as SS on the drawing. Since MP is a body in pure rotation, we know that V_M must be directed perpendicular to MP, along line TT on the drawing. Now, since V_M is a vector equal to the sum of the 2 vectors V_L and $V_{M/L}$, V_M will be the closing side of the triangle whose other two sides are V_L and $V_{M/L}$. The velocity of L is known, the velocity of M relative to L is along line SS

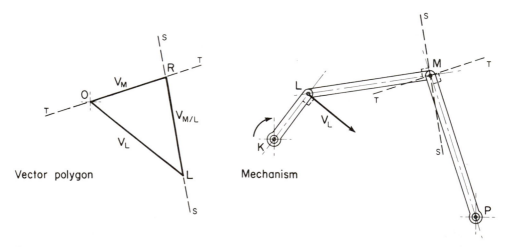

Vector polygon   Mechanism

Figure 3.27 Velocities on a four-bar linkage

and must terminate at point R, where line SS meets line TT, so that the closing side of the triangle (equal to the velocity of M) will lie along TT, which is the correct direction of V_M. Line OR is then the vector of the desired V_M, and is directed from O to R. Its magnitude may be scaled from the drawing.

This vector triangle is called a *velocity polygon*. Since it is a free vector diagram, separated from the mechanism, it may be drawn at large scale to yield accurate results. The notation may be simplified by labeling as shown in Figure 3.28. The point O is the origin of all absolute velocities so that they will all be directed away from point O. The letters at the other end of each vector will designate which velocity is represented. For example, line OL is the vector of the velocity of L (directed from O towards L) and OM is the velocity of M. The line joining L and M is the vector of the velocity of M relative to L (directed from L towards M). When reading relative velocities, the vector is directed away from the point to which the velocity is relative. Fixed points like K and P on the mechanism have zero velocity so these may be placed at the origin also. They may be thought of as vectors of zero length.

Figure 3.28 Large scale velocity polygon

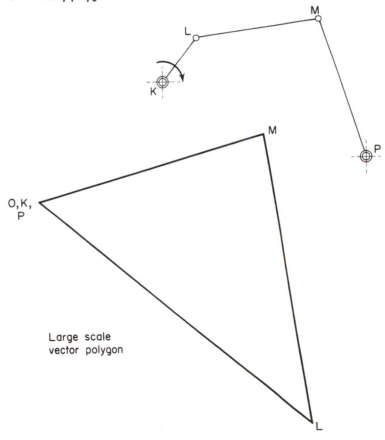

Large scale
vector polygon

This is consistent, as vector LK would be read $V_{L/K}$ (directed down to the right) and would therefore be equal to the absolute V_L, since $V_K = 0$. Note that the vectors are related in inclination to members of the mechanism, but the scale of the velocity polygon may be enlarged to increase the accuracy of the solution.

3.23 Polygon for Velocities of Unknown Direction

The linkage shown in Figure 3.29(a) is like that shown in Figure 3.16. The velocities of the blocks at R and T are given, and the velocity of pin S is to be determined in both magnitude and direction.

Let us construct a velocity polygon for this linkage. From a convenient origin O, we first lay out the given velocity of R at a suitable scale. This is shown as the line OR in Figure 3.29(b). On the rigid body RS, $V_S = V_R + \rightarrow V_{S/R}$, so to obtain the velocity of S, we must add the velocity of S relative to R to the known velocity of R. We do not know the magnitude of $V_{S/R}$, but we do know that its inclination must be perpendicular to the bar RS, since S can have no motion relative to R in the direction of RS. A line a of this inclination is drawn through R in the velocity diagram of Figure 3.29(b). The vector $V_{S/R}$ starts at point R and terminates somewhere on this line a.

Now let us lay out the given velocity of T on the diagram. This is the line OT shown in Figure 3.29(c). On the rigid body ST, $V_S = V_T + \rightarrow V_{S/T}$, so the velocity of S may be obtained by adding the velocity of S relative to T to the known velocity of T. Again, we do not know the magnitude of $V_{S/T}$, but we know that it is directed perpendicular to bar ST. This inclination of $V_{S/T}$ is shown as a line b through T and perpendicular to ST on the drawing. The $V_{S/T}$ starts as point T and terminates somewhere on this line b.

Since $V_S = V_T + \rightarrow V_{S/T}$, the vector of V_S would be the closing side of a triangle whose other 2 sides are V_T and $V_{S/T}$. The vector of V_S would originate at O and terminate at point S, which is located somewhere on line b.

If point S on the velocity polygon must then lie on line a and also on line b, this point must be at the intersection of lines a and b, as shown in Figure 3.29(d). We can now draw the vector OS, which has the correct magnitude and the exact inclination of the velocity of pin S. Like all absolute velocity vectors, it is directed from O towards S. Its magnitude may be scaled from the drawing.

3.24 Instant Centers and Velocity Polygons

It has previouly been shown (in Article 3.15) that a rigid body in combined motion may be considered at any instant to be in pure rotation about some point on the body which, at that instant, has zero velocity. This axis of

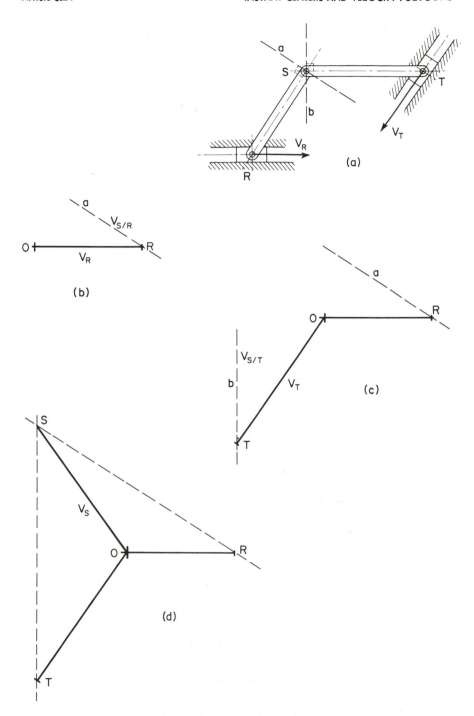

Figure 3.29 Determining direction of the velocity of pin S

rotation is called the instant center, and, once located, it simplifies determining the magnitude and direction of the velocity of any point on the body. This concept can be verified by the velocity-polygon method and is just as useful when that technique is employed.

Suppose we have a rigid body W, as shown in Figure 3.30, and suppose we know the velocities of two points, A and B. Let us assume a third point I, which is also on this same body W and is so located that it lies at the intersection of line PP, perpendicular to V_A through A, and line NN, perpendicular to V_B through B. Now, to find the velocity of point I we must draw a velocity polygon. Starting at origin O, lines OA and OB represent the given velocities of A and B. We know that $V_I = V_A + \to V_{I/A}$, where $V_{I/A}$ is perpendicular to line IA on W, whereas IA is perpendicular to V_A and OA in the polygon.

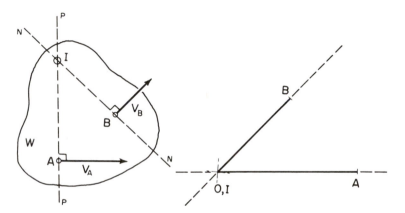

Figure 3.30 Instant center of rotation

So the vector representing $V_{I/A}$ is parallel to V_A and therefore lies along OA in the polygon. We do not know its length.

Also, $V_I = V_B + \to V_{I/B}$, where $V_{I/B}$ is perpendicular to line IB, or parallel to V_B on the rigid body. So in the vector polygon, the vector IB representing $V_{I/B}$ lies along OB, the vector of V_B.

In the polygon, the velocity of I would be shown as a vector OI which, according to the above equation, would be the closing side of a triangle composed of OA (V_A) and IA ($V_{I/A}$). This same velocity of I (OI) is the closing side of another triangle composed of OB (V_B) and IB ($V_{I/B}$). Therefore the point I on the polygon must lie on both OA and OB. Point I, then, must lie at O, the only point that OA and OB have in common, their intersection. If points I and O coincide, then the line OI has no length; the velocity of I is zero; and I is the instant center of body W.

This is significant because it describes the motion of a rigid body in a new way. *If a body is in motion and one point on the body is fixed, the body can only be in pure rotation about that fixed point*, like a wheel turning about its axis. So, in this example, body W may be considered at this instant to be in pure rotation about I. This means that the velocity of any point on W must be directed perpendicular to a line connecting it with I and that the velocities of all points on W must be proportional to their distances from I.

Instant centers of rotation, like I, can be located if we know the inclination of the velocities of two points on a body. It is only necessary to draw two perpendiculars to the velocity vectors through the points. The instant center will lie at their intersection. As a body moves in combined motion, the velocities of moving points change their inclinations, so points like I will be located in different positions for each position of the body. Thus they are properly called *instant centers* of rotation.

To find out how instant centers are used to determine the direction of velocity vectors, let us examine Figure 3.31. In the linkage shown, bar BE is a continuous rigid member. The angular velocity of AB is given, and it is required that we find the velocity of point E.

The velocity of B can be calculated ($V_B = \omega_{AB} \times AB$), and its vector OB can be laid out perpendicular to crank AB. Then we know that:

$$V_E = V_B + \rightarrow V_{E/B}$$

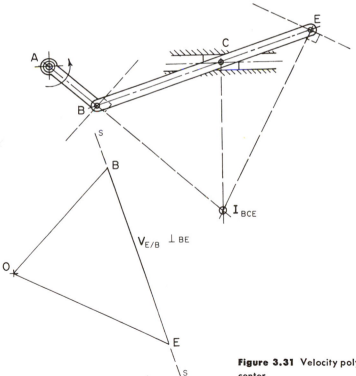

Figure 3.31 Velocity polygon and instant center

The direction of $V_{E/B}$ is perpendicular to bar BE, and its vector will pass through point B on the polygon. We do not know the magnitude of $V_{E/B}$, so we can only show that line SS has indefinite length. The closing side of the vector triangle would be OE, but since we do not know the inclination of V_E, we cannot locate point E.

Let us find the direction of V_E by using the instant center of rotation of bar BE. This will lie at the intersection of a line through B perpendicular to V_B and a line through C perpendicular to V_C. Pin C is constrained to move along the center line of the fixed guides, and B moves perpendicular to AB. The location of the instant center I is shown in Figure 3.31. Now we can draw IE and claim that V_E is perpendicular to IE.

In the vector polygon we then can draw a vector through O, parallel to V_E as determined above, to intersect line SS, thus locating point E. Vector OE represents the velocity of E and can be measured.

3.25 Translation-Rotation Components

The velocities of points A and B are known on the bar shown in Figure 3.32, and the velocity of C is required. In the direction AC, $ecV_A = ecV_C$. The direction of V_C is needed and would normally be found by locating the instant center of the bar. In this case, however, if V_A and V_B are nearly parallel, we find that perpendiculars to these velocities will intersect at a distant point, making I inaccessible unless the drawing is very large.

Let us consider other means of determining the velocity of C. Combined motion is simultaneous translation and rotation. We have noted that these motions can be considered to take place separately for convenience of analysis.

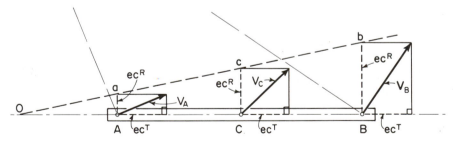

Figure 3.32 Effective components of translation and rotation

The bar AB (Figure 3.33) can move from the original position to $A_1 B_1$ by first rotating about point O (the intersection of AB and $A_1 B_1$ extended) to position $A_0 B_0$, and then translating along line $A_0 B_0$ to $A_1 B_1$. Let us examine the

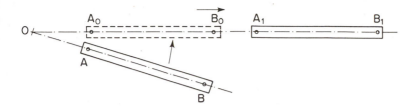

Figure 3.33 Rotation and translation in combined motion

velocities of A and B with reference to these separate motions. We may replace V_A by a pair of effective components along, and perpendicular to, AB. V_B may be similarly represented, as in Figure 3.32. The components along AB are called *translation components* (symbol: ec^T), since they define velocities of A and B as the body translates along AB. The components perpendicular to AB are known as *rotation components* (symbol: ec^R), because they define velocities of A and B as the body rotates.

The translation components along AB are known to be equal by the rigid-body principle. The rotation components also are definitely related to one another. We have observed that the velocities of points on a rotating body are proportional to their distances from the axis of rotation. In moving from AB to A_1B_1 (Figure 3.33), rotation of the body about fixed point O was required. Therefore, as the bar turns, the velocities of rotation (ec^R) of A, B, and C are proportional to OA, OB, and OC, respectively:

$$\frac{ec^R \text{ of } C}{ec^R \text{ of } A} = \frac{OC}{OA} \quad \text{and} \quad \frac{ec^R \text{ of } A}{ec^R \text{ of } B} = \frac{OA}{OB}$$

Graphically, if we draw a line from a to b, we will find that ec^R of C terminates at c on this line, as shown in Figure 3.32. In the similar triangles OAa and OCc thus formed,

$$\frac{Cc}{Aa} = \frac{ec^R \text{ of } C}{ec^R \text{ of } A} = \frac{OC}{OA} \quad \text{(as stated above)}$$

Since, when I is remote, O is also remote, this graphical method is effective as it does not require location of point O.

Now that we have determined ec^R and ec^T of V_C, perpendiculars from this mated pair of rectangular components will define the resultant V_C. The example in Figure 3.34 will illustrate the use of this method. Crank HL turns about fixed shaft H with a known angular velocity. The velocity of point S is required.

Since $V = \omega r$,

$$V_L = \omega_{HL} \times HL$$

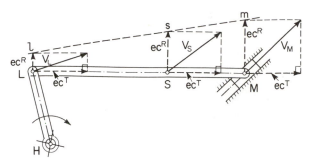

Figure 3.34 Rotational components are proportional

and is directed perpendicular to *HL*. Along *LM* the $ecV_L = ecV_M$ (labeled ec^T here) by the rigid-body rule. Since the resultant V_M is along the fixed guides, we can readily determine V_M by a perpendicular from ec^T of *M*.

Now we break V_L and V_M into rotation and translation components, ec^R and ec^T. The ec^T of $S = ec^T$ of *L* and *M*. Line *lm* defines the length of ec^R of *S*. The resultant V_S is determined as the resultant of ec^T and ec^R of *S*.

3.26 Pure Rolling Contact

When one body rolls on another without slipping, they are said to be in *pure rolling contact*. If the bodies slip at contact, the contact point on one body will move relative to that on the other. We are perhaps most familiar with the contact between a tire and the pavement in the case of an automobile. Figure 3.35 shows wheel *W* in contact with the surface *R* of the road. Point P_W on the wheel contacts P_R of the road. If slipping or "skidding" occurs, P_W will slide along the road surface so that it has velocity, while P_R, being a point on the road, remains stationary. In terms of relative velocity, we may say that if there is slipping the velocities of P_W and P_R are different, whereas if there is no slip the velocities of these contact points are identical. "No slip" means "no relative velocity between contact points," which is the case with pure rolling contact. If there is pure rolling contact between *W* and *R*, P_W has the same velocity as P_R, which is zero. Since the instant center of rotation is a point of zero velocity on a body, we note that I_W is at P_W, the contact point, at all times. Therefore, all points on *W* will have velocities perpendicular to lines joining them with I_W and proportional to their distances from I_W.

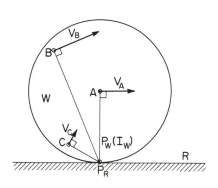

Figure 3.35 Instant center of a rolling cylinder

Wheels *D* and *F* in Figure 3.36 turn on fixed shafts *A* and *B* in pure rolling contact at C. The velocity of C on body D equals $\omega_D \times AC$ $(V = \omega r)$. The velocity of C on F equals $\omega_F \times BC$. V_C on D equals V_C on F, since there is no slip. Therefore we can write a ratio of angular velocities:

$$V_C^F = V_C^D$$

And substituting:

$$\omega_F \times BC = \omega_D \times AC \quad \text{or} \quad \frac{\omega_F}{\omega_D} = \frac{AC}{BC}$$

(dividing both sides by ω_D and BC)

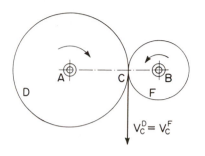

Figure 3.36 Velocities in rolling contact

Since AC equals the radius of D (R_D) and BC the radius of F (R_F):

$$\frac{\omega_F}{\omega_D} = \frac{R_D}{R_F}$$

This is called the *speed ratio* of the rolling cylinders, and it may be stated that the angular speeds are inversely as the radii. The sense of rotation (shown by the sense of V_C) is opposite on D and F.

The cam with roller follower in Figure 3.37 illustrates that all rolling bodies are not circular. The angular velocity of the cam C is given, and the angular velocity of follower arm ST is to be determined for the position shown. It is specified that there is no slip between cam C and roller F.

The contact point P relates the motion of the cam and follower. The velocity of P on C can be readily found. Cam C is a body in pure rotation about O, so V_P on $C = \omega_C \times OP$ and is perpendicular to OP. Since there is no slip, V_P on $C = V_P$ on roller F. To determine the angular velocity of ST we need to know the velocity of S. P and S are on the same rigid body, so, along the line PS, $ecV_P^F = ecV_S$. V_S is known to be perpendicular to ST (since ST rotates about a fixed axis T), so a perpendicular to ecV_S will define V_S. $\omega_{ST} = V_S/ST$, and the sense of this angular velocity must be consistent with V_S.

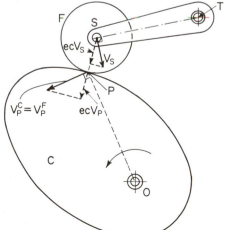

Figure 3.37 Velocities on cam with roller follower

3.27 Velocity Polygons for Pure Rolling Contact

The application of the velocity-polygon method to rolling-contact problems offers the same advantages of efficiency and precision as it does when applied to pin-connected linkages. The principles of relative velocity and the use of instant centers combine to provide an alternate method to the somewhat cumbersome effective components

In pure rolling contact, all contact points have identical velocities and therefore have no relative velocity one to the other. As we have seen, the velocity of one point on a rigid body relative to another point is directed perpendicular to the line connecting these two points, since there can be no relative velocity in the direction connecting them. When a body is in combined motion, a line drawn from any point on the body to its instant center must be perpendicular to the velocity vector of that point—since the point can have no velocity component along this line. The instant center is a point on a body which has zero velocity, so it has no component of velocity in any direction.

With these facts in mind, let us look at an example involving pure rolling contact. An *epicyclic wheel train* is shown in Figure 3.38. Wheel D turns clockwise about fixed axis B. The triangular arm A also turns about B independently of D and is designed to keep wheel J in contact with wheels D and G. Wheel G also maintains continuous contact with the fixed ring F, which is concentric with D and only partially shown here. It is required to determine the angular velocity of arm A for a given angular velocity of wheel D, assuming there is no slip at all rolling-contact points.

The velocity of point P, the contact point between wheels D and J, can be calculated, since on D, $V_P = \omega_D \times BP$, and it will be perpendicular to BP. Point P on wheel J will have the same velocity. We can draw the line OP on the vector polygon representing V_P. It is perpendicular to BP on the mechanism.

Now, on wheel J, point R will be the contact point between J and G, and the velocity of R will be the same on both wheels J and G. Again, applying the relative velocity expression, we see that $V_R = V_P + \rightarrow V_{R/P}$, where $V_{R/P}$ will be perpendicular to line PR on wheel J but its magnitude will be unknown. This is shown as line TT of indefinite length through P on the polygon.

In the vector triangle representing the above expression, one side will be V_P (line OP), a second side will be $V_{P/R}$, and the closing side will be V_R (line OR). If we could determine just the direction of V_R, we could complete this triangle. We note that R is a point on wheel G as well as on J. Point I on wheel G has the same velocity as the corresponding point I on the stationary ring F, namely zero. *A point of zero velocity on a moving body is an instant center of rotation.* Therefore, wheel G can be considered to be turning about I at this instant—establishing the direction of V_R as perpendicular to the line IR on wheel G. So, on the vector polygon, we draw a vector from O, perpendicular to IR (on the mechanism), to meet line TT at R. This defines the vector OR, which is the velocity of the contact point R on both G and J.

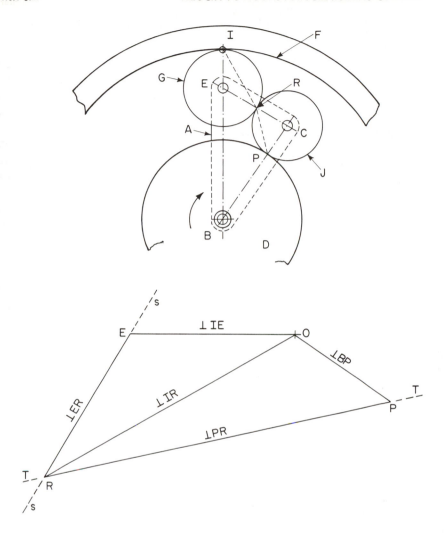

Figure 3.38 Velocity polygon for epicyclic train

To determine the required angular velocity of arm A, we need to know the linear velocity of some point on A (other than the fixed point B). Point E is probably the easiest to investigate. All points on wheel G have velocities proportional to their distances from I, the instant center. So we can compute the velocity of E from the velocity of R:

$$\frac{V_E}{V_R} = \frac{IE}{IR} \quad \text{and} \quad V_E = \frac{IE}{IR} \times V_R$$

This velocity of E is shown as the vector OE on the polygon, perpendicular to IE on wheel G and directed from right to left.

An alternate method of determining the velocity of E will check the calculated value and the sense of vector OE:

$$V_E = V_R + \ \to V_{E/R}$$

The velocity of R is known (OR on the polygon), so to this we add the vector of the velocity of E relative to R. This $V_{E/R}$ is perpendicular to line RE on the mechanism drawing. So through point R on the polygon, we draw a line SS of indefinite length, perpendicular to RE on the mechanism (we do not know the magnitude of $V_{E/R}$). With the direction of the velocity of E determined from the instant center of wheel G (above), we can draw a line through O, perpendicular to IE, which is the correct direction of the velocity of E. This line will intersect SS at point E, closing the vector triangle ORE on the polygon. This point E at the intersection should make OE equal to the calculated value used above and should verify that the velocity of E (vector OE) is directed toward the left.

Since E is a pin on the arm A as well as on wheel G, the angular velocity of A can now be calculated. As A turns about the fixed axis B,

$$\omega_A = \frac{V_E}{BE}$$

The sense will be counterclockwise, consistent with the velocity of E.

3.28 Velocities in Sliding Contact

We have studied mechanisms in which blocks were designed to slide in fixed guides. When the guides are stationary, they serve only to define the direction of motion of the sliding member and thus present no problems in velocity analysis. In some mechanisms the guides for the sliding member are themselves in motion. To determine the motion of the moving guides, we must study the displacement relationship between the sliders and the guides.

In Figure 3.39, a slotted bar M contains a slider S. The slider is free to move horizontally along the slot, but is restrained from vertical motion by the slot. If the bar translates upward to M_1, the position shown above, and at the same time the slider moves along the slot to position S_1, the displacement of M is the vertical distance d, between the center lines of the slot in the two positions. The displacement of S is the diagonal distance e, measured between the two positions of the center of the slider at C and C_1. If we consider the vertical component of e, we note that it is equal to d, the displacement of M. Since the center of S cannot depart from the center line of the slot, this equality will always hold true for any displacement of M and S.

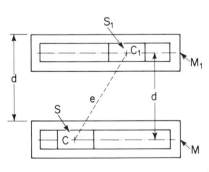

Figure 3.39 Displacement in sliding contact

In any given time interval, the displacements of M and S in the direction perpendicular to the center line of the slot will be equal. Since $V = \Delta S/\Delta T$, the velocities of M and S, in the direction perpendicular to sliding, will always be equal. Stated in more general terms, this fact may be applied to all sliding members as follows: *In the direction perpendicular to sliding, the velocities of two members in sliding contact are equal.* Armed with this principle, we are equipped to analyze velocities on bodies in sliding contact.

A well-known mechanism called the *Scotch yoke* is shown in Figure 3.40. It consists of a slotted yoke Y restrained in vertical bearings to translate up and down. The yoke is driven by a crank in rotation about fixed shaft O, on which point P is pinned to a block B sliding freely in the slot in Y.

With the angular velocity of OP known, it is required to determine the velocity of Y for the position shown. Pin P is attached to crank OP, which is in pure rotation. The velocity of P on OP is given by $V = \omega r$:

$$V_P^{OP} = \omega_{OP} \times OP$$

Pin P is also attached to block B, therefore,

$$V_P^{B} = \omega_{OP} \times OP$$

We must now relate the velocity of P on the block with the velocity of some point on Y in order to determine the velocity of Y. If the yoke Y is constructed as shown in Figure 3.40(a), so that the slot in Y has a solid back, there is a point on the back of Y lying on the center line of the pin P directly in back of point P on the block. Let us compare the velocity of this point P^Y with the velocity of the point P^B on the block. In Figure 3.40(b) these two points will lie one over the other. The sliding velocity rule states that, in the direction perpendicular to sliding, or perpendicular to the center line SS of the slot, points P^B and P^Y have the same velocity. Now

$$V_P^{B} = V_P^{OP}$$

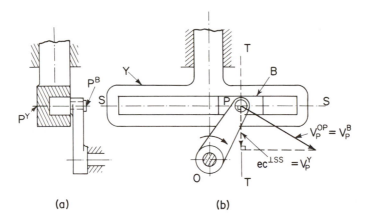

(a)   (b)

Figure 3.40 Velocities on a Scotch yoke

which is perpendicular to OP, so the velocity of P^B in the direction perpendicular to SS is the effective component of V_P^B along line TT.

Therefore, ecV_P^B along $TT = V_P^Y$ along TT. Since Y is in translation, all points on Y have identical velocities and these velocities are in a vertical direction. Thus, V_P^Y along TT is the resultant velocity of point P on Y or, in fact, the resultant velocity of the entire yoke Y.*

$V_Y = ecV_P^B$ along TT, and the sense will be downward, as shown in Figure 3.40(b).

Figure 3.41 shows a crank AB turning with a given angular velocity about fixed axis A. A slider S is pinned to AB at B and slides freely in the slot cut in the bent bar N, which turns about fixed pin C. It is required to find the angular velocity of bar N.

Since AB is in rotation,

$$V_B^{AB} = \omega_{AB} \times AB \quad \text{(with direction perpendicular to } AB\text{)}$$

But pin B is a pin on block S as well as on AB, so

$$V_B^{AB} = V_B^S$$

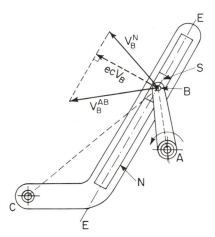

Figure 3.41 Velocities on a rotating slider

since pin B is common to both members. As before, let us consider a second point B, located on bar N, directly beneath point B on the slider S. Points B on S and B on N have identical amounts of velocity in the direction perpendicular to sliding (or perpendicular to EE). Since N is in rotation, V_B^N must be perpendicular to CB. Neither V_B^S nor V_B^N is perpendicular to the center line EE of the slot, but the rule states that the effective velocities of B^N and B^S perpendicular to EE are equal. Since V_B^S has been obtained, the ecV_B^S perpendicular to EE can be determined as shown. This $ecV_B^S = ecV_B^N$, and, with the direction of V_B^N known, we can find V_B^N by erecting a perpendicular from ecV_B^N to the line of action of V_B^N, as shown. Now the angular velocity of N may be found by $\omega = V/r$:

$$\omega_N = \frac{V_B^N}{CB}$$

with sense counterclockwise as dictated by the sense of V_B^N.

A cam with flat follower offers another example of sliding-contact velocity analysis. Cam C turns counterclockwise about fixed shaft A in Figure 3.42, with a given angular velocity, driving the flat follower F through contact at P. The velocity of F is desired for the position shown.

*When the angular velocity of the crank (OP) is constant, the motion of the yoke (Y) is described as simple harmonic motion.

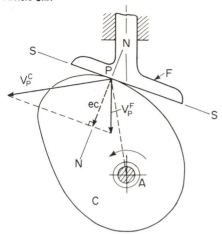

Figure 3.42 Velocities on a cam with flat follower

As before, we relate the velocities of point P on C and P on F. Cam C is in rotation, so

$$V_P^C = \omega_c \times AP$$

and is perpendicular to AP. Sliding takes place along surface SS of F. Follower F is in translation in a vertical direction, as dictated by the fixed bearings in which it slides. Points P^C and P^F have the same amount of velocity in the direction NN, perpendicular to SS, so $ecV_P^C = ecV_P^F$ along NN. Point P^F, like all points on F, is constrained to move only in a vertical path, so ecV_P^F defines the resultant V_P^F, which equals V_F.

3.29 Velocity of Sliding

When a body A slides on another body B (Figure 3.43), A moves along the surface of B. In order to do this, A must have motion relative to B. If A and B had identical velocities, they would move together at the same speed and there would be no sliding between them. If A is to slide on B, their absolute velocities must be different in magnitude, inclination, or sense, or any combination of these. The difference of absolute velocities has been defined as

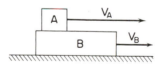

Figure 3.43 Velocity of sliding on translating bodies

relative velocity. This relative velocity of two bodies which move in contact is called the velocity of sliding. It is a very significant velocity in many machine design problems.

The velocity of sliding between A and B is the velocity of A relative to B, or $V_A - V_B$. Since A and B have parallel velocities, $V_{A/B}$ is an algebraic difference in this case.*

   *$V_{A/B} = V_{B/A}$ in magnitude and differs only in sense. With velocities of sliding we are usually interested only in the magnitude, or possibly inclination, of the relative velocity and need not distinguish as to which body slides on which.

3.30 Rolling or Sliding?

When two moving bodies touch one another, they are either in pure rolling or sliding contact, and the first step in a velocity study is to identify the nature of this contact. This is easily done if we examine the velocities at the contact point. If both bodies have identical velocities at the contact point, they are in pure rolling contact. If the velocities of the contacting points on each body differ in any respect (magnitude, inclination, or sense), they are in sliding contact.

3.31 Determining Velocity of Sliding

In Figure 3.44, arm A turns clockwise about fixed shaft D, causing body B to turn about fixed axis E, through contact at C. The center of the circular portion of B is at O. The velocity of contact point C on arm A is perpendicular to DC and equal to $\omega_A \times DC$ (see vector V_C^A). The velocity of contact point C on body B is perpendicular to EC (see vector V_C^B). These two contact velocities have different directions, so, whatever their magnitudes, we have sliding contact at C.

Let us find the velocity of sliding. V_C^A and V_C^B have the same effective component along NN, perpendicular to sliding (ec on the drawing). From this

Figure 3.44 Velocity of sliding on rotating bodies

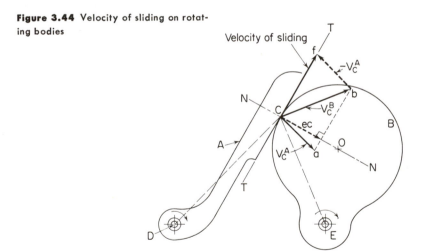

common component, we establish the magnitude of V_C^B in the usual manner. The velocity of sliding is the velocity of C on B relative to C on A (or vice versa). This is the *vector difference*, $V_C^B - \rightarrow V_C^A$, since these velocities are in different directions.

To find this relative velocity we lay out a negative V_C^A from the head end of V_C^B. (A negative V_C^A has the opposite sense to the actual V_C^A). The vector drawn from C to the head of $-V_C^A$ is the relative velocity of sliding. Note that, as V_C^A is parallel and equal to $-V_C^A$, the figure $Cabf$ is a parallelogram.

Therefore, f lies on the line TT which is an extension of the contacting surface of A. TT is tangent to body B at C, so it is established that the velocity of slip lies along the tangent to the contacting surfaces. This will always hold true, whatever the size of the contact point velocities.

In summary, the velocity of sliding is the relative velocity of the contact points and is directed along the common tangent.

3.32 Velocity Polygons for Sliding Contact

The advantages of the simplicity and large-scale precision which are provided by the vector-polygon method may be gained by the adaptation of this technique to sliding-contact problems. In this application we exploit the velocity of slip, which is the same as the velocity of sliding. This relative-velocity concept replaces the effective component perpendicular to sliding.

When two bodies are in sliding contact, the contact points have different velocities on each body. The velocity of slip is the difference between the two contact-point velocities, but this difference is equal to the velocity of one contact point minus the velocity of the other contact point. If the contact points are A and B:

$$V_B - \rightarrow V_A = V_{\text{slip}}$$

and solving for the velocity of B:

$$V_B = V_A + \rightarrow V_{\text{slip}}$$

In Figure 3.45, a cam C turns counterclockwise about fixed shaft A and maintains continuous contact with flat follower F. The follower slides up and down in fixed guides. Let it be required to determine the velocity of follower F when the cam is in the position shown. (This mechanism is similar to that of Figure 3.42.)

The contact point P on cam C (P^C) has a velocity directed perpendicular to line AP and equal to $\omega_C \times AP$. This is shown on the vector diagram as the line OP^C from a convenient origin O. The corresponding point P on the follower F (P^F) has a velocity parallel to the vertical follower slide, so the velocities of the contact points differ and sliding takes place.

The direction of sliding is tangent to the cam's surface at P, or along the flat surface of the follower (see line SS). In this case the relative-velocity expression for the two contact points (P^C and P^F) is:

$$V_P^F = V_P^C + \rightarrow V_{\text{slip}}$$

The velocity of slip is along the common tangent SS, through P.

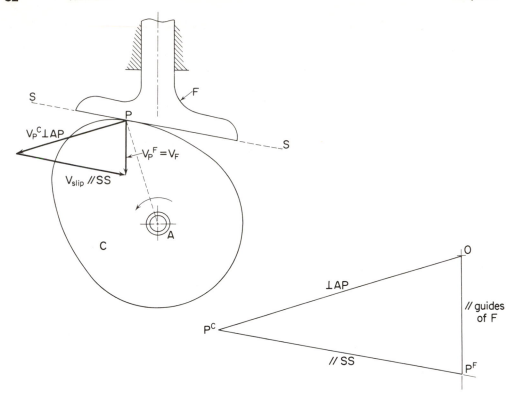

Figure 3.45 Velocity polygon for a cam

On the vector diagram we have the V_P^C shown as OP^C, laid out to scale. We can add to this a line through P^C parallel to SS. This represents the direction of the V_{slip}, whose magnitude is not yet determined. The sum of these two velocities (V_P^C and V_{slip}) equals V_P^F, according to the above expression. So the vector OP^F must be the third side of the vector triangle. Since the velocity of P on F is parallel to the follower guides, this vector is drawn in this direction through O to meet the V_{slip} vector at P^F. (This vector triangle is also shown on the mechanism for explanatory purposes, but this is not necessary to the solution.) The magnitude of the velocity of F (V_P^F) equals the length of the vector OP^F. The velocity of slip may also be scaled from the polygon by measuring the vector $P^C P^F$, which is the relative velocity of the two points P.

A second example of a sliding-contact mechanism (similar to Figure 3.41) is shown in Figure 3.46. Crank AB turns counterclockwise at known speed about fixed axis A. Slider S is pinned to the crank at B and slides in the slot in the bent arm N. This slotted arm turns about a fixed axis at C. Let it be required to find the angular velocity of arm N when N is in the position shown.

The velocity of pin B on crank AB can be calculated: $V_B^{AB} = \omega_{AB} \times AB$. Since this pin B connects crank AB to slider S, the velocity of B on AB is equal to the velocity of B on S. See vector OB^S, drawn from origin O in the

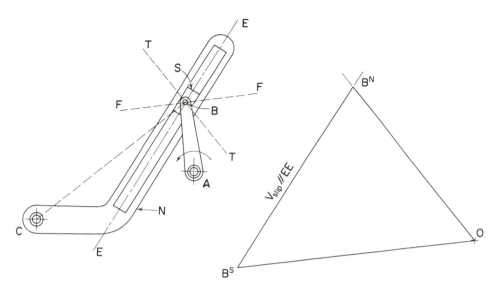

Figure 3.46 Velocity polygon for sliding contact

direction perpendicular to crank *AB* (parallel to line *FF* on the mechanism drawing).

Let us now consider the velocity of a second point *B* which lies directly beneath the pin *B* on the slider, but which is a point on the arm *N*. The velocity of pin *B* on *S* relative to the coincident point *B* on *N* is called the velocity of slip. It is equal to the difference of the velocities of the two points *B* and in a direction parallel to the centerline of the slot *EE*.

We can rewrite $V_{\text{Slip}} = V_B^N - \rightarrow V_B^S$ in the following form:

$V_B^N = V_B^S + \rightarrow V_{\text{Slip}}$, since it is V_B^N that we wish to find.

Now, following this expression in the vector diagram, we may add to the V_B^S (vector OB^S) a vector representing the velocity of slip. We do not know the magnitude of V_{Slip} but we do know that it must be parallel to line *EE* which is the direction of sliding.

The closing side of this vector triangle will be V_B^N, the sum of the V_{Slip} and V_B^S. Again we do not know the magnitude of this vector V_B^N, but we do know its direction. Like all points on *N*, the velocity of *B* on *N* will be perpendicular to a line connecting B^N to the fixed axis of rotation, *C*. The vector representing V_B^N will be called OB^N on the vector triangle and will be drawn through *O* perpendicular to *CB* (or parallel to line *TT*) on the mechanism. This will intersect the V_{Slip} vector at point B^N, thus establishing the length of both OB^N (V_B^N) and $B^S B^N$ (V_{Slip}).

This vector OB^N may be scaled to determine the velocity of point *B* on arm *N*. To find the required angular velocity of *N*, we divide V_B^N by the distance *CB*.

$$\omega_N = \frac{V_B^N}{CB}$$

PROBLEMS

Many of these problems are designed to be solved either by calculation or by graphical methods. In all cases the graphical solution is faster and shorter and is recommended. The accuracy of the graphical layout is dependent upon the precision of linework, measurement, and a wise choice of scales for the vectors, as stated in Chapter 1. If calculations are used, most of the triangles to be solved are right triangles and the Pythagorean theorem* will apply. In many cases these triangles involve common angles or integral ratios of sides, as shown in Figure 3.47. The student should seek to recognize and take advantage of this simplification. Calculations should always be guided by free-hand sketches which point out the relationships.

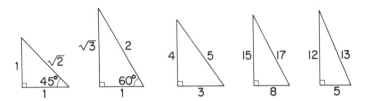

Figure 3.47 Ratios of sides of common right triangles

If the problems are done graphically, the vectors may be drawn in place on the mechanism or the vector-polygon method may be used. Since the latter method employs a free-vector diagram, separated from the mechanism, larger velocity scales may be used, which will contribute to the accuracy of solution. The choice of method may be designated by the instructor.

The author suggests that it is practicable in initial assignments, where the emphasis is on learning the method rather than obtaining accurate solutions, to sketch the vector systems entirely free-hand, establishing the scale, relative sizes, and inclinations of the vectors by eye. Answers determined by sketching are of practically no value, but the method may be mastered very efficiently if scaling and instrument manipulation is eliminated. Precision drawing can be postponed until the student is competent to attempt more ambitious problems. If precision instrumental drawings are used, much layout time may be eliminated by "skeletonizing" the mechanism as has been suggested earlier in the text.

One detail is extremely important whatever technique of solution is used: Careful and thorough labeling of vectors, components, and instant centers is as essential to the student doing the problem as it is to the instructor correcting it.

*This theorem states that the square of the hypotenuse of a right triangle equals the sum of the squares of the other two sides.

3.1. A jet plane flew nonstop from New York to Madrid, 3330 miles, in 6 hr 48 min. What was its average speed for the trip? The return trip was not flown over the same course, in an effort to avoid bad weather, and required 7 hr 36 min. If the average speed on the return trip is estimated at 450 mph, how many miles were traveled?

3.2. Wheel W, 12 in. in diameter, turns clockwise at constant speed about fixed axis O at its center (Figure P3.2). The velocity of point C is 84 in./min. What is the angular velocity of W, the surface speed of the circular part of the wheel, and the angular velocity of the flat side AB. Dimensions are given on the drawing. Report angular velocities in radians/sec.

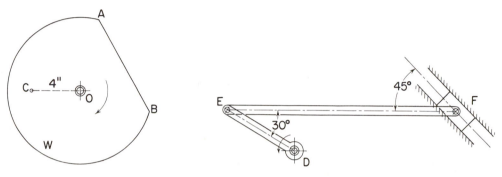

Figure P3.2   **Figure P3.3**

3.3. Crank DE, 2 in. long turns counterclockwise about fixed shaft D at 10 radians/sec. Block F slides freely in the fixed guides. What is the velocity of F in the position shown in Figure P3.3? Show vector to indicate sense.

3.4. In the linkage shown in Figure P3.4, A and E are fixed axes. DBC is one rigid bar. $AB = DB = BC = BF = FE = 3$ in. If the velocity of D in the position shown is 40 in./sec, toward A, determine the linear velocity of C and the angular velocity of FE.

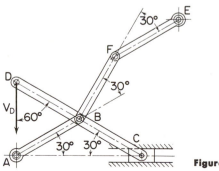

Figure P3.4

3.5. The linkage in Figure P3.5 is driven by crank *OR* turning counterclockwise about fixed shaft *O* at 70 radians/min. Member *STL* turns about fixed axis *T*. Dimensions are given on the drawing. Determine the angular velocities of *STL* and *MP* in the position shown.

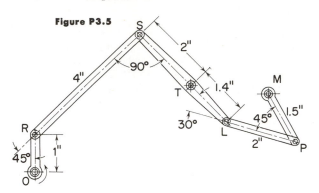

Figure P3.5

3.6. *AB* turns clockwise about fixed axis *D* at 1 radian/sec. $AD = BD = 1$ in., $BC = 5$ in., and the slopes of *AE* and *CE* are shown in Figure P3.6. Determine the linear velocity of *E* for the position shown.

Figure P3.6

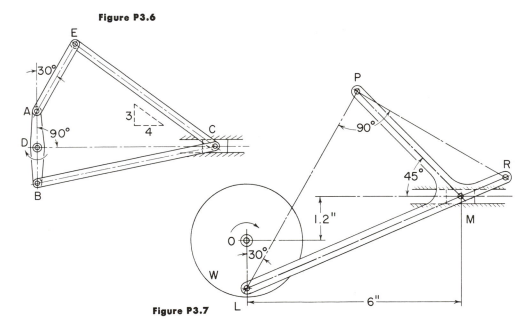

Figure P3.7

3.7. Wheel *W* turns clockwise 10 radians/sec about fixed axis *O*. Points *L, P, R,* and *M* are all on one rigid body, which is pinned to *W* at *L* and to the block sliding in fixed guides at *M*. $OL = 1.3$ in., $LM = 6.5$ in., and other dimensions and angles are given in Figure P3.7. Find the linear velocities of points *P* and *R*.

3.8. In the linkage shown in Figure P3.8, arm *ABC* turns clockwise about fixed axis *A* at 0.5 radians/sec. $AB = 2$ in., $BC = 3$ in., $CD = 3$ in., $BE = 2$ in., F is the mid point of bar DE, $FG = 3\frac{1}{2}$ in., $AG = 4\frac{1}{8}$ in., *G* is a fixed axis. Find the linear velocity of *F* and the angular velocity of *GF*.

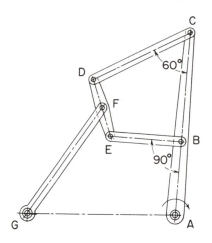

Figure P3.8

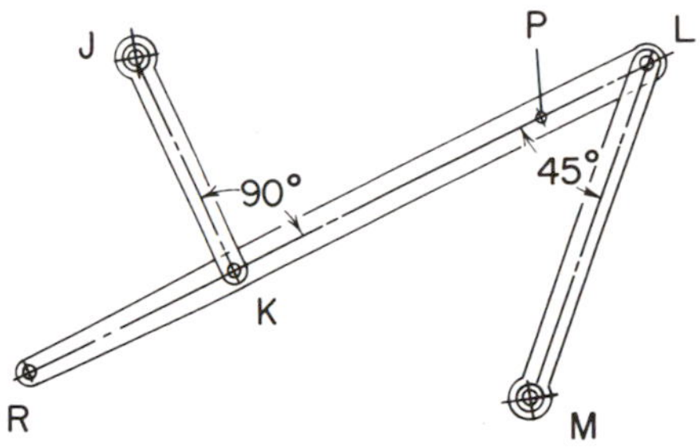

Figure P3.9

3.9. In Figure P3.9, *J* and *M* are fixed axes, *RL* is one continuous bar, $JK = 2$ in., $LM = 3$ in., $RK = 2$ in., $KP = 3$ in., $PL = 1$ in. Locate the instant center of rotation of *RL*. Find the linear velocity of points *P* and *R* (in the position shown) if the angular velocity of *JK* is 10 radians/sec.

3.10. In Figure P3.10 *ST* turns about fixed axis *S*. *RT* is prependicular to *ST* and *RW*. The block at *W* slides in fixed guides perpendicular to *RW* with a velocity of 3 in./sec, downward, when in the position shown. $ST = 1$ in., $TR = 2$ in., and $RW = 6$ in. Locate the instant center of rotation of *RW* and find the angular velocity of *RW*.
(a) When *ST* turns clockwise at 1 radian/sec.
(b) When *ST* turns counterclockwise at 1 radian/sec.

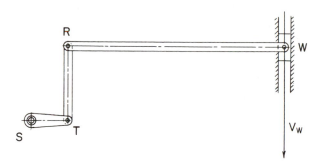

Figure P3.10

3.11. A four-bar linkage in Figure P3.11 has fixed axes at K and G. $HJ = 5$ in., $GH = 2.5$ in., and $JK = 3$ in. Angles are shown on the drawing. If the angular velocity of $GH = 4$ radians/sec, find the angular velocities of KJ and HJ when in the position shown.

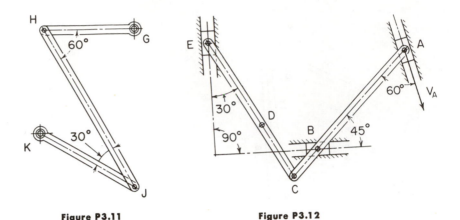

Figure P3.11   Figure P3.12

3.12. Two rigid bars, ABC and CDE, are pinned together at C and pinned to blocks sliding in fixed guides at A, B, and E, as shown in Figure P3.12. $AB = 5.5$ in., $BC = 1.5$ in., $CD = 2.5$ in., $DE = 4$ in. Angles are given on the drawing. In the position shown, the velocity of $A = 2$ in. per sec. Locate the instant center of rotation of bar CDE and determine the velocity of point D.

3.13. In Figure P3.13 MR and ST are rigid bars. M is a fixed axis, and the block at S slides in fixed guides. $MQ = QR = SQ = QT = RP = TP = 3$ in. If MR turns counterclockwise at $\frac{1}{3}$ radian/sec, find the velocity of R relative to T and the velocity of P relative to Q.

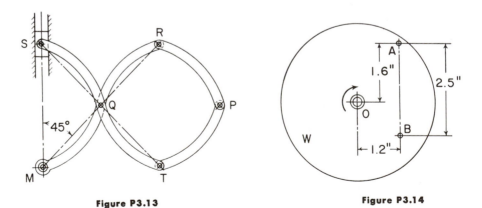

Figure P3.13   Figure P3.14

3.14. Wheel W in Figure P3.14 turns about fixed center O. If the velocity of point A relative to $B = 5$ in. per sec, find the angular velocity of W and the velocity of B.

3.15. Crank *OL* turns clockwise about fixed axis *O* at 10 radians/min. *LR* is a rigid bar pinned to a block at *R* which slides in fixed vertical guides. *PK* is a rigid bar pinned to *RL* at *M* and to horizontal crank *JK* at *K*. *JK* turns about fixed axis *J*. *OL* = 1.5 in., *LM* = 4 in., *MR* = 1 in., *PM* = 2 in., *MK* = 4 in. and *KJ* = 2.8 in. Other dimensions and angles are shown in Figure P3.15. Determine the linear velocity of *P* and the angular velocity of *PK* when crank *OL* is in the vertical position shown.

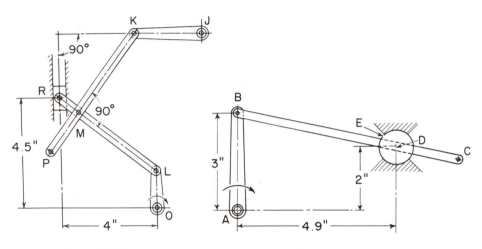

Figure P3.15   Figure P3.16

3.16. In Figure P3.16 vertical crank *AB* turns clockwise about fixed axis *A* at 1 radian/sec. Rod *BC* is pinned to *AB* at *B* and slides freely through a hole in cylinder *E*. *D* is the fixed axis of *E* which rotates freely in the stationary bearings shown. *AB* = 3 in., and *BC* = 7 in. Other dimensions are given on the drawing. Determine the linear velocity of point *C* and the angular velocity of rod *BC*.

3.17. The wheel train shown in Figure P3.17 is a prototype of a gear train. All wheels are mounted on fixed shafts. There is pure rolling contact between *A* and *B* and between *C* and *E*. *B* and *C* are fastened together so that they turn at the same speed. The diameters of the wheels are as follows: *A* = 9 in., *B* = 3 in., *C* = 7 in., and *E* = 2 in. If *A* is the driver and *E* the follower, what is the ratio of ω_E to ω_A (called *speed ratio*)? If *A* turns 100 rpm clockwise, what is the angular speed and sense of *E* (give answer in rpm)?

Figure P3.17

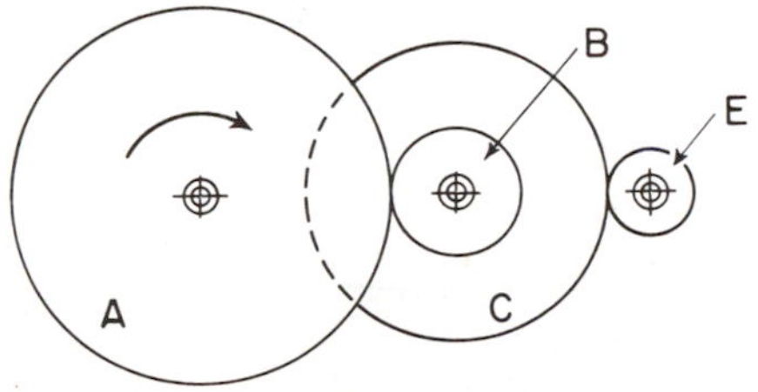

3.18. An *epicyclic wheel train* is shown in Figure P3.18. Wheel W and arm A turn independently about axis O, located at the center of the fixed ring R. The inside diameter of R is 8 in., the diameter of W is 5 in., and the diameter of N is 1.5 in. Wheel N rolls without slip on W and R. If W turns 100 rpm, clockwise, determine the angular speed and sense of arm A (give answer in rpm).

Figure P3.19

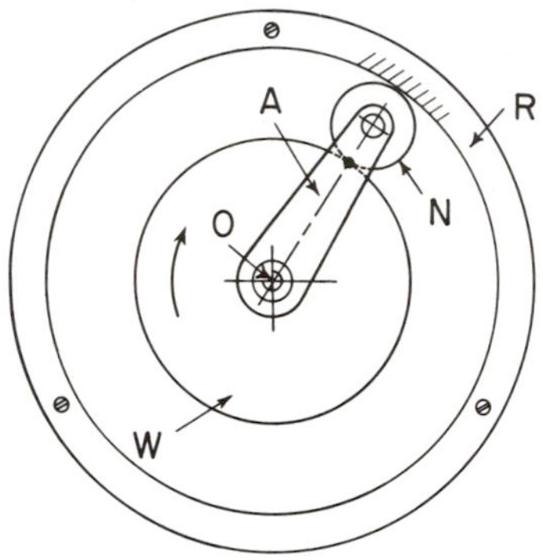

Figure P3.18

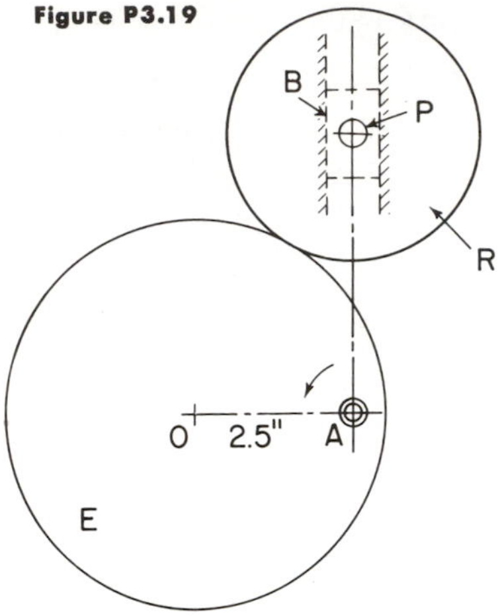

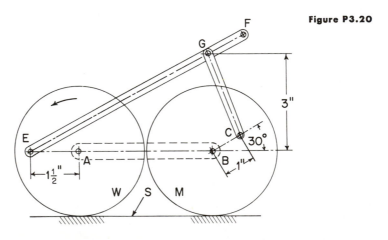

3.19. Eccentric E, 6 in. in diameter with center at O, turns about fixed axis A at 0.5 radian/sec, counterclockwise. Roller R, 4 in. in diameter, rotates freely about pin P in block B, shown dotted in Figure P3.19. Pure rolling contact is maintained between E and R by a spring not shown. Block B slides in fixed vertical guides (dotted) whose center line passes through A. Determine the linear velocity of P and the angular velocity of R when in the position shown.

3.20. Rod AB, $4\frac{1}{8}$ in. long, is pinned to the centers of two 4 in. diameter wheels, W and M, which roll without slip on fixed surface S. Bar EF ($7\frac{1}{2}$ in. long) is

Figure P3.20

pinned to W at E. EG is $6\frac{1}{4}$ in. long. Bar CG is pinned to M at C and to EF at G, as shown in Figure P3.20. If the angular velocity of W is 0.5 radian/sec counterclockwise, determine the linear velocity of pin F.

3.21. In Figure P3.21, M is a rigid member turning freely about fixed axis O. Crank AB (2 in. long) turns about fixed axis A at 1.5 radians/min counterclockwise. CD (3.6 in. long) turns about fixed axis D. Sleeves sliding freely on M are pinned to AB and CD at B and C. Determine the angular velocities of M and CD for the position shown.

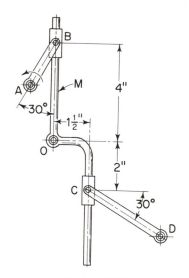

Figure P3.21

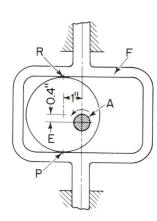

Figure P3.22

3.22. Eccentric E, 4 in. in diameter, turns about fixed axis A at 1 radian/sec counterclockwise. E contacts frame F at R and P, as shown in Figure P3.22. F slides in vertical bearings as shown. Determine the velocity of F and the velocity of sliding between E and F at contact points R and P.

3.23. Cam C turns clockwise about fixed axis O at 1 radian/sec. Pointed follower F slides in vertical fixed guides and is held in contact with the cam by a spring not shown. Determine the velocity of F in the position shown in Figure P3.23 and the velocity of slip between F and C.

Figure P3.23

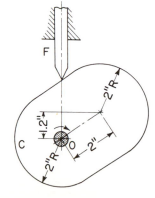

3.24. Wheels W and X are in pure rolling contact. W turns 2 radians/min, counterclockwise, about fixed axis O. X turns about fixed axis R. Slotted bar M is pinned to W at A. Block B turns freely on pin P on X and slides in the slot in M. Determine the angular velocity of bar M and the linear velocity of point S for the position shown in Figure P3.24. $AS = 8.35$ in.

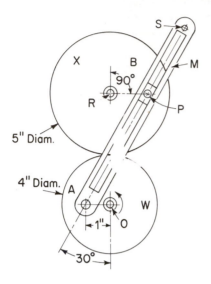

Figure P3.24

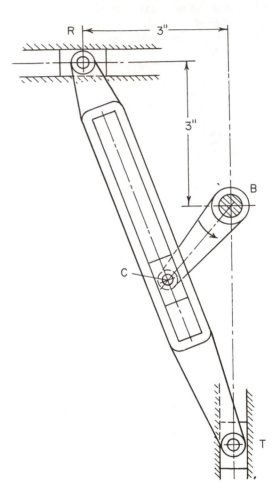

Figure P3.25

3.25. In the *sliding-beam mechanism* shown in Figure P3.25 crank BC is 2 in. long and turns about fixed axis B at 1 radian/sec. Pin C carries a block which slides freely in the slot in the beam. $RT = 8\frac{1}{2}$ in. The fixed guides for blocks at R and T are at right angles.

(a) Determine the velocities of R and T when BC is in the position shown.

(b) Determine the velocities of R and T when R is in the position shown but BC is in the alternate position.

3.26. A *variable-speed friction drive mechanism* is shown in Figure P3.26. The input shaft turns at constant speed of 1725 rpm driving cone 1. The idler wheel between the cones makes pure rolling contact in all positions with both cones. It may be adjusted to different positions along its shaft by the screw and control knob. (The idler turns freely on a bearing that is threaded internally and constrained against turning by a lever not shown.) The output shaft turns with cone 2. Determine the speed of the output shaft when the idler is in positions *A*, *B*, and *C* shown on the illustration. State the direction of rotation of the output relative to the input shaft.

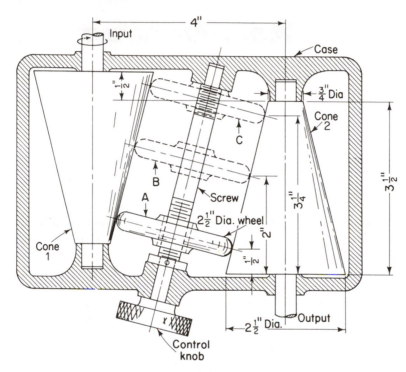

Figure P3.26

4

Acceleration

Acceleration is a very important property of motion to the machine designer. A basic formula of dynamics states that force equals mass times acceleration ($F = ma$). While kinematics is not fundamentally concerned with forces, the study of acceleration becomes vitally important in preparation for the work in dynamics which follows.

Acceleration is the rate of change of velocity with respect to time. It is the rate at which we "speed up" an automobile from a velocity of 10 to 40 mph, for example. The change in speed divided by the time required to make that change is the acceleration. The rate of "slowing down" is technically called *deceleration*. This is negative acceleration, and the distinction of terms is not very strictly observed.

Acceleration is more difficult to visualize and predict than velocity. It demands that strict mental discipline and confidence in sound analytical reasoning so typical of engineering training.

4.1 Linear Acceleration   (Symbol: a)

The change of linear velocity per unit of time is called *linear acceleration*. It is a property of the motion of a point on a body in translation. Common units of linear acceleration are inches per second per second or feet per second per second. Since it has both magnitude and direction, it is a vector quantity; so vectors are freely used to define accelerations and to avoid lengthy calculations.

4.2 Uniform Linear Acceleration

If a point moves along a straight path so that its velocity changes the same amount in successive equal intervals of time, it is said to have *uniform* or *constant* acceleration. A freely falling body has constant acceleration, for example. The acceleration equals the change in velocity (ΔV) divided by the time interval (ΔT) in which that change takes place:

$$a = \frac{\Delta V}{\Delta T}$$

If we designate the velocity at the start of the interval as V^0 and the velocity at the end of the interval as V^f, then

$$\Delta V = V^f - V^0 \quad \text{and} \quad a = \frac{\Delta V}{\Delta T} = \frac{V^f - V^0}{\Delta T}$$

If a body, dropped from a stationary position, acquires a velocity of 16 ft per sec at the end of $\frac{1}{2}$ sec and a velocity of 32 ft per sec after the next $\frac{1}{2}$ sec, etc., the acceleration equals:

$$a = \frac{\Delta V}{\Delta T} = \frac{V^f - V^0}{\Delta T} = \frac{32 - 16}{\frac{1}{2}} = 32 \text{ ft/sec/sec}$$

When linear acceleration is constant we can devise equations for velocity and displacement which are helpful in the analysis of mechanisms.

Since $a = (V^f - V^0)/\Delta T$, we can solve for a final velocity (V^f) which will be attained when a point having an initial velocity (V^0) travels for a time ΔT with a given constant acceleration:

$$V^f = V^0 + a\,\Delta T \text{ (solving for } V^f \text{ in the equation above)}$$

The displacement of a point during any time interval ΔT equals the average velocity multiplied by the time ΔT:

$$S = V^{av}\,\Delta T$$

The average velocity, however, equals the sum of the velocity at the start of the interval (V^0) and the velocity at the end of the interval (V^f) divided by 2:

$$V^{av} = \frac{V^0 + V^f}{2}$$

If we substitute $V^f = V^0 + a\,\Delta T$

$$V^{av} = \frac{V^0 + (V^0 + a\,\Delta T)}{2} = V_0 + \frac{a\Delta T}{2}$$

But $S = V^{av}\,\Delta T$, so

$$S = \left(V^0 + \frac{a\,\Delta T}{2}\right)\Delta T = V^0\,\Delta T + \frac{a(\Delta T)^2}{2}$$

Final velocity can also be expressed in terms of acceleration and displacement without using the time factor ΔT:

$$V^f = V^0 + a \, \Delta T \quad \text{(above)}$$

$$\therefore \ (V^f)^2 = (V^0)^2 + 2V^0(a \, \Delta T) + (a \, \Delta T)^2 \quad \text{(squaring both sides of equation)}$$

$$S = V^0 \Delta T + \frac{a(\Delta T)^2}{2} \quad \text{(above)}$$

$$\therefore \ (2a)S = (2a)V^0 \, \Delta T + \frac{(2a)a(\Delta T)^2}{2} \quad \text{(multiplying by } 2a\text{)}$$

or
$$2aS = 2V^0 a \, \Delta T + (a \, \Delta T)^2$$

Note that the last two terms of the squared equation above are equal to $2aS$. If we substitute $2aS$ for these last two terms of the equation:

$$(V^f)^2 = (V^0)^2 + 2aS$$

4.3 Variable Linear Acceleration

If the change in velocity of a moving point is not the same in each successive equal interval, the acceleration is *variable* and will have a different value at each instant. If a point travels a straight path, starting from rest, so that its velocity is 5 ft/sec at the end of $\frac{1}{10}$ sec, 15 ft/sec at the end of the second $\frac{1}{10}$ sec interval, and 30 ft/sec at the end of the third $\frac{1}{10}$ sec, we observe that it has variable acceleration. In the first interval the *average acceleration is:*

$$a_1^{av} = \frac{\Delta V}{\Delta T} = \frac{V^f - V^0}{\Delta T} = \frac{5 - 0}{\frac{1}{10}} = 50 \text{ ft/sec/sec}$$

In the second interval the average acceleration is:

$$a_2^{av} = \frac{V^f - V^0}{\Delta T} = \frac{15 - 5}{\frac{1}{10}} = 100 \text{ ft/sec/sec}$$

In the third interval:

$$a_3^{av} = \frac{30 - 15}{\frac{1}{10}} = 150 \text{ ft/sec/sec}$$

These average accelerations are largely of statistical interest, as the actual acceleration varies continually throughout each interval. The actual acceleration at a given instant is usually required in a design study.

The approach to instantaneous accelerations follows the same pattern, except that a very small time interval (ΔT) and the corresponding change in velocity (ΔV) are used. As these small time increments approach zero we obtain the instantaneous acceleration:

$$a = \frac{\Delta V}{\Delta T} \quad \text{(as } \Delta T \text{ approaches zero)}$$

In the language of calculus, this definition becomes $a = dV/dT$, the first derivative of velocity with respect to time. (Since $V = dS/dT$, acceleration may also be expressed in terms of calculus as $a = d^2S/dT^2$, the second derivative of displacement with respect to time.)

Where a point travels a straight path, its velocity is at all times directed along that path. Changes in velocity are changes in magnitude and possibly sense, but the inclination remains constant. Velocity differences are therefore *algebraic*, and vectors need not be used. Since all velocities are of the same inclination, all accelerations are likewise directed along the straight path of motion.

4.4 Linear Acceleration Along Curved Paths

If a point travels a curved path, its velocity changes direction as well as magnitude. The velocity change ΔV is therefore a *vector difference* between the original (V^0) and final (V^f) velocity at the start and end of a time interval ΔT:

$$a = \frac{\Delta V}{\Delta T} = \frac{V^f - \rightarrow V^0}{\Delta T}$$

In Figure 4.1(a), point A travels the curved path shown. At the beginning of a time interval ΔT, the point is at A^0 and has a velocity at that instant of V^0, directed along or tangent to the curve at A^0. At the end of time ΔT the point has reached position A^f and has attained an increased velocity of V^f, again tangent to the curve at A^f. The change in velocity (ΔV) during time ΔT is therefore the vector difference ($V^f - \rightarrow V^0$) in the free vector diagram shown in Figure 4.1(b). This takes into account the change in both magnitude and direction.

(a)   (b)

Figure 4.1 Change in velocity of a point on a curved path

The direction of the linear acceleration in all cases will be the same as the direction in which the velocity is changing or the same as the vector ΔV.

A body in translation may be considered to have linear acceleration. Since the velocities of all points on the body are identical at any given instant, changes in velocities of these points will be equal in any given interval. The linear acceleration of the body as a whole will therefore equal that of any point on the body.

4.5 Angular Acceleration   [Symbol: α (alpha)]

The change in angular velocity per unit of time is called *angular acceleration*. It is a property of the motion of bodies or lines and does not apply to points, as their angular motion has no significance. Common units of angular acceleration are radians per second per second.

4.6 Uniform Angular Acceleration

If a body rotates so that its angular velocity changes the same amount in equal successive time intervals, its angular acceleration is described as *uniform* or *constant*. Just as with linear acceleration, the change in angular velocity ($\Delta\omega$) taking place in a given time interval, divided by that time interval, equals the angular acceleration (α):

$$\alpha = \frac{\Delta\omega}{\Delta T}$$

If we designate the angular velocity at the beginning of the interval as ω^0 and at the end as ω^f,

$$\Delta\omega = \omega^f - \omega^0 \quad \text{and} \quad \alpha = \frac{\Delta\omega}{\Delta T} = \frac{\omega^f - \omega^0}{\Delta T}$$

This equation will yield an expression for the final angular velocity (ω^f) attained by a line which has an original angular velocity of ω^0 and which rotates for a time ΔT, with a constant angular acceleration:

$$\omega^f = \omega^0 + \alpha\Delta T \quad \text{(solving for } \omega^f \text{ in the equation above)}$$

The angular displacement (θ) of a line during any time interval ΔT equals the average angular velocity during that interval multiplied by the time ΔT:

$$\theta = \omega^{\text{av}} \times \Delta T$$

This average angular velocity equals the sum of the initial and final velocities divided by 2:

$$\omega^{\text{av}} = \frac{\omega^0 + \omega^f}{2}$$

and substituting $\omega^f = \omega^0 + \alpha\Delta T$:

$$\omega^{\text{av}} = \frac{\omega^0 + (\omega^0 + \alpha\,\Delta T)}{2} = \omega^0 + \frac{\alpha\,\Delta T}{2}$$

Since $\theta = \omega^{\text{av}} \times \Delta T$,

$$\theta = \left(\omega^0 + \frac{\alpha\,\Delta T}{2}\right)\Delta T = \omega^0\Delta T + \frac{\alpha(\Delta T)^2}{2}$$

Final angular velocity can be expressed in terms of angular acceleration and angular displacement by eliminating the time factor.

Since $\omega^f = \omega^0 + \alpha\,\Delta T$, squaring both sides:

$$(\omega^f)^2 = (\omega^0)^2 + 2\omega^0\alpha\,\Delta T + (\alpha\,\Delta T)^2$$

$$\theta = \omega^0\,\Delta T + \frac{\alpha(\Delta T)^2}{2} \quad \text{(above)}$$

and multiplying by 2α:

$$2\alpha\theta = 2\alpha\omega^0\,\Delta T + (\alpha\Delta T)^2$$

In the $(\omega^f)^2$ equation, we substitute $2\alpha\theta$ for the last two terms and

$$(\omega^f)^2 = (\omega^0)^2 + 2\alpha\theta$$

Note that these equations are similar to those involving uniform linear acceleration in Article 4.2, except that α is substituted for a, ω for V, and θ for S.

4.7 Variable Angular Acceleration

If the changes in angular velocity of a rotating line are not the same in successive equal time intervals, the angular acceleration is *variable* and will have a different value each instant. In a given time interval ΔT the *average* angular acceleration will equal the change in angular velocity during that interval divided by the time ΔT:

$$\alpha^{\text{av}} = \frac{\Delta\omega}{\Delta T} = \frac{\omega^f - \omega^0}{\Delta T}$$

This average will not be the correct acceleration throughout the interval, as the acceleration will be continually changing.

As before, we approach the true value of instantaneous acceleration by using a very small time interval ΔT and the corresponding change, $\Delta\omega$. As the small time increment approaches zero, we obtain the instantaneous value of angular acceleration:

$$\alpha = \frac{\Delta\omega}{\Delta T} \quad \text{(as } \Delta T \text{ approaches zero)}$$

In the language of calculus this definition becomes

$$\alpha = \frac{d\omega}{dT}$$

the first derivative of angular velocity with respect to time or, since $\omega = d\theta/dT$,

$$\alpha = \frac{d^2\theta}{dT^2}$$

the second derivative of angular displacement with respect to time.

Since we are limiting our study to motion in a single plane or in parallel planes, the changes in angular velocities involve only magnitude and possibly sense, so the velocity differences are algebraic and not vector differences.*

4.8 Accelerations on Bodies in Translation

Bodies in translation move without turning. We have observed that the velocities of all points on translating bodies are identical (Article 3.1) for any given instant. If these velocities are changing from one instant to the next, they must all change at the same rate or they could not maintain equality at all times.

Since acceleration equals rate of change of velocity and all points on the body have the same rate of change of velocity, it follows that *all points on a body in translation have identical accelerations at any given time.* The direction of these accelerations will be the same as the direction of motion or along the (parallel) paths of the points, as shown in Figure 4.2.

Figure 4.2 Accelerations on a translating body

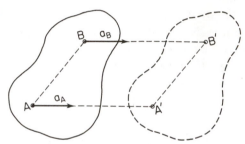

4.9 Linear Accelerations on Bodies Rotating at Constant Speed

Let us first consider the case of a body in pure rotation with a *constant angular velocity.* Any point on such a body will have a velocity equal to the angular velocity multiplied by its distance from the axis of rotation ($V = \omega r$) directed perpendicular to the radial line joining the point and the axis. Since both ω and r remain constant, this linear velocity will also remain constant in magnitude, but will change inclination as the body rotates. Thus there will be a change in velocity and therefore acceleration.

The wheel W, in Figure 4.3(a), turns clockwise with a constant angular

---

*Angular velocities can be represented vectorially, and this is necessary where axes of rotation are not parallel, but we do not have need for such vectors in plane kinematics.

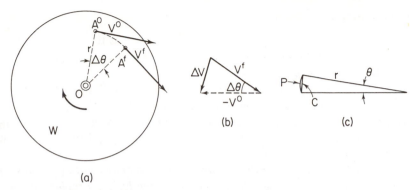

Figure 4.3 Velocity changes on body rotating at constant speed

velocity ω_w about a fixed axis O, at the center. A typical point on W, such as A, located at distance r from O, has a linear velocity $V_A = \omega_w r$, perpendicular to OA. Let us assume that, in a small interval of time (ΔT), line OA turns through a small angle $\Delta\theta$ as A moves from A^0 to A^f. Since ΔT is a very small interval of time, we may consider the acceleration to be uniform throughout the interval without introducing a serious error.

Then, by definition:

$$a_A = \frac{\Delta V_A}{\Delta T} \quad \text{where} \quad \Delta V_A = V^f - \to V^0$$

This vector subtraction is shown in the free vector diagram in Figure 4.3(b). Since V^f is drawn perpendicular to OA^f and V^0 perpendicular to OA^0, the angle between V^f and V^0 will be equal to the angle between OA^f and OA^0, which is the angle $\Delta\theta$. We have shown (in Article 2.3) that the length of arc (P) subtended by an angle (θ) is equal to the radius of the arc (r) multiplied by the angle in radians ($P = r\theta$). When the angle is very small, as in Figure 4.3(c), the subtended arc and its chord (C) are very nearly equal, so the chord is essentially equal to the radius multiplied by the angle in radians ($C = r\theta$).

In Figure 4.3(b), V^f equals V^0, so ΔV may be considered the chord of a circle of radius V subtended by an angle of $\Delta\theta$. Since $\Delta\theta$ is very small, we may state that $\Delta V = V\,\Delta\theta$ as $\Delta\theta$ approaches zero.

The acceleration of A then becomes:

$$a_A = \frac{\Delta V_A}{\Delta T} = \frac{V_A \Delta\theta}{\Delta T}$$

If we substitute $\omega = \Delta\theta/\Delta T$:

$$a_A = V_A \omega_w$$

and substituting $V_A = \omega_w r$:

$$a_A = (\omega_w)^2 r *$$

Since $\Delta\theta$ was in radians in the equation $\Delta V = V \Delta\theta$, ω in the equation above must be in *radians per unit of time*.

The direction of this acceleration will be the same as that of ΔV. As the angle $\Delta\theta$ approaches zero, the angle between ΔV and V_A will approach $90°$, in which case ΔV would be parallel to OA, and have a sense towards axis O. Since r and ω are both constants, the magnitude of the a_A will remain constant as the body turns.

In conclusion, *the acceleration of any point on a body rotating at constant speed equals the square of the angular velocity of the body (in radians per unit of time) multiplied by the distance of the point from the axis of rotation:*

$$a = \omega^2 r \quad \text{(directed toward the axis of rotation)}$$

Since this acceleration is normal (perpendicular) to the path of the point at all times, and therefore radial with respect to the axis, it is designated as the *normal* or *radial* acceleration (a^N or a^R). This acceleration is due to the change in inclination of velocity alone, as the magnitude remains constant when the body has uniform angular speed.

4.10 Linear Accelerations on Bodies Rotating with Variable Angular Velocity

When the angular velocity of a body is changing, the linear velocity of a point on the body changes in magnitude as well as in direction.

Let us consider the acceleration due to the changes in *magnitude* of velocity separately and ignore changes in direction for the moment.

In Figure 4.4(a), wheel W turns about fixed axis O with variable angular velocity. In position A^0, point A has a velocity V^0, as shown. After W turns a small angle $\Delta\theta$, point A moves along an arc to A^f and has a velocity V^f (larger than V^0). Since we are ignoring changes in direction of velocity, the circular path A^0A^f may be considered to be straight, as shown in Figure 4.4 (b). The acceleration of A due to changes in magnitude of velocity alone would be:

$$a_A = \frac{\Delta V}{\Delta T} \quad \text{(using the equation for straight line motion in Article 4.3)}$$

$$\Delta V = V^f - V^0 \quad \text{(algebraic difference)}$$

*This equation can assume another form:

$$a_A = \frac{(V_A)^2}{r}$$

if we substitute $\omega_W = V_A/r$ in the equation $a_A = V_A\omega_W$. In some instances this form is more convenient.

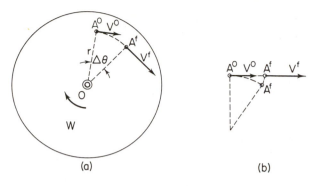

(a)   (b)

Figure 4.4 Velocity changes due to angular acceleration

Since the angular velocity is changing, ω^0 of line OA^0 will differ from ω^f of line OA^f.

$$V^f = \omega^f r \quad \text{and} \quad V^0 = \omega^0 r$$

Substituting:

$$\Delta V = \omega^f r - \omega^0 r = r(\omega^f - \omega^0)$$

But

$$\omega^f - \omega^0 = \Delta\omega \quad \text{(the change in } \omega\text{)}$$

Therefore

$$a = \frac{r\,\Delta\omega}{\Delta T}$$

And since $\Delta\omega/\Delta T = \alpha$

$$a = r\alpha$$

This acceleration will be constant only if the angular acceleration α is constant; otherwise it is an instantaneous value for the position A^0 only. The direction of this acceleration is the same as that of ΔV, which (since we are considering changes in magnitude only) is tangential to the path (perpendicular to OA), and it is logically called *tangential acceleration* (a^T).

Point A also has an acceleration due to change in inclination of V_A. This normal acceleration $a^N = \omega^2 r$, as shown in Article 4.9. Since the angular acceleration is not involved in this equation, it is permissible to consider acceleration due to changes in *direction* of velocity separately from that due to changes in *magnitude*. a^N is not constant in this case.

Since ω is continually changing, the normal acceleration is an instantaneous value and the ω^2 term employs the ω of line OA^0.

The *resultant acceleration* of A is the resultant of two components: a^N, directed toward O, and a^T, perpendicular to OA. This resultant acceleration may be found graphically (Figure 4.5) as the sum of the two vectors, or, since

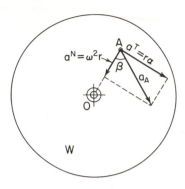

Figure 4.5 Resultant of normal and tangential accelerations

a^N and a^T are at 90° with one another, we can calculate it by using the Pythagorean theorem: Since $a^2 = (a^N)^2 + (a^T)^2$,

$$a = \sqrt{(a^N)^2 + (a^T)^2}$$

The angle β between the resultant acceleration a_A and the radial line OA is determined by the ratio of a^T to a^N or $\alpha r/\omega^2 r$, which becomes α/ω^2 (canceling r). In trigonometry, the *tangent** of angle β equals α/ω^2 and β may be found in the tables.

It may be noted here that, since at any instant the entire body has the same angular velocity and the same angular acceleration, the angle β is the same for all points on the body in any given position.

This is illustrated in Figure 4.6.

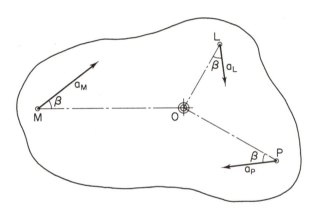

Figure 4.6 Inclination of accelerations with radial lines

Example 1

In the position shown, the bell-crank lever in Figure 4.7 is turning about fixed axis A with an angular velocity of 2 radians/min and an angular acceleration of 3 radians/min/min, both clockwise. $AB = 2$ in., $AC = 3$ in., and $AD = 4$ in. The linear accelerations of points B, C, and D are required.

The normal component of the acceleration of B equals $\omega^2 r$:

$$a_B^N = 2^2 \times 2 = 8 \text{ in./min/min} \quad \text{(toward } A\text{)}$$

The tangential component of the acceleration of B equals αr:

$$a_B^T = 3 \times 2 = 6 \text{ in./min/min} \quad \text{(perpendicular to } AB \text{, in the same sense as } \alpha\text{)}$$

*In terms of trigonometry, the tangent of an angle of a triangle equals the side opposite the angle divided by the side adjacent to the angle.

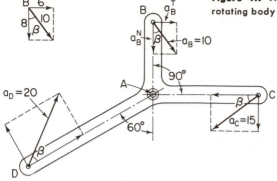

Figure 4.7 Accelerations of points on a rotating body

The resultant acceleration of B is the vector sum of a_B^N and a_B^T, and since the angle between them is $90°$ (see small diagram in Figure 4.7):

$$a_B = \sqrt{(a_B^N)^2 + (a_B^T)^2} = \sqrt{64 + 36}$$

$$= \sqrt{100} = 10 \text{ in./min/min}$$

Angle β, between a_B and AB has a tangent equal to $\alpha/\omega^2 = \frac{3}{4}$. (In a right triangle having sides of 3, 4, and 5 units, β is the angle opposite the 3-unit side.)

Similarly:

$$a_C^N = \omega^2 r = 4 \times 3 = 12$$

$$a_C^T = \alpha r = 3 \times 3 = 9$$

$$a_C = \sqrt{12^2 + 9^2} = \sqrt{225}$$

$$= 15 \text{ in./min/min}$$

Angle β between a_C and AC equals angle β between a_B and AB.

$$a_D^N = \omega^2 r = 4 \times 4 = 16$$

$$a_D^T = \alpha r = 3 \times 4 = 12$$

$$a_D = \sqrt{16^2 + 12^2} = 20 \text{ in./min/min}$$

The same angle β lies between a_D and AD.

4.11 Relative Accelerations on a Rigid Body

The acceleration of one point on a rigid body relative to another point on the same body is very useful in the analysis of acceleration on bodies in combined motion. The rigid body imposes restraints on accelerations just as it does upon velocities.

Figure 4.8 shows two points, A and B, on the same rigid body. First, let us consider the motion of B relative to A. This can be visualized if we imagine ourselves standing on A and watching the motion of B. From that vantage point the motion we observe will be the motion of B relative to A.

The concept of the rigid body guarantees that, whatever the motion of the body, B will never approach closer to A nor move farther away from A. Then the only motion that B can have relative to A is that of moving in a circular path (of radius AB) about A. Relative to A, the line AB is then in *rotation about A*. The acceleration of B relative to A must then be the acceleration of a point on a body in pure rotation about a fixed axis at A, for when we stand upon A, from our viewpoint, A does not move.

We have already investigated the accelerations of points on bodies in pure rotation and have found that such points have two components of acceleration: a normal component $\omega^2 r$ (directed towards the axis) and a tangential component αr (perpendicular to a line to the axis.) The ω and α in these terms are the angular velocity and angular acceleration of the line joining the point to the axis of rotation.

Since the motion of B relative to A is that of a point on a body turning about A, we can state that the acceleration of B relative to A has two similar components: the normal acceleration of B relative to A,

$$a^N{}_{B/A} = (\omega_{AB})^2 \times AB \quad \text{(towards } A\text{)}$$

and the tangential acceleration of B relative to A,

$$a^T{}_{B/A} = \alpha_{AB} \times AB \quad \text{(perpendicular to } AB\text{)}.$$

At a given instant, a rigid body can have but one value of ω and α. These values are the same no matter what point we select as an axis of reference. Therefore, the absolute values of ω and α are the same values employed in relative acceleration. The total acceleration of B relative to A is the vector sum of these two components, and B can have no other acceleration relative to A.

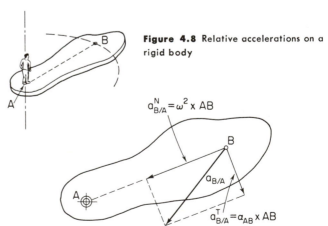

Figure 4.8 Relative accelerations on a rigid body

Note that, in this exploration, we have not specified the class of motion of the body concerned, so these findings are true whatever kind of motion takes place.

If the body is in translation, it does not turn, so both ω and α equal zero. Therefore, $\omega^2 r = 0$ and $\alpha r = 0$, and there is no relative acceleration of one point on a body to another. This verifies the statement in Article 4.3 that *all points on a body in translation have identical accelerations*. Any relative acceleration of one point to another is due to rotation of the body alone.

If the body is in rotation or in combined motion, the absolute values of ω and α are used to obtain the relative acceleration of one point to another.

Just as in the case of velocities, if the absolute accelerations of two points on a body are known, the relative acceleration of one to the other is the difference of their absolute accelerations. In Figure 4.9 the accelerations of points E and F are known (a_E and a_F), and the acceleration of F relative to E is required:

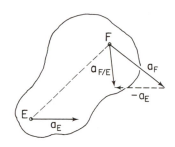

Figure 4.9 Relative acceleration equals the difference of absolute accelerations

$$a_{F/E} = a_F - \!\!\rightarrow a_E$$

This vector subtraction is performed in the same manner as with velocities, by adding a negative a_E to the given a_F. $a_{F/E}$ is the resultant of this operation.

We have pointed out above that this relative acceleration is due to rotation of the body. It therefore provides a means of calculating the ω and α of the body. We know that the acceleration of F relative to E is the resultant of two components: $\omega^2 r$, along FE, and αr, perpendicular to FE. Let us resolve $a_{F/E}$ into two components along and perpendicular to FE, as shown in Figure 4.10:

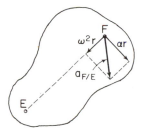

$$a_{F/E}^N = \omega^2 \times FE \qquad \text{so} \qquad \omega^2 = \frac{a_{F/E}^N}{FE}$$

$$\text{and} \qquad \omega = \sqrt{\frac{a_{F/E}^N}{FE}}$$

$$a_{F/E}^T = \alpha \times FE \qquad \text{so} \qquad \alpha = \frac{a_{F/E}^T}{FE}$$

Figure 4.10 Normal and tangential components of relative acceleration

4.12 Accelerations on Bodies in Combined Motion

In velocity analysis we used the concept of relative velocities on a rigid body as a basic tool. We noted that, in the direction connecting two points on a body, there was no relative velocity. This was the basis for claiming that the effective components were equal along the line between two points. (If there is no relative velocity, then the velocities must be equal.)

In acceleration analysis, we similarly employ the concept of relative acceleration on a rigid body. The only difference is that (except on a body in translation) the relative accelerations are a bit more complex.

We have observed, in Figure 4.9 above, that the acceleration of F relative to E equals the difference between their absolute accelerations:

$$a_{F/E} = a_F - \rightarrow a_E$$

If we solve this equation for a_F we have:

$$a_F = a_{F/E} + \rightarrow a_E$$

This equation affords a means of determining the acceleration of F when the acceleration of another point E is known and the relative acceleration of F to E can be found.

For example, two points, C and D, on a body M are shown in Figure 4.11. The angular velocity and angular acceleration of M and the absolute linear

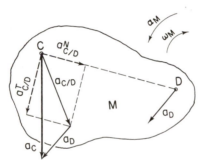

Figure 4.11 Acceleration of a point on a body in combined motion

acceleration of D are known. The acceleration of C is required.

Following the relative acceleration equation above:

$$a_C = a_{C/D} + \rightarrow a_D$$

The acceleration of C relative to D is composed of two parts:

$$a_{C/D}^N = \omega_M^2 \times CD \quad \text{and} \quad a_{C/D}^T = \alpha_M \times CD$$

These are calculated and laid out from C as shown. The sense of the αr component is made consistent with the sense of α_M (counterclockwise with respect to D).

Next we determine the resultant of these two components, which is $a_{C/D}$, as shown. Our formula states that we must add the acceleration of D to this $a_{C/D}$ to obtain a_C. This is shown in Figure 4.11, in which a_D is added at the head end of $a_{C/D}$ and the resultant a_C drawn. The magnitude and direction of this vector are correct as determined by its components, $a_{C/D}$ and a_D.

4.13 Effective Components of Acceleration

Let us apply this *relative acceleration method* to a mechanism. The crank and slider linkage in Figure 4.12 is typical. Crank AB turns counterclokwise about fixed axis A with an ω and α given. Let us find the acceleration of C.

AB is in pure rotation, so a_B has two components:

$$(\omega_{AB})^2 \times AB \quad \text{and} \quad \alpha_{AB} \times AB.$$

Their resultant is the a_B, as shown. BC is in combined motion, so we use the relative acceleration equation for a_C:

$$a_C = a_{C/B} + \rightarrow a_B$$

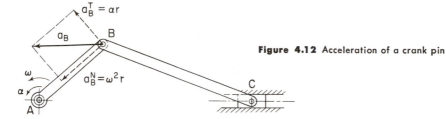

Figure **4.12** Acceleration of a crank pin

We can predict that a_C is directed along the center line of the guides. Since V_C is always along this center line, any change in velocity will be in the same direction.

Again, the acceleration of C relative to B has two components:

$$a_{C/B} = (\omega_{CB})^2 \times CB + \rightarrow \alpha_{CB} \times CB$$

The length of CB is given, and we can find ω_{CB}, but α_{CB} is not readily available, so we try another approach.

In Figure 4.13 a resultant vector R is shown to be the sum of two components, J and K. Let us find the effective components of J, K, and R along any line, such as SS. Dropping perpendiculars to SS, we determine:

$$OM = ec_J, \qquad MP = ec_K, \qquad OP = ec_R$$

Figure **4.13** Sum of effective components

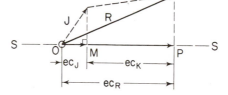

We note:

$$OP = OM + \rightarrow MP \quad \text{or} \quad ec_R = ec_J + \rightarrow ec_K$$

In any given direction then, the effective component of a resultant vector equals the sum of the effective components of its component vectors.

Applying this to the relative acceleration equation ($a_C = a_{C/B} + \rightarrow a_B$), it follows that in any given direction:

$$ec\ a_C = ec\ a_{C/B} + \rightarrow ec\ a_B$$

In the problem in Figure 4.12, since we know the *direction* of the a_C, we need but one effective component of the acceleration of C to determine the *magnitude* of a_C. (A perpendicular from this $ec\ a_C$ to the center line of the guides will define a_C.)

Let us get an effective component of the acceleration of C along BC. Along BC:

$$ec\ a_C = ec\ a_{C/B} + \rightarrow ec\ a_B$$

The $ec\ a_{C/B}$ along BC equals $(\omega_{BC})^2 \times BC$ because the αr component, being perpendicular to BC, has no effect along BC. We find ω_{BC} by locating I_{BC}, as shown in Figure 4.14. Since IB has the same ω as BC (both lines are on the same body)

$$\frac{V_B}{IB} = \omega_{BC}$$

But since $V_B = \omega_{AB} \times AB$,

$$\omega_{BC} = \omega_{AB} \times \frac{AB}{IB}$$

We next square this ω_{BC} and multiply it by BC, and we have the $ec\ a_{C/B}$ along BC, shown in Figure 4.14.

Next we need the $ec\ a_B$ along BC, which is easily obtained at B. This ec is shown added to $ec\ a_{C/B}$ in Figure 4.14. These two ec's together comprise the $ec\ a_C$ along BC. A perpendicular to this $ec\ a_C$ at d intersects the center line of the guides at f, defining Cf as the absolute acceleration of C.

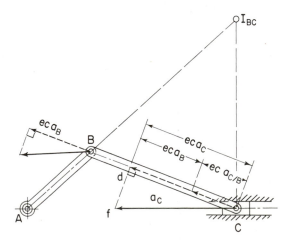

Figure 4.14 Accelerations on a crank and slider

Effective components of acceleration can thus be used to obtain resultant accelerations in the same manner that we employed effective components of velocity to obtain resultant velocities. Every effective component of acceleration along a line connecting two points on a rigid body is composed of two parts:

1. The acceleration component of the first point relative to the second. (This is due to rotation of the body, causing the first point to travel about the second, and therefore is an $\omega^2 r$ component.)
2. The acceleration component of the second. (This is due to the translation of the body and therefore is the same for both points.)

4.14 Determining Directions of Linear Acceleration

When the direction of an acceleration is not defined by straight guides, we employ two effective components just as we did with velocities.

In the linkage in Figure 4.15, the driving crank OM turns clockwise about fixed axis O with a constant angular velocity ω. L is a fixed axis, and bar MKP is a single rigid body. The acceleration of K and P are required.

First we find the acceleration of M. Points on a body in rotation, such as OM, have two components of acceleration: $\omega^2 r$ and αr. In this case, however, OM has a constant ω, and therefore α equals zero. The resultant acceleration of M is equal to $a_M^N = (\omega_{OM})^2 \times OM$ and is directed toward O.

Second, let us consider the acceleration of K. We can find one effective component of the acceleration of K along MK:

$$ec\ a_K = ec\ a_{K/M} + \rightarrow ec\ a_M \quad \text{(all in direction } MK\text{)}$$

$$ec\ a_{K/M} = (\omega_{MK})^2 \times MK \quad \text{(since the } \alpha r \text{ component of } a_{K/M} \text{ has no}$$
effect along MK, being perpendicular to MK)

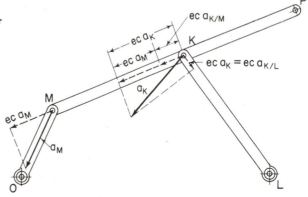

Figure 4.15 Accelerations on a four-bar linkage

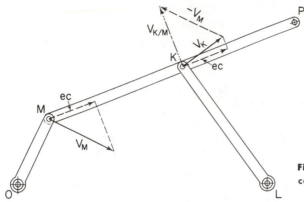

Figure 4.16 Relative velocities on a connecting rod

The angular velocity of MK may be calculated by locating I_{MK}, as in Article 4.13, or, if I_{MK} is inaccessible, the $V_{K/M}$ may be determined and $\omega_{MK} = V_{K/M}/MK$. This requires a separate velocity study, which is shown separately in Figure 4.16. The total $ec\ a_K$ along MK is shown in Figure 4.16, by the addition of $ec\ a_{K/M}$ to $ec\ a_M$.

Since the direction of a_K is not known in this problem, we must find another $ec\ a_K$ to determine a_K completely. A second $ec\ a_K$ can be found in the direction KL. Applying the equation along KL:

$$ec\ a_K = ec\ a_{K/L} + \rightarrow ec\ a_L$$

Here

$$ec\ a_{K/L} = (\omega_{KL})^2 \times KL \quad \text{(again the } \alpha r \text{ component of } \alpha_{K/L}$$
$$\text{has no effect along } KL)$$

$$\omega_{KL} = V_K/KL \quad \text{(Figure 4.16 shows } V_K)$$

$ec\ a_L = 0$ in this case, since L is a fixed axis and has no acceleration. The total effective component of a_K, along KL is simply:

$$ec\ a_{K/L} = (\omega_{KL})^2 \times KL$$

[as shown in Figure 4.15]

We now have two effective components of a_K, one along MK and one along KL. Perpendiculars from the head ends of these vectors will intersect to determine the resultant a_K in magnitude and direction, as illustrated in Figure 4.15.

Now let us find the acceleration of P. Our basic equation states:

$$a_P = a_{P/M} + \rightarrow a_M$$

$$a_{P/M} = (\omega_{MP})^2 \times MP + \rightarrow \alpha_{MP} \times MP$$

$$\omega_{MP} = \omega_{MK} \quad \text{(which has already been determined)}$$

α_{MP} can now be determined, since we already know the resultant accelerations of two points, M and K, shown in Figure 4.17. The acceleration of K relative to M equals the difference of their absolute accelerations:

$$a_{K/M} = a_K - \to a_M$$

This vector subtraction is shown in Figure 4.17, yielding $a_{K/M}$. Since K moves in a circular path relative to M, the acceleration of K relative to M has two components: a normal conponent ($\omega^2 r$) and a tangential component (αr), or more specifically:

$$a_{K/M} = (\omega_{MK})^2 \times MK + \to \alpha_{MK} \times MK$$

We are only interested in the tangential component, involving α_{MK}. This is an effective component of $a_{K/M}$, perpendicular to MK, as shown in Figure 4.17.

Since this component equals $\alpha_{MK} \times MK$,

α_{MK} equals this component divided by MK

$$\therefore a_{MK} = \frac{a_{K/M}}{MK}$$

Since α_{MK} equals α_{MP}, we can now lay out:

$$a_{P/M} = (\omega_{MP})^2 \times MP + \to \alpha_{MP} \times MP$$

This is shown in Figure 4.17. If we add a_M to $a_{P/M}$ we have a_P since, as stated above:

$$a_P = a_{P/M} + \to a_M$$

This vector addition, yielding a_P, is shown in Figure 4.17.

The relative acceleration equation can be adapted to determine linear accelerations of any point on a body in combined motion. The procedure outlined in the above example is typical. Effective components may be used to obtain absolute accelerations of two points on the body, then the angular acceleration can be found and the equation applied to resultant accelerations to study other points.

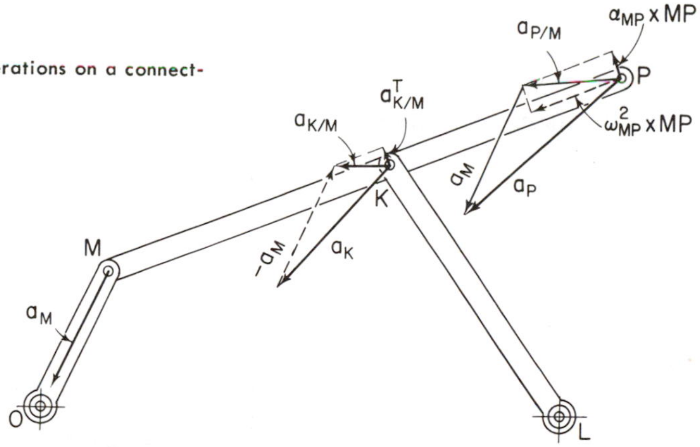

Figure 4.17 Accelerations on a connecting rod

4.15 Accelerations by the Vector-Polygon Method

In the study of velocity, the vector-polygon method has been offered as an alternative to the effective-component technique. A similar system of analysis can be used in the investigation of accelerations, one which provides the same advantages of economy of drafting time and increased accuracy of graphical solutions. The use of free vector polygons, separate from the mechanism drawing, permits the use of larger acceleration-vector scales and less complicated layouts, both of which enhance the accuracy of the solutions. In the case of acceleration analyses, the relative-acceleration concept is the guiding principle in both the effective-component and polygon techniques, so there is a simpler transition from one method to the other. The simplification of the graphical work and the more direct method of attack leads the author to recommend the polygon system over the effective-component technique, especially for accelerations on bodies in combined motion. The reader may make his own choice.

4.16 Constructing an Acceleration Polygon

As the fundamental properties of acceleration and the relative accelerations of points on a rigid body have been presented, we can best explain the polygon method by applying this technique to a typical problem, step by step. In Figure 4.18 a four-bar linkage is shown. Points A and D are fixed axes with the coupler BC extended through C to point F and a projecting strut carrying point E so that $BCFE$ is one rigid body. Let us assume that AB has an angular acceleration (α) of 0.4 radian/sec/sec, counterclockwise and an angular velocity (ω) of 0.6 radian/sec, clockwise, when in the given position.

Let us determine the acceleration of points B, C, F, and E and the angular accelerations of CD and $BCFE$.

1. *Acceleration of B*

    First, we must determine the acceleration of B. Crank AB is in pure rotation, so the absolute acceleration of B is the sum of two component

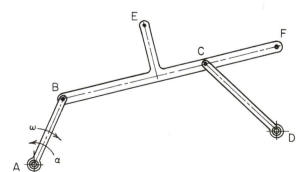

Figure 4.18 Four-bar linkage with extended coupler

vectors: a normal acceleration component equal to $\omega_{AB}^2 \times AB$ directed parallel to AB down toward A, and a tangential component equal to $\alpha_{AB} \times AB$, which is perpendicular to AB and has sense up to the left, consistent with the counterclockwise sense of α. The magnitude of these components can be calculated from the given data.

We start the vector polygon from any convenient origin O, laying out B's normal acceleration vector at a convenient scale (such as 1 in. = $\frac{1}{3}$in./sec/sec) and drawing it parallel to AB directed downward from O. (See Figure 4.19). From the head end of this vector we add the tangential component of B's acceleration directed perpendicular to AB and with sense up to the left. The head end of this second vector is labeled B. The absolute acceleration of B will then be the sum of these two components—a vector drawn from O to B and with sense toward B. The magnitude of this acceleration of B can be determined by measuring OB at the designated scale. The arrowheads shown in Figure 4.19 may be omitted if we establish a rule that the acceleration of B is always directed toward the point labeled B on the polygon, etc.

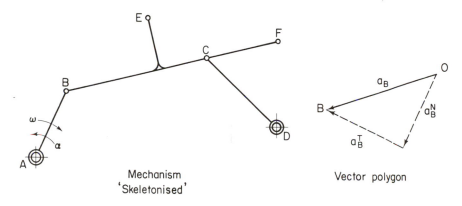

Mechanism
'Skeletonised'

Vector polygon

Figure 4.19 Acceleration polygon—step 1

2. *Acceleration of C*

Pin C is a point on both the coupler BC and the crank CD. Its acceleration can be found by observing its motion relative to points B and D. A relative acceleration is the vector difference between two absolute accelerations. For example, the acceleration of C relative to B equals the absolute acceleration of C minus (vectorially) the absolute acceleration of B, which can be stated in symbolic form:

$$a_{C/B} = a_C - \rightarrow a_B$$

Since we are seeking a_C, this equation can be rearranged:

$$a_C = a_{C/B} + \rightarrow a_B$$

Pins C and B are on the same rigid body, so the only motion pin C can have relative to B will be that of traversing a circular path of radius BC about B. If the reader were to imagine himself standing on pin B, observing the motion of the coupler BC, he would see BC swinging about him as a pivot, as if B were a fixed point. All points on the coupler would travel in circular arcs about his position at B (Figure 4.20). If one point on a rigid body is assumed fixed, the only motion the body can have relative to the fixed point is that of pure rotation. Thus, if BC swings about B, the acceleration of C relative to B will be that of a point on a body in pure rotation relative to its fixed axis. Relative to B then, C will have a normal component of acceleration equal to $\omega^2 r$ and a tangential component of αr, where ω is the angular velocity of BC (in the given position) and r is the length of BC and α is the angular acceleration of BC. In symbolic form:

$$a_{C/B} = (\omega^2_{BC} \times BC) + \rightarrow (\alpha_{BC} \times BC)$$

[The vector $(\omega^2_{BC} \times BC)$ would be parallel to BC with sense from C toward B. The vector $(\alpha_{BC} \times BC)$ would be perpendicular to BC.] Following the equation above

$$a_C = a_B + \rightarrow a_{C/B}$$

Since we have already laid out the vector a_B, we may add the two vectors of $a_{C/B}$ to the vector diagram if we can compute their magnitude. To compute the normal component $(\omega^2_{BC} \times BC)$, we must determine ω_{BC} when it is in the given position. This can be found from the velocities of points B and C as follows:

$$V_B = \omega_{AB} \times AB$$

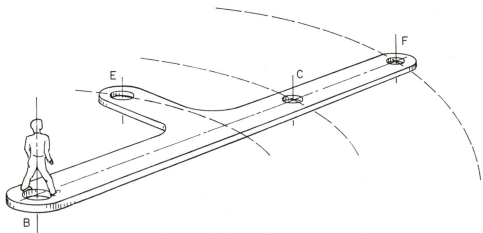

Figure 4.20 Motion of coupler relative to B

Locate *I*, the instant center of *BC* as shown in Figure 4.21. Then on rigid body *IBC*, all lines have the same angular velocity.

$$\omega_{BC} = \omega_{IB} = \frac{V_B}{IB} = \frac{\omega_{AB} \times AB}{IB}$$

Thus we can measure *AB* and *IB*, and we can calculate ω_{BC}. We then square it and multiply it by *BC*, and we have the value of the normal component of $a_{C/B}$. This vector will be laid out from point *B* on the polygon directed parallel to *BC* with sense from *C* toward *B* on the mechanism layout. (See line *B*1 in Figure 4.22).

The tangential component of $a_{C/B}$ or $\alpha_{BC} \times BC$ cannot be computed as easily since α_{BC} is not readily available. However, we know that its inclination is perpendicular to *BC*, so we can draw a line in this direction through point 1. We know only that the point *C*, designating the head end of the acceleration vector of pin *C*, must lie somewhere on this line. (See Figure 4.22.)

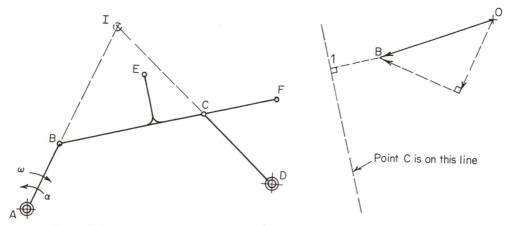

Figure 4.21 Instant center of coupler   **Figure 4.22** Acceleration polygon—step 2

To determine the acceleration of pin *C* completely, we can consider its acceleration relative to another point like *D*. Since *C* and *D* both lie upon the crank *CD*, it follows that:

$$a_C = a_{C/D} + \rightarrow a_D$$

The acceleration of *D* is zero (it is a fixed point), so we need only lay out the $a_{C/D}$ from the origin *O*. This, like all points on bodies in pure rotation, has two components:

$$a_{C/D} = (\omega_{CD}^2 \times CD) + \rightarrow (\alpha_{CD} \times CD)$$

$$\uparrow \qquad\qquad\qquad \uparrow$$

$$(\parallel \text{ to } CD) \qquad (\perp \text{ to } CD)$$

The angular velocity of crank CD can be found by finding the V_C, using I, the instant center of BC, as follows:

$$V_B = \omega_{AB} \times AB$$

$$V_C = V_B \times \frac{IC}{IB} = \omega_{AB} \times AB \times \frac{IC}{IB}$$

Then we know that:

$$\omega_{CD} = \frac{V_C}{CD} = \omega_{AB} \times \frac{AB}{CD} \times \frac{IC}{IB}$$

This value can be computed, squared, and multiplied by CD to obtain the normal component. The magnitude of this vector can be laid out to scale from O on the vector polygon, parallel to CD, and in the sense from C toward D on the mechanism drawing. (See line $O2$ in Figure 4.23.)

To this we must add the other component of $a_{C/D}$, which is $\alpha_{CD} \times CD$. Here again we cannot easily obtain α_{CD}, but we can predict that the direction of this αr vector is perpendicular to CD. If we lay out a line in this inclination through point 2 we can predict that the point C, designating the head end of the resultant acceleration vector of pin C, must lie somewhere on this perpendicular.

Now we have two lines which contain point C, the αr component of $a_{C/B}$ (through point 1) and the αr component of $a_{C/D}$ (through point 2). The intersection of these lines is the required point C, and the vector OC can be drawn, determining the absolute acceleration of pin C. By convention this vector is directed from O towards point C on the polygon (Figure 4.23).

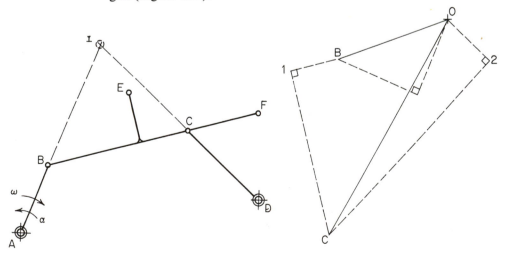

Figure 4.23 Acceleration of C—step 3

3. *Angular Acceleration of CD*

The tangential component of *C*'s acceleration relative to *D* (line 2*C*) is now determined in length as well as in direction and may be measured at the assigned scale. Since vector *OC* is the sum of vectors *O*2 and 2*C*, *O*2 is directed towards 2 and 2*C* is directed toward *C*. This component 2*C* is equal to $\alpha_{CD} \times CD$, so α_{CD} may be found by dividing the measured value of 2*C* by *CD*. The sense of α_{CD} may be determined by observing the sense of vector 2*C* with respect to *CD* on the mechanism. In this case α_{CD} is counterclockwise.

4. *Angular Acceleration of Body BCFE*

At any given instant, all lines on a rigid body have the same angular velocity and the same angular acceleration. Therefore *BC*, *BF*, and *BE* will all have the same angular acceleration. Let us determine this α_{BC} next, since it will be useful in finding the acceleration of points *F* and *E*. The angular acceleration of *BC* can readily be determined from the previous layout in Figure 4.23. In this polygon, line 1*C* is the tangential component of $a_{C/B}$ which is equal to $\alpha_{BC} \times BC$ in magnitude. We can divide the measured length of this component by *BC* and obtain the value of the angular acceleration of *BC*, which equals that of *BF* and *BE* or of any other line on body *BCFE*.

5. *Acceleration of F*

To find the acceleration of pin *F*, we observe its acceleration relative to pin *B*.

$$a_F = a_{F/B} + \rightarrow a_B$$

The acceleration of *B* has already been found, so we must add the relative acceleration vectors to *OB* in the polygon. (See Figure 4.23.) As in the case of pin *C*, *F*'s motion relative to *B* is that of pure rotation about *B* so that

$$a_{F/B} = (\omega_{BF}^2 \times BF) + \rightarrow (\alpha_{BF} \times BF)$$

$$\uparrow \qquad\qquad\qquad \uparrow$$

$$(\| \text{ to } BF) \qquad (\perp \text{ to } BF)$$

Since the coupler *BCF* is a rigid body, note that $\omega_{BF} = \omega_{BC}$ and $\alpha_{BF} = \alpha_{BC}$. (All lines on a rigid body have the same α and ω in any given position.) ω_{BC} has already been computed, so we can square this value and multiply it by *BF*. This vector is laid out to scale from point *B* on the polygon parallel to *BF* and in a sense from pin *F* toward pin *B* in the mechanism drawing (line *B*3 in Figure 4.24).

The value of α_{BC} already determined in 4 above is identical for α_{BF}, so we can multiply this value by *BF* to obtain the tangential component of *F* relative to *B*. This is now added to the normal component ($\omega_{BF}^2 \times BF$) laid out above. This αr component must be perpendicular to

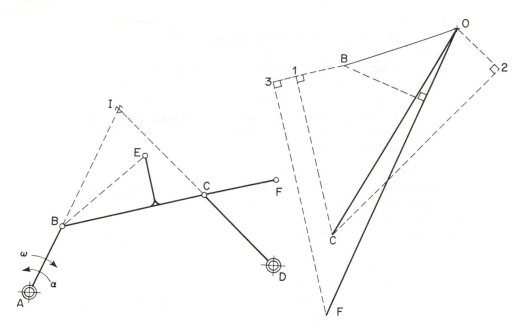

Figure 4.24 Acceleration of F—step 4

BF and in the same sense as the $\alpha_{BC} \times BC$ component (downward from 3 and parallel to $1C$). Point *F* will fall at the head end of this vector (line $3F$), and the absolute acceleration of *F* will be equal to the vector *OF*, directed from *O* towards *F*.

6. *Acceleration of E*

We find the acceleration of *E* by observing its acceleration relative to *B*:

$$a_E = a_{E/B} + \rightarrow a_B$$

in which the a_B is already known and

$$a_{E/B} = (\omega_{EB}^2 \times EB) + \rightarrow (\alpha_{EB} \times EB)$$

$$\uparrow \qquad\qquad\qquad \uparrow$$

(‖ to *EB* toward *B*)   (⊥ to *EB*)

But since *E*, *B*, and *C* are all on the same rigid body, that body has but one α and one ω in this position, so $\omega_{BC} = \omega_{EB}$ and $\alpha_{BC} = \alpha_{EB}$. We have already computed this ω and α. It is therefore easy to compute the magnitude of the normal component ($\omega_{EB}^2 \times EB$) and the tangential component ($\alpha_{EB} \times EB$) of *E*'s acceleration relative to *B*.

We have already determined the acceleration of *B* as line *OB*, so on the polygon we lay out from *B* the $\omega^2 r$ component (normal acceleration of *E* relative to *B*) parallel to line *EB* and directed from *E* toward *B* on the mechanism. (See line $B4$ in Figure 4.25(a).) To this we now add the αr component (tangential acceleration of E relative to B), perpendicular to EB (and to $B4$) and directed downward, consistent with the other αr components of accelerations relative to B. (See line $4E$.) The head end of this vector is labeled E, and the line OE is the vector of the acceleration of pin E, with the sense of O towards E.

In the process of this study, the vector polygon has been in a series of progressive steps in its growth. In practice of course only one polygon need be drawn for the entire investigation, each new vector being added to the system already drawn. This labor-saving technique results in a compact diagram which graphically illustrates relative values.

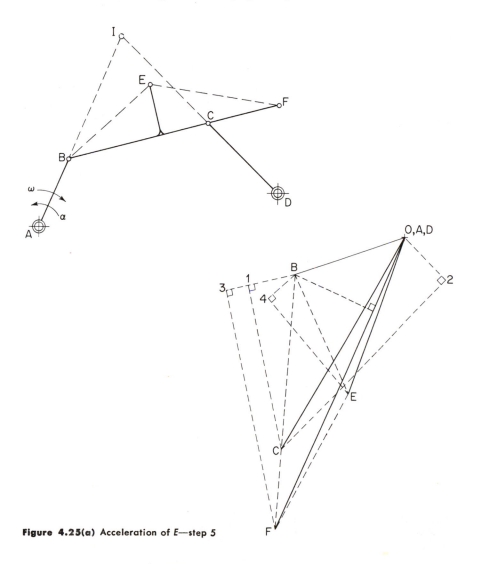

Figure 4.25(a) Acceleration of E—step 5

4.17 Reading Relative Accelerations

Point O on the vector polygon may also be labeled A or D since these are fixed points on the mechanism. (Point O may be thought of as an acceleration vector of zero length.) The vector BO (or BA) is the absolute acceleration of B or the acceleration of B relative to A, which has no motion. In like manner, we may "read" other relative accelerations directly from the vector polygon now laid out. For example, the vector CF represents the acceleration of pin C relative to pin F, directed from F toward C on the vector polygon. Similarly, the vector FE is the acceleration of pin F relative to pin E (directed toward F). These vectors are dotted in Figure 4.25(a).

4.18 Use of the Vector Image

If we compare the geometry of the triangle BEF on the coupler with the vector triangle labeled BEF, we will see that these two triangles are similar (the vector triangle is rotated clockwise through an angle of approximately $107°$). We will also see that point C falls on line BF in each triangle. (See Figure 4.25(b).)

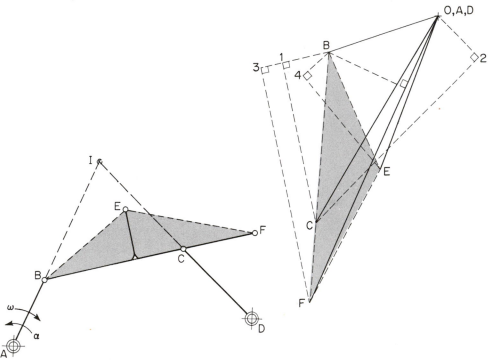

Figure 4.25(b) Vector image of coupler accelerations

This relationship is due to the fact that relative accelerations of points on the same rigid body ($\omega^2 r + \rightarrow \alpha r$) vary with r, the distance between them, and therefore all such accelerations make the same angle with their connecting line on the mechanism. This reveals a quick way of finding the accelerations of any new points on a coupler once the accelerations of any two points on that body have been determined. We first locate the new points on the vector image in positions corresponding to those on the mechanism layout; we then draw the required vectors from these points to the origin; and finally we measure off the magnitude at scale. This method is usually shorter than the relative-acceleration process since it eliminates all computations.

4.19 Accelerations on Bodies in Rolling Contact

When two bodies are connected by a pin, the acceleration of the pin defines the acceleration of one point on each body. By determining the acceleration of each successive pin in turn, we can make a complete analysis of a complex linkage.

If motion is transmitted from one member to another by other means, we focus our attention on the accelerations of the contact points in order to proceed from one member to another.

Let us consider accelerations on bodies in pure rolling contact. Wheel D (Figure 4.26) turns about fixed axis A with a given angular velocity ω_D and angular acceleration α_D, both clockwise, in the position shown. D drives wheel F, which turns about fixed axis B, with pure rolling contact at point P. The angular acceleration of F and the linear acceleration of point C on F is required.

First we determine the linear acceleration of point P on body D. Since D turns about a fixed axis,

$$a_P^D = \omega_D^2 \times AP + \rightarrow \alpha_D \times AP$$

As ω_D, α_D, and AP are all known, these components can be calculated and their resultant (a_P^D) drawn, as shown in Figure 4.26.

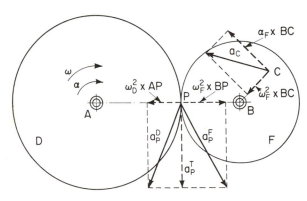

Figure 4.26 Accelerations in pure rolling contact

Now we turn to the contact point on body F (P^F). We know that $V_P^D = V_P^F$ at all times. If this velocity is changing, the change in magnitude of velocity during any time interval must be equal for both bodies:

$$\Delta V_P^D = \Delta V_P^F$$

The acceleration components of both points P must always remain identical in this tangential direction, so perpendicular to AB:

$$a_P^T \text{ on } D = \frac{\Delta V_P^D}{\Delta T} = \frac{\Delta V_P^F}{\Delta T} = a_P^T \text{ on } F$$

Therefore, these *tangential components* of the acceleration of the two points P are equal. Since F is in pure rotation, any point on F has two components of acceleration: a normal component toward B, equal to $\omega^2 r$, and a tangential component, perpendicular to AB, equal to αr. The resultant acceleration of P on F is the sum of these two:

$$a_P^F = \omega_F^2 \times BP + \rightarrow \alpha_F \times BP$$

The tangential component of the acceleration of P on F is the αr component:

$$a_P^T = \alpha_F \times BP$$

And, solving for α_F:

$$\alpha_F = \frac{a_P^T}{BP}$$

The sense is counterclockwise so as to be consistent with the vector a_P^T, as referred to wheel F.

Note that with rolling cylinders like D and F:

$$a_P^T = \alpha_D \times AP = \alpha_F \times BP$$

Or, rearranging:

$$\frac{\alpha_F}{\alpha_D} = \frac{AP}{BP} = \frac{\text{radius of } D}{\text{radius of } F}$$

So, like angular velocities, angular accelerations on two rolling cylinders are inversely proportional to their radii.

The linear acceleration of C (like all points on F) has two components:

$$a_C = \omega_F^2 \times BC + \rightarrow \alpha_F \times BC$$

From velocity analysis (Art 3.26)

$$\frac{\omega_F}{\omega_D} = \frac{\text{radius of } D}{\text{radius of } F}$$

Solving for ω_F:

$$\omega_F = \omega_D \times \frac{AP}{BP}$$

The two components of a_C can now be calculated and the resultant a_C defined, as shown in Figure 4.26.

Another example involving pure rolling contact is shown in Figure 4.27. Wheel W rolls without slip on fixed surface S. The angular velocity and angular acceleration of W are given in the position shown. Both are clockwise. The linear accelerations of center A and contact point B on W are required.

First, let us consider the motion of W. The wheel turns about A while A translates, so W is in combined motion. Furthermore, the path of A is known to be a straight horizontal line, parallel to S. We known from velocity study that B on W is the instant center of rotation for wheel W. This means that B has zero velocity, but it does not mean that B has zero acceleration. As the name implies, all instant centers are locations of a rotation center good only for one instant or one position of a body. An instant later the center has a different position. No matter what position it occupies, it is just about to change that position or about to acquire a velocity. This impending change in velocity proves that an instant center has acceleration.

We can predict, in this case, that B on W can only have acceleration in a vertical direction, toward A. Both points B, on W and S, have identical velo-

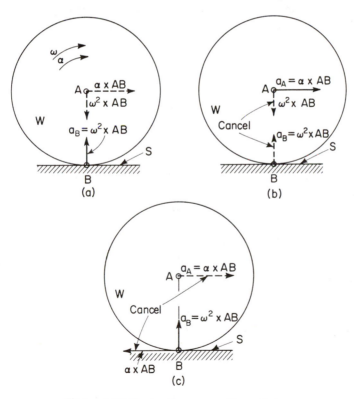

Figure 4.27 Accelerations on a rolling cylinder

cities along surface S because there is no slip. Since B^s has zero velocity, B^W also has zero velocity. Surface S imposes complete restraint upon B^W in every direction except vertically upward. Therefore, as W turns, any velocity which B^W acquires must be in this vertical direction. A change in the velocity of B^W from zero to a real value causes acceleration, so it follows that the acceleration of B^W in the contact position can only be directed upward toward A.

The relative-acceleration equation states:

$$a_A = a_{A/B} + \rightarrow a_B$$

And since

$$a_{A/B} = \omega_W^2 \times AB + \rightarrow \alpha_W \times AB$$

it follows that:

$$a_A = \omega_W^2 \times AB + \rightarrow \alpha_W \times AB + \rightarrow a_B$$

The $\omega_W^2 \times AB$ component is directed toward B, whereas we know that the resultant a_A is horizontal. Since the $\alpha_W \times AB$ component is known to be horizontal, the only component which can cancel the $\omega_W^2 \times AB$ component is a_B. Therefore, a_B must have a component toward A, equal to $\omega_W^2 \times AB$. We have shown above that a_B is entirely vertical, with no horizontal component along surface S, This proves that:

$$a_B = \omega_W^2 \times AB \text{ directed toward } A \quad \text{[Figure 4.27(a)]}$$

If we substitute this value in the equation for a_A:

$$a_A = (-\omega_W^2 \times AB) + \rightarrow \alpha_W \times AB + \rightarrow (+\omega_W^2 \times AB)$$

The $\omega_W^2 \times AB$ component of $a_{A/B}$ is downward toward B, while the $\omega_W^2 \times AB = a_B$ is upward toward A, so the component of $a_{A/B}$ is given a minus sign in the above equation.* These two components will cancel, leaving $a_A = \alpha_W \times AB$ directed parallel to S toward the right [Figure 4.27(b)].

The a_B can now be verified, since $a_B = a_{B/A} + \rightarrow a_A$

$$\therefore a_B = \omega_W^2 \times AB + \rightarrow (-\alpha_W \times AB) + \rightarrow (+\alpha_W \times AB)$$

To be consistent with α, the $\alpha_W \times AB$ component of $a_{B/A}$ is toward the left, therefore negative above, while a_B is positive. These will cancel, leaving $a_B = \omega_W^2 \times AB$ [Figure 4.27(c)].

An epicyclic wheel train is shown in Figure 4.28. Arm A and wheel D turn independently about fixed axis O. Wheel B, carried on pin M on A, rolls without slip on D and the fixed ring C. The angular velocity and angular acceleration of A are both counterclockwise and known. The angular acceleration of D is required. Pin M, on rotating body A, has a tangential component

---

*Assume that accelerations directed toward the right or upward are positive (+) and those directed downward or toward the left are negative (−).

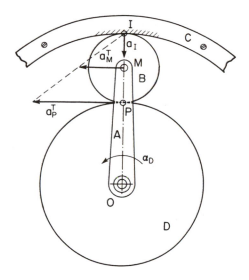

Figure 4.28 Accelerations on epicyclic rolling cylinders

of acceleration (a_M^T) and a normal component (a_M^N). Since we are seeking α_D here, we are mainly concerned with tangential components, which alone involve α.

$$a_M^T = \alpha_A \times OM \quad \text{(Figure 4.28)}$$

This tangential acceleration of M on A also equals the tangential acceleration of M on B. On wheel B, point I (the instant center of rotation) has no acceleration tangential to fixed ring C, as we learned in the former example ($a_I^T = 0$).

$$\alpha_B = \frac{a_{M/I}^T}{IM} = \frac{a_M^T - a_I^T}{IM}$$

But since $a_I^T = 0$;

$$\alpha_B = \frac{a_M^T}{IM}$$

The tangential component of the acceleration of P equals:

$$a_P^T = \alpha_B \times IP \text{ (shown in Figure 4.28)}$$

This tangential acceleration of P is the same for both bodies B and D, since there is no slip at P.

$$\alpha_D = \frac{a_{P/O}^T}{PO} = \frac{a_P^T - a_O^T}{PO}$$

But $a_O^T = 0$, therefore

$$\alpha_D = \frac{a_P^T}{PO}$$

4.20 Choice of Methods for Acceleration Study

The relative-acceleration method, whether executed by effective components or by vector polygons, is quite as analytical as it is graphical. Vectors are used to picture the calculated values and to determine resultants without resorting to unwieldy trigonometry. Once the student becomes familiar with the fundamentals of the relative-acceleration concept, the vector-polygon method is recommended as the most efficient and accurate graphical technique. It can be employed on rolling-contact problems just as effectively as on pin-connected mechanisms.

As we consider the acceleration analysis of mechanisms involving sliding along moving guides, the relative-acceleration theory becomes much more complex. Since this text is written especially for freshmen and sophomores and is dedicated to the exploitation of graphical methods, we will first undertake a simple and very enlightening graphical technique for sliding-contact acceleration studies before attempting a more analytical and sophisticated method.

The *velocity-difference method*, which we will employ in the next article, not only offers an excellent indoctrination in sliding-contact accelerations but may just as effectively be used on pin-connected linkages, rolling-contact devices, or, in fact, on any kind of mechanism in which acceleration is to be investigated. This method is so clear and graphic that we would use it for all acceleration studies but for one disadvantage—there are instances in which it does not yield sufficiently accurate results. *All graphical techniques are dependent not only on the theory but on the configuration of the geometry*, and, in some instances, accelerations are so sensitive that the degree of graphical precision that can ordinarily be employed does not prove adequate for reliable results. The educational value of the velocity-difference method far outweighs this occasional fault, however, so for that reason it is offered here as an introductory technique before we undertake a more precise and more complex solution.

4.21 The Velocity-Difference Method for Accelerations

This method has the advantage of simplicity, since it is based upon fundamental definitions rather than involved formulas and uses graphical solutions to avoid lengthy calculations. It employs velocity analysis directly in determining accelerations, logically relating these two studies instead of depending upon separate special techniques.

Linear acceleration is defined as the rate of change of velocity with respect to time:

$$a = \frac{\Delta V}{\Delta T}$$

If we use a small finite change in velocity ΔV and represent the corresponding interval of time (during which the velocity change took place) as ΔT, the definition may be expressed as the limit of the ratio of ΔV to ΔT as ΔT approaches zero. ΔV is the difference between the final velocity at the end of the time interval and the initial velocity at the start of the interval:

$$\Delta V = V_f - V_0$$

So, to determine the average acceleration for the interval ΔT, we must find this ΔV and divide it by the ΔT in which the velocity change occurred. Applications of this method are shown in the examples that follow.

Example 1

In the linkage in Figure 4.29, crank AB turns counterclockwise about fixed axis A at constant speed of 1 radian/sec. The acceleration of slider C in the position shown is required.

$$a_C = \frac{\Delta V_C}{\Delta T}$$

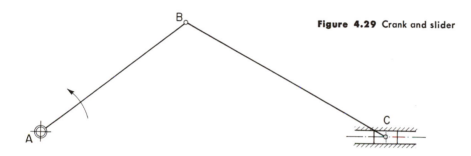

Figure 4.29 Crank and slider

We must determine ΔV_C for a small interval of time ΔT. The size of ΔT is dependent upon the speed of the mechanism and can most easily be expressed in terms of angular displacement of AB, since ω_{AB} is specified. Experience shows that the time during which AB turns $\frac{1}{10}$ radian is usually a reasonable and convenient value to use for ΔT. (As shown in Figure 4.30, an angle of $\frac{1}{10}$ radian may be accurately laid out by measuring a $\frac{1}{2}$-in. chord on an arc of 5 in. radius. For $\frac{1}{20}$ radian a chord of $\frac{1}{4}$ in. may be used on the same radius of arc.) Rather than calculate this ΔT in sec, it may be expressed in terms of angular velocity and angular displacement.

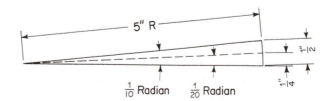

Figure 4.30 Layout of $\frac{1}{10}$ radian

By definition:

$$\omega = \frac{\Delta\theta}{\Delta T} \quad \text{so} \quad \Delta T = \frac{\Delta\theta}{\omega}$$

The acceleration obtained by this method is an average for the time interval used. To get accurate results for acceleration in the given position, we should consider this given position to be halfway between the initial and final positions of the linkage at the start and end of time ΔT. The initial position of AB will then be AB_0, $\frac{1}{20}$ radian clockwise from the given position AB which was shown in Figure 4.29. The final position AB_f will be $\frac{1}{20}$ radian counterclockwise from the given position AB. Figure 4.31 shows these displacements of the crank.

To determine ΔV_C, we first lay out the linkage in position AB_0C_0 and find the velocity of C_0 by the usual methods; using instant center I_0:

$$V_{C_0} = V_B \times \frac{I_0C_0}{I_0B_0} = \omega_{AB} \times AB \times \frac{I_0C_0}{I_0B_0}$$

(substituting $V_B = \omega_{AB} \times AB$)

Next we lay out the linkage in position AB_fC_f (with AB displaced $\frac{1}{10}$ radian from AB_0C_0) and determine V_{C_f}.

$$V_{C_f} = \omega_{AB} \times AB \times \frac{I_fC_f}{I_fB_f}$$

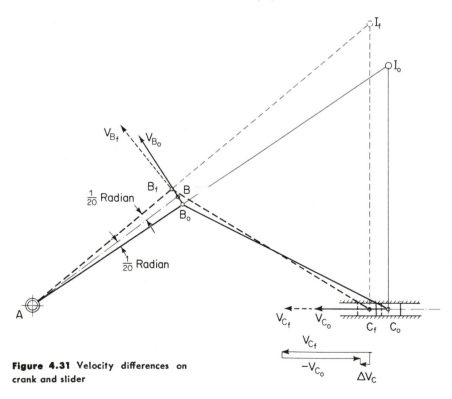

Figure 4.31 Velocity differences on crank and slider

The velocity difference $\Delta V_C = V_{C_f} - \rightarrow V_{C_0}$. Note that this is ordinarily a vector difference, but in this case, where both velocities have the same inclination, an algebraic difference is also correct.

The acceleration of C in the given position shown in Figure 4.29 will be:

$$a_C = \frac{\Delta V_C}{\Delta T} = \frac{\Delta V_C}{\Delta \theta_{AB}/\omega_{AB}} = \frac{\Delta V_C \times \omega_{AB}}{\Delta \theta_{AB}} = \frac{\Delta V_C \times 1}{\frac{1}{10}} = \Delta V_C \times 10$$

The sense of a_C is the same as that of the vector ΔV_C, provided that ΔV_C is obtained by subtracting V_{C_0} from V_{C_f}.

Note that this method involves only definitions and proficiency in velocity analysis. No special acceleration techniques are employed.

4.22 Accelerations Involving Sliding on Moving Guides

The complications usually attendant upon acceleration analysis of mechanisms with sliding contact (Coriolis) are not encountered when the *velocity-difference method* is used, as illustrated in the next example.

Example 2

Crank AB turns counterclockwise about fixed axis A at a constant speed of 10 radians per second. B is pinned to a sleeve which slides freely on arm CD. Arm CD swings about fixed axis D.

Find the acceleration of C and the angular acceleration of CD when the mechanism is in the position shown in Figure 4.32.

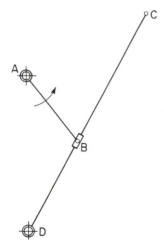

Figure 4.32 Crank and slider with extended connecting rod

As in the previous example:

$$a_C = \frac{\Delta V_C}{\Delta T} = \frac{\Delta V_C \times \omega_{AB}}{\Delta \theta_{AB}}$$

Selecting a $\Delta \theta_{AB} = \frac{1}{10}$ radian, we first draw the mechanism in the position AB_0C_0D in which AB is displaced $\frac{1}{20}$ radian clockwise from the given position. This is shown Figure 4.33. We now determine V_{C_0} in this position by the usual velocity vector method.

Next, we lay out the mechanism in position AB_fC_fD, in which AB_f is displaced $\frac{1}{10}$ radian from AB_0, counterclockwise, and find V_{C_f} in this position (Figure 4.33).

V_{C_f} and V_{C_0} differ in both inclination and magnitude in this case, so ΔV_C must be obtained by subtracting vectorially:

$\Delta V_C = V_{C_f} - \rightarrow V_{C_0}$ as shown in the vector diagram in Figure 4.34. As this ΔV_C is small, the scale for this diagram should be much larger than that used for the velocity analysis. As is the case in all graphical solutions, this method demands drafting precision of high quality if results are to compare favorably with analytical methods.

The acceleration of C in the given position in Figure 4.32 will be:

$$a_C = \frac{\Delta V_C \times \omega_{AB}}{\Delta \theta_{AB}} = \frac{\Delta V_C \times 10}{\frac{1}{10}} = \Delta V_C \times 100$$

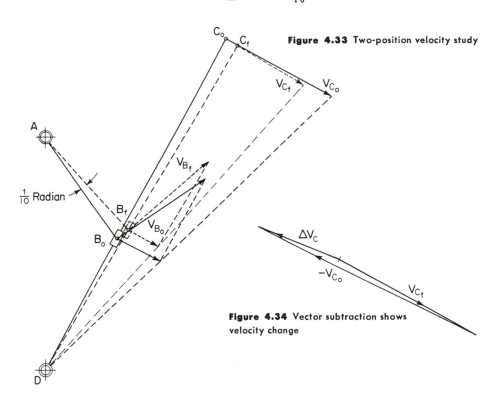

Figure 4.33 Two-position velocity study

Figure 4.34 Vector subtraction shows velocity change

The inclination and sense of the a_C are the same as those of the velocity difference vector ΔV_C in Figure 4.34.

The angular acceleration of CD is, by definition,

$$\alpha_{CD} = \frac{\Delta \omega_{CD}}{\Delta T}$$

When the mechanism is in the initial position AB_0C_0D, the V_{C_0} has been determined. Since, by definition, $V = \omega r$, the angular velocity of C_0D in this position will be:

$$\omega_{C_0 D} = \frac{V_{C_0}}{CD}$$

Similarly, we can determine the angular velocity of C_fD when the mechanism is in the final position AB_fC_fD, which will be:

$$\omega_{C_f D} = \frac{V_{C_f}}{CD}$$

The change in ω_{CD} during the time ΔT (while AB turned through $\frac{1}{10}$ radian) is the angular velocity difference:

$$\Delta \omega_{CD} = \omega_{C_f D} - \omega_{C_0 D}$$

This is an algebraic difference, since all angular velocities are in the same plane.

The angular acceleration of CD for the given position shown in Figure 4.32 will then be:

$$\alpha_{CD} = \frac{\Delta \omega_{CD}}{\Delta T} = \frac{\Delta \omega_{CD} \times \omega_{AB}}{\Delta \theta_{AB}} = \frac{\Delta \omega_{CD} \times 10}{\frac{1}{10}} = \Delta \omega_{CD} \times 100$$

The sense of α_{CD} is counterclockwise, the same as the sense of $\Delta \omega_{CD}$ ($\Delta \omega_{CD}$ is counterclockwise here since ω_{C_0} is greater than ω_{C_f} and both are in the clockwise direction).

In each of the above examples, the driving member turned at constant angular speed. This is the usual condition, as most mechanisms are powered by constant speed motors. There are exceptions in gravity driven devices, etc., in which the driving crank has angular acceleration.

Example 3

Crank AB turns about fixed axis A with a constant angular acceleration of 10 radians/sec/sec, clockwise. In the position shown in Figure 4.35, the angular velocity of AB is 5 radians/sec, clockwise.

The acceleration of point E is required in the given position. As before, $a_E = (\Delta V_E / \Delta T)$ but ΔT cannot be expressed as $\Delta \theta / \omega$ since ω is not constant for all positions of AB. However, since α_{AB} is known, we may express ΔT in terms of α and ω.

By definition:

$$\alpha = \frac{\Delta \omega}{\Delta T} \quad \text{or} \quad \Delta T = \frac{\Delta \omega}{\alpha}$$

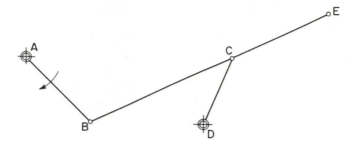

Figure 4.35 Four-bar linkage

Then:

$$a_E = \frac{\Delta V_E}{\dfrac{\Delta \omega_{AB}}{\alpha_{AB}}} = \frac{\Delta V_E \times \alpha_{AB}}{\Delta \omega_{AB}}$$

Taking ΔT as the time in which AB is displaced $\frac{1}{10}$ radian, we first draw the mechanism in position $AB_0C_0DE_0$ with AB_0 rotated $\frac{1}{20}$ radian counter-clockwise from the given position (see Figure 4.36). We must now find ω_{AB_0} which, owing to α_{AB}, is somewhat slower than ω_{AB} in the given position shown in Figure 4.35.

Figure 4.36 Two-position velocity study

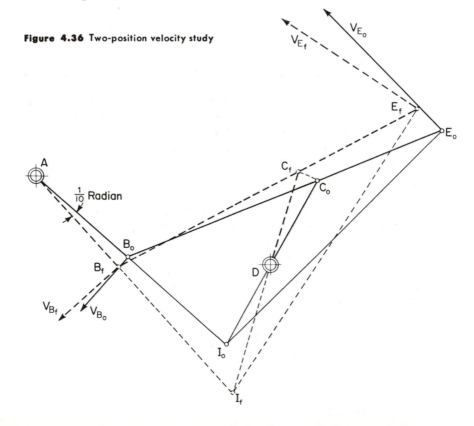

To do this we may apply the familiar constant acceleration formula (from Article 4.6):

$$\omega_f^2 = \omega_0^2 + 2\alpha\theta$$

which here reads:

$$(\omega_{AB})^2 = (\omega_{AB_0})^2 + 2\alpha\,\Delta\theta \quad (\Delta\theta = \tfrac{1}{20}\ \text{radian})$$

From this:

$$(\omega_{AB_0})^2 = (\omega_{AB})^2 - 2\alpha\,\Delta\theta = 5^2 - 2 \times 10 \times \tfrac{1}{20} = 24$$

$$\omega_{AB_0} = \sqrt{24} = 4.9$$

Using this value of ω_{AB_0} we now determine V_{E_0} by the usual methods:

$$V_{E_0} = V_B \times \frac{I_0 E_0}{I_0 B_0} = \omega_{AB_0} \times AB_0 \times \frac{I_0 E_0}{I_0 B_0}$$

Next we must find V_{E_f}, so we lay out the mechanism in position $AB_f C_f DE_f$ in which AB_f is displaced $\tfrac{1}{20}$ radian clockwise from the given position AB. It has been shown above that in displacing AB $\tfrac{1}{20}$ radian, the change in ω_{AB} was 0.1 radian/sec $(5 - 4.9 = 0.1)$. Since α is constant, we can predict that, in turning from AB to AB_f, the change in ω_{AB} will also be 0.1 radian per second. The value of ω_{AB_f} will therefore be:

$$5 + 0.1 = 5.1\ \text{radian/sec}$$

Using this value of ω_{AB_f} we find

$$V_{E_f} = \omega_{AB_f} \times AB \times \frac{I_f E_f}{I_f B} \quad \text{(Figure 4.36)}$$

Now

$$\Delta V_E = V_{E_f} \longrightarrow V_{E_0}$$

This is shown in magnitude, inclination and sense by the large scale vector subtraction diagram in Figure 4.37.

$$\Delta\omega_{AB} = \omega_{AB_f} - \omega_{AB_0} = 5.1 - 4.9$$

$$= 0.2\ \text{from the values obtained above}$$

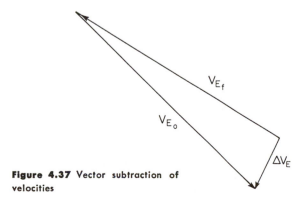

Figure 4.37 Vector subtraction of velocities

Therefore:

$$a_E = \frac{\Delta V_E \times \alpha_{AB}}{\Delta \omega_{AB}} = \frac{\Delta V_E \times 10}{0.2} = \Delta V_E \times 50$$

This a_E has the same inclination and sense as the vector ΔV_E in Figure 4.37.

Although the mechanisms used here are rather simple examples, this method is never more complex than the velocity analysis upon which it depends. Drafting precision and large scales for vector diagrams are essential to accuracy, as is true of all graphical methods. In cases where the velocity difference is very small, this method may prove to be quite inaccurate. Where velocity differences are significant (i.e., where accelerations are large enough to be important), it is reasonably accurate. The simplicity of exploiting definitions and expanding velocity analysis, as compared to learning special and involved techniques for acceleration study, justifies the velocity-difference method as a valuable tool.*

4.23 Coriolis Acceleration

The velocity-difference method offers a clear exposition of the basic velocity and acceleration relationships in which there is sliding along moving guides. With this introduction, we can now pursue a more rigorous analysis which is more complicated in principle but which has the advantage of yielding more consistently accurate solutions.†

Let us start with the simple mechanism shown in Figure 4.38. A circular disk M turns with a constant angular velocity ω about a fixed axis at its center O. A straight radial slot is cut in the face of the disk. In the slot is a block which slides radially outward. Consider two coincident points A and B at distance r from O. Point A is on disk M, directly beneath point B, which is on the sliding block. Let it be required that we determine the absolute acceleration of B.

*Reference: John A. Hrones, "A Job Shop Approach to Mechanism Analysis," *Machine Design*, February, 1954.

†If he has mastered the velocity-difference method, the student has at his command one tool with which to determine accelerations on mechanisms where there is sliding contact in moving guides. This method is general and can be applied to all mechanisms. For this reason, the use of Coriolis acceleration is not absolutely essential in order to solve the special problems described above. It is educationally challenging and provides a degree of accuracy necessary in certain instances, but it can be omitted if time is limited. While it is necessary to study instantaneous accelerations when mechanisms are in one given position, in practice it is usually required to make a comprehensive study of the mechanism throughout its complete motion cycle. Competence in this is our ultimate practical objective.

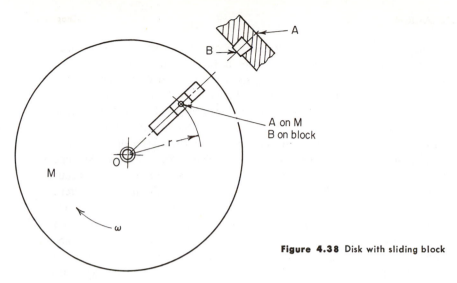

Figure 4.38 Disk with sliding block

1. The velocity of A (on M) is equal to $V_A = \omega r$ and is perpendicular to OA. The acceleration of A is equal to $a_A = \omega^2 r$ and is directed radially towards O.
2. Now, assuming that B (on the block) is moving radially outward in the slot with a velocity relative to A of $V_{B/A}$, the absolute $V_B = V_A + \rightarrow V_{B/A}$, following the relative-velocity principle.
3. If this $V_{B/A}$ is not constant, then B has an acceleration relative to A and the absolute $a_B = a_A + \rightarrow a_{B/A}$ (Figure 4.39). At first thought, this appears to be a proper and complete expression for the absolute acceleration of B.

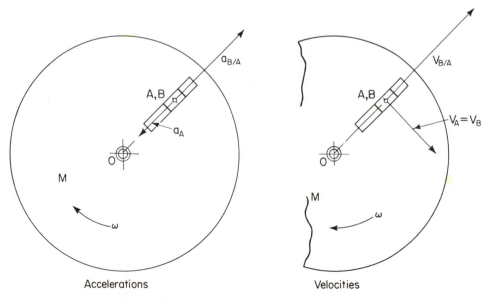

Accelerations   Velocities

Figure 4.39 Incomplete acceleration of block B

This, however, is not true, because this equation fails to take into account two other changes which are taking place in the velocity of point B: first, the change in direction of $V_{B/A}$ due to the rotation of the disk M, and second, the change in magnitude of the tangential velocity of B (V_B^T) due to the change in its distance from O. $(V_B^T = \omega r$, and r is increasing.) The complete equation for the absolute a_B must contain terms to express these two velocity changes.

First, consider only the first change. As M turns, the direction of $V_{B/A}$ changes with the rotating line OA, to position OA'. (See Figure 4.40(a).) Since B is not moving radially along OA in this instance, points A and B are the same and points A' and B' are also coincident. In a small time increment ΔT, this change in direction will be $\Delta \theta$ and the path ΔS from A to A' will be equal to $r\Delta\theta$. For small angles, this arc $r\Delta\theta$ may be considered equal to the chord AA'. In this same time interval ΔT, the vector $V_{B/A}$ will also turn through $\Delta\theta$. The change in $V_{B/A}$ due to this rotation is shown in the vector subtraction in Figure 4.40(b) as vector ΔV. The triangle OAA' in Figure 4.40(a) is similar to the vector triangle 123 (corresponding sides are parallel), so it follows that:

$$\frac{r\Delta\theta}{r} = \frac{\Delta V}{V_{B/A}}$$

so: $\Delta V = V_{B/A}\Delta\theta$

The acceleration of B due to this direction change will be:

$$a_B = \frac{\Delta V}{\Delta T} = \frac{V_{B/A}\Delta\theta}{\Delta T}$$

But

$$\frac{\Delta\theta}{\Delta T} = \omega$$

So $a_B = V_{B/A}\omega$, due to the first change.

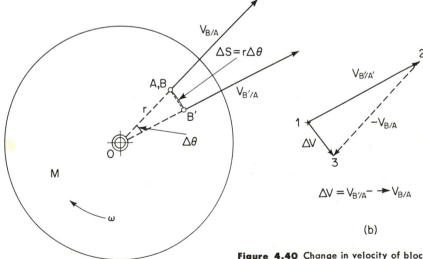

Figure 4.40 Change in velocity of block due to rotation

The direction of this acceleration is perpendicular to OA (parallel to ΔV) or tangential and consistent with the sence of ω.

Now, consider only the second change. As B moves out radially from O, the radius OB increases to OB' and, since $V = \omega r$, the tangential $V_{B/A}$ will increase (Figure 4.41). In a small time increment ΔT, the radius OB will change an amount equal to Δr and the change in $V_{B/A}$ will be $\omega \Delta r$. The acceleration of B due to this change in r will be:

$$a_B = \frac{\Delta V}{\Delta T} = \frac{\omega \Delta r}{\Delta T} = \omega V_{B/A} \quad \left(\frac{\Delta r}{\Delta T}, \text{ like } \frac{\Delta S}{\Delta T}, = V \right)$$

This acceleration is also perpendicular to OA or tangential and consistent with the sense of ω.

Since these two additional terms for the acceleration of B relative to A are equal and in the same direction, they may be added together into one term:

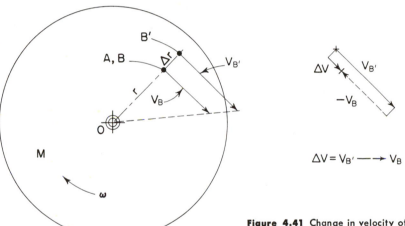

Figure 4.41 Change in velocity of block due to sliding

$2V_{B/A}\omega$. This term is called the *Coriolis acceleration*, after its discoverer. *This vector is always directed perpendicular to $V_{B/A}$, and it is pointed as if it had been rotated about its tail end through an angle of $90°$ with $V_{B/A}$ in the same direction as ω.*

This Coriolis term $2V_{B/A}\omega$ must be added to the incomplete equation in 3 above to give the absolute acceleration of B:

$$a_B = a_A + \rightarrow a_{B/A} + \rightarrow 2V_{B/A}\omega$$

In this equation, B is a point on the sliding block. A is a point on disk M, coincident with B. The term $a_{B/A}$ refers to the acceleration of B in a radial direction along the slot, relative to point A. (It could be described as "the acceleration of B when disk M is stationary.") The Coriolis term $2V_{B/A}\omega$ is directed perpendicular to the slot and is due to the angular velocity of M and the change in position of the block in the slot. A diagram of the system of vectors which determine the absolute acceleration of B (the vector polygon) is shown in Figure 4.42. (Relative size of the vectors is arbitrary.)

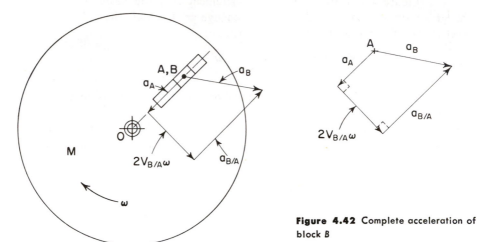

Figure 4.42 Complete acceleration of block B

In the above example, disk M is turning with a constant angular velocity. If M also had angular acceleration (Figure 4.43), the same considerations would hold true, except that the first term (a_A) would have tangential as well as normal acceleration components and the acceleration of A would be:

$$a_A^N + \rightarrow a_A^T \quad \text{or} \quad \omega^2 r + \rightarrow \alpha r$$

The other terms of the equation would remain unchanged, so the correct equation for these conditions would be:

$$a_B = a_A^N + \rightarrow a_A^T + \rightarrow a_{B/A} + \rightarrow 2V_{B/A}\omega$$

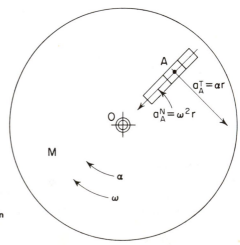

Figure 4.43 Disk with angular acceleration

The two examples above apply to points, like B, that move along straight lines on a rotating body. If the slot that guides B is curved with center of curvature at C (Figure 4.44) and the disk M has both ω and α, the basic equation again holds true, with further modifications:

1. The first term (a_A) would become $a_A^N + \rightarrow a_A^T$ because of the angular acceleration α.
2. The second term $a_{B/A}$ is now clearly described as $a_{B/M}$ and will include a normal and a tangential component, since the curved slot changes the direction of $a_{B/M}$ as well as its magnitude. The $a_{B/M}$ would then have two components:

$$a_{B/M}^N + \rightarrow a_{B/M}^T \quad \text{or} \quad \omega_{BC}^2 \times BC + \rightarrow \alpha_{BC} \times BC$$

(The ω and α in this case are the ω and α of the line BC on the disk M).
3. The Coriolis component remains unchanged, so the a_B in this general case would be:

$$a_B = a_A^N + \rightarrow a_A^T + \rightarrow a_{B/M}^N + \rightarrow a_{B/M}^T + \rightarrow 2V_{B/A}\omega$$

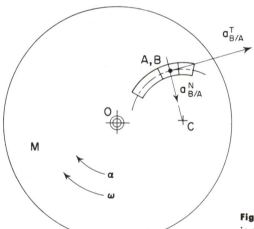

Figure 4.44 Accelerating disk with block in curved slot

Let us try a problem involving Coriolis' acceleration. In Figure 4.45 the slotted crank M turns about fixed axis O with a constant angular velocity of 5 radians/sec, clockwise. Arm DB turns about fixed axis D. A block which slides in the slot in M is pinned to DB at B (so B is a point on the block and on DB). Point A is a point on M directly beneath pin B. Arm DB is 3 in. long. In the position shown, points A and B are 4 in. from O. Centerline DO is $2\frac{1}{2}$ in.

Let it be required that we find the absolute acceleration of pin B and the angular acceleration of arm DB. The equation for the acceleration of B is:

$$a_B = a_A + \rightarrow a_{B/A} + \rightarrow 2V_{B/A}\omega_M$$

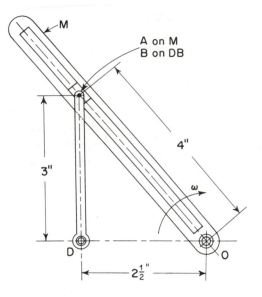

Figure 4.45 Slotted crank mechanism

We can compute the $a_A = \omega_M^2 \times OA = 5^2 \times 4 = 100$ in./sec^2, and we know it is directed from A toward O. Only the direction of $a_{B/A}$ is known. Since, relative to M, pin B can only move along the slot, its acceleration relative to M must be in the direction OB. We can compute the Coriolis component if we first find $V_{B/A}$. This is determined graphically as shown in Figure 4.46, where V_A is subtracted from V_B, yielding $V_{B/A} = 16.3$ in./sec (which, incidentally, is the velocity of slip).

$$2V_{B/A}\omega_M = 2 \times 16.3 \times 5 = 163 \text{ in./sec}^2$$

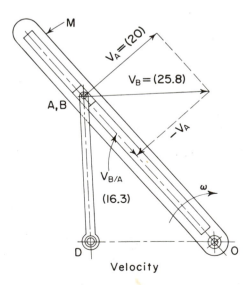

Figure 4.46 Velocity of block B relative to A on M

Velocity

We know the direction of this vector: It is rotated 90° from $V_{B/A}$ (which is along OA) about its tail end in the direction of ω_M (clockwise).

If we determine the angular velocity of DB, we can compute the normal component of the acceleration of B ($\omega_{DB}^2 \times DB$).

$$\omega_{DB} = \frac{V_B}{DB} = \frac{25.8}{3} = 8.6 \text{ rad/sec} \quad \text{so} \quad a_B^N = 8.6^2 \times 3 = 222 \text{ in./sec}^2$$

and is directed from B towards D.

Since we know only the direction of $a_{B/A}$, let us rewrite the equation so that we may solve for that term:

$$a_{B/A} = a_B - \rightarrow a_A - \rightarrow 2V_{B/A}\omega_M$$

$$\text{(in which } a_B = \omega_{DB}^2 \times DB + \rightarrow \alpha_{DB} \times DB)$$

We can now draw an acceleration polygon since 3 of the 5 vectors are completely known in magnitude and direction and the directions of the other 2 are also known. This polygon is shown in Figure 4.47. Starting from origin O, we lay out $-a_A$ (opposite in sense from $+a_A$), then $-2V_{B/A}\omega$, then $+a_B^N$, then a line in the direction of $a_B^T(\perp$ to $a_B^N)$, and finally, a line through O in the direction of $a_{B/A}$ ($\parallel$ to OA). These lines of a_B^T and $a_{B/A}$ intersect at a point B, thus determining the length of both of these vectors. The vector $a_{B/A}$ is the resultant of the others (see equation above), so this one is directed away from the origin, towards the intersection B.

If we examine this polygon, we will note that a_B^N and a_B^T are two adjacent (and $\perp$) vectors. Since BD is a rotating body, these two components may be added vectorially to yield the resultant a_B. This vector is shown in Figure 4.47. It measures 225 in./sec² and is directed down to the left of the arm BD.

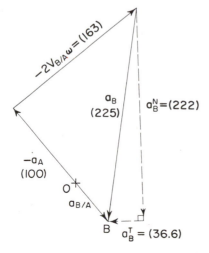

Acceleration

Figure 4.47 Vector polygon for acceleration of block B

The tangential component of B's acceleration (a_B^T) is equal to $\alpha_{BD} \times BD$ and is 36.6 in./sec². Therefore,

$$\alpha_{BD} = \frac{a_B^T}{DB} = \frac{36.6}{3} = 12.2 \text{ radians/sec}^2$$

Coriolis acceleration is not limited to linkages with sliding in moving guides. It applies to all mechanisms in which a point follows a straight or curved path on a rotating body.

PROBLEMS

4.1. In a "drag" race, a hot rod starting from a standstill reaches a speed of 50 mph in 10 sec. What is the average acceleration during this time? If the acceleration of the car was uniform, how far does the car travel before reaching 50 mph?

4.2. A sprinter in a track meet runs the 100-yd dash in 10 sec flat. If his acceleration was uniform, what was his maximum speed in ft/sec?

4.3. A body dropped from rest has a constant acceleration of 32 ft/sec/sec. Determine the distance it falls during each successive period of 1 sec for 4 sec. What is the ratio of the distance covered in the first period to the second period, to the third, and to the fourth? (This ratio of displacements during each successive equal period of time is the same ratio for all bodies moving with constant linear acceleration).

4.4. Upon being started, a motor accelerates uniformly until it reaches its running speed of 3600 rpm. If it requires $\frac{3}{10}$ sec to accelerate from 1000 rpm to 1900 rpm, what is its acceleration and how many revolutions must it make before it reaches its running speed?

4.5. In the four-bar linkage shown in Figure P4.5, crank LM turns about fixed axis L with a constant angular velocity of 100 radians/min. $LM = TR = 4$ in., $MR = LT = 6$ in., $MS = 3.5$ in., and $TX = 2$ in. Determine the accelerations of points S and X in the position shown.

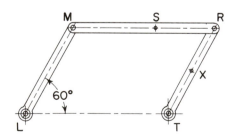

Figure P4.5

4.6. Wheel W turns about fixed axis O with an angular velocity of 3 radians/sec clockwise and an angular acceleration of 12 radians/sec/sec counterclockwise in the position shown in Figure P4.6. Determine the linear accelerations of points A, B, and C on the wheel.

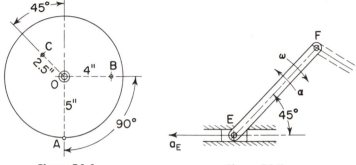

Figure P4.6   **Figure P4.7**

4.7. The absolute acceleration of pin E on the bar EF is 2 in./sec/sec toward the left. The angular acceleration of EF is 3 radians/sec/sec counterclockwise, and the angular velocity of EF is 1 radian/sec clockwise, in the position shown. Find the absolute acceleration of F (Figure P4.7).

4.8. In Figure P4.8, AB turns about fixed axis A. The mechanism is driven by slider C, which has a constant velocity, right to left, of 2 in./sec. $AB = 2$ in. and $BC = 4$ in. Determine the linear acceleration of B and the angular acceleration of AB in the position shown.

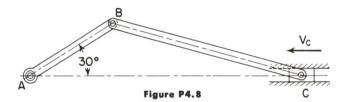

Figure P4.8

4.9. Links AB, BC, and CD of the linkage shown in Figure P4.9 are each 3 in. long. A and D are fixed axes. AB turns counterclockwise with a constant angular velocity of 2 radians/sec. Determine the acceleration of point C.

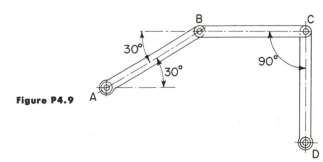

Figure P4.9

4.10. In Figure P4.10 crank DA turns about fixed axis D with an angular velocity of 2 radians/sec counterclockwise and an angular acceleration of 4 radians/sec/sec clockwise. $DA = AB = BC = CA = 3$ in. Determine the acceleration of slider C, the angular acceleration of the body F, and the acceleration of point B.

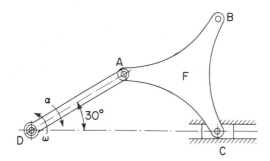

Figure P4.10

4.11. Crank DE in Figure P4.11 has a constant angular velocity of 1 radian/sec. $DE = 2$ in., $EG = 4$ in., $FG = 2.5$ in., and $DF = 6$ in. H is the midpoint of EG. D and F are fixed axes. Find the absolute acceleration of G and H and the angular acceleration of FG.

Figure P4.11

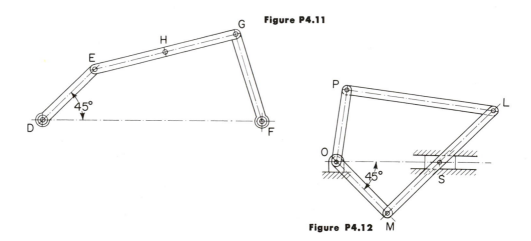

Figure P4.12   M

4.12. Links OM and OP turn independently about fixed axis O in Figure P4.12. Rigid bar ML is pinned at S to the block sliding in fixed guides. $OM = OP = 2$ in., $ML = PL = 4$ in., and $MS = 2$ in. Driving crank OM has an angular velocity of 1 radian/sec and an angular acceleration of 2 radians/sec/sec, both clockwise. Determine the acceleration of P in the position shown.

4.13. The winch shown in Figure P4.13 is composed of two drums, A and E, and two gears, B and C (which operate as wheels turning on one another without slip). Gear B is secured to A so that they turn together, and gear C is secured to E in the same manner. A and B turn on fixed axis G. C and E turn on fixed axis H. Weight W is attached to a cable wound around drum E, and the lifting force is applied (downward) at P, the end of a second cable wound around drum A. The diameters are as follows: $A = 12$ in., $B = 4$ in., $C = 10$ in., and $E = 6$ in. If the linear acceleration of P is 10 in./sec/sec, find the linear acceleration of W.

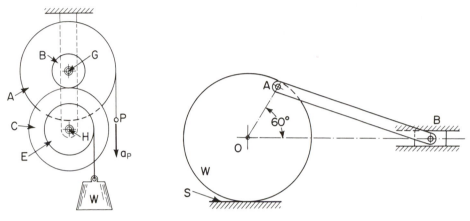

Figure P4.13   Figure P4.14

4.14. Wheel W rolls without slip on fixed surface S with an angular velocity of 1 radian/sec and an angular acceleration of 2 radians/sec/sec, both counter-clockwise (Figure P4.14). Rod AB is pinned to W at A and to the block sliding in fixed horizontal guides at B. W is 4 in. in diameter. AB is 5 in. long. OA $= 1.8$ in. Determine the linear accelerations of points A and B in the position shown.

4.15. An epicyclic wheel train is shown in Figure P4.15. Arm A turns about fixed axis O with an angular velocity of 10 radians/sec and an angular acceleration of 5 radians/sec/sec. Wheel B is stationary and is 4 in. in diameter. Wheel C is carried on the pin on A and is 2 in. in diameter. Ring E turn sfreely about center O guided by rollers L, M, Q, and R. There is no slip between B, C, and E. Determine the angular acceleration of ring E.

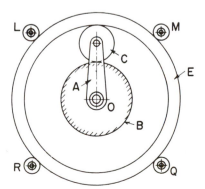

Figure P4.15

4.16. The Scotch yoke in Figure P4.16 is driven by crank *AB*, 2 in. long, which turns counterclockwise about fixed axis *A* with a constant angular velocity of 5 radians/sec. Determine the acceleration of yoke *C* in the position shown.

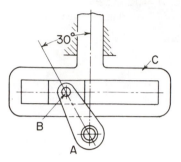

Figure P4.16

4.17. Solve problem 4.16 by the velocity difference method and compare answers with the solution of problem 4.16 by the method originally used.

4.18. In Figure P4.18 *OP* turns clockwise about fixed axis *O* with a constant angular velocity of 2 radians per sec. Sleeve *S* turns freely on pin *P* and slides along rod *R*. *M* is the fixed axis of *R*. Distance *MO* = 3 in., *ML* = 9 in. Find the linear acceleration of *L* and the angular acceleration of *R*.

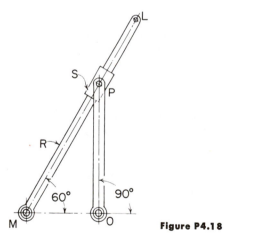

Figure P4.18

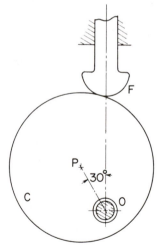

Figure P4.19

4.19. Circular cam *C* turns counterclockwise about fixed shaft *O* with a constant angular velocity of 20 radians/min. Diameter of *C* = 6 in. Eccentricity *OP* = 2 in. Radius of the contacting surface of follower *F* = 1 in. Find the linear acceleration of the mushroom follower *F* in Figure P4.19.

4.20. In Figure P4.20 block D has a linear velocity of 10 in./sec and a constant linear acceleration of 20 in./sec/sec along the fixed guides from left to right in each case. AC is a rigid bar 5 in. long. $AB = 3$ in., $BD = 8$ in., and $CE = 3$ in. E is pinned to a sleeve which slides freely on BD. Determine the linear acceleration of point E on CE in the position shown.

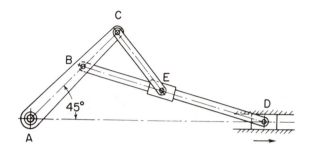

Figure P4.20

5

Graphical
Analysis of Motion

In previous chapters we have studied methods of determining displacements, velocities, and accelerations of mechanisms. In most cases these studies have been limited to the analysis of motion of a mechanism in one position only, and the velocities and accelerations obtained have been instantaneous values, good only for a single position in a complete cycle of motion.

In machine design it is generally necessary to make a complete study of a mechanism throughout its entire program of motion, so as to determine maximum velocities and accelerations. If this were attempted by a large number of vector studies in various positions, the process would be very slow and laborious. Our next objective is to apply the basic knowledge gained in these single-position studies in an effort to develop more rapid and efficient techniques for the survey of complete motion cycles.

5.1 Graphs to Represent Changing Values

We have used vectors to picture quantities and directions, so that we can visualize velocities, for example, without using numbers and measured angles. A graph is also a picture drawn to represent the relative size or continuous change in a quantity from one instant to the next. It has the advantage of showing comparisons between two or more related quantities, but lacks the vector's ability to represent direction as well as magnitude.

If we examine the velocity of a point traveling along a straight line at 1-sec intervals, we can draw a picture which represents the variation of velocity in relation to time. Each second of time is represented by a distance (called an *abscissa*) measured along a horizontal (*X*) axis, where some convenient unit of

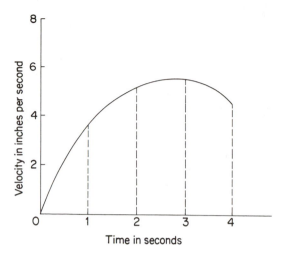

Figure 5.1 A velocity-time graph

length is equal to 1 sec. The velocity at the end of each time interval is also represented as a length of line (called an *ordinate*) laid out parallel to a vertical axis (Y) from the time mark at which that velocity was attained. This layout is shown in Figure 5.1, with velocities plotted at four 1-sec intervals.

A smooth continuous curve, drawn through the top ends of the velocity lines, shows the trend of velocity variation and indicates approximate values between those actually plotted.

5.2 Graphical Calculus

The value of graphs is not limited to visual representation of the variation of two related quantities. By using plotted curves, instead of numbers and complex equations, we can reduce a calculus problem to the level of simple geometry and thus bring its solution within the reach of those who have not studied more advanced mathematics. The graphical techniques are quickly learned and easily remembered. Skill in precision drafting is the only additional requirement. Calculus and higher forms of mathematics are not things to be avoided, but the study of mechanisms frequently precedes mathematics courses of that caliber. The graphical form of calculus fills this gap and, in addition, offers a practical introduction to analytical work to follow.

Graphical calculus has even further value in that it affords a means of solution to problems which, were they attempted analytically, would demand the talent of a genius. Some fairly simple linkages have motions which are very difficult to express as mathematical equations, whereas their graphical representation presents no special problems.

5.3 Graphical Integration

If a point moves with uniform velocity, it will have the same speed at each successive time interval. This motion can be pictured by a velocity-time graph. The velocity "curve" is a straight horizontal line, as shown in Figure 5.2(a).

Suppose we wish to know the distance traveled by the point in a given time interval. The displacement equation for uniform motion is $S = V \times T$. If we examine the graphic representation of this problem [Figure 5.2(b)], we note that the area beneath the constant velocity line for a time interval T is a rectangle, whose altitude is V and base T. This area is equal to $V \times T$. The equation states that $V \times T = S$, the displacement of the moving point. Therefore, we conclude that, in graphical terms, the area beneath a velocity curve during a given time interval equals the displacement of a point moving at speed V for a time T.

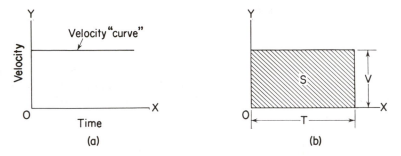

Figure 5.2 Displacement due to constant velocity

This relationship also holds true for a point moving with uniform acceleration. Figure 5.3(a) shows a velocity-time graph for such motion. Acceleration is the ratio of velocity to time or

$$a = \frac{\Delta V}{\Delta T}$$

If a is uniform, then the ratio $\Delta V/\Delta T$ must at all times be the same. Thus, the graph of velocity will be a line of constant slope, and only a straight line can maintain constant slope. So, the velocity graph is a straight line at angle θ with the X axis, where the size of angle θ is determined by a "run" of ΔT and a "rise" of ΔV (or, in terms of trigonometry, the tangent of $\theta = \Delta V/\Delta T$).

Now, let us determine the displacement during time T in this case. The equation for displacement in uniformly accelerated motion is

$$S = V^{\mathrm{av}} \times T \quad \text{(Article 4.2)}$$

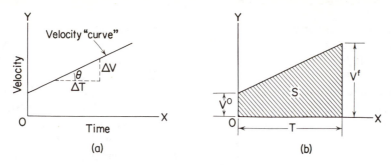

Figure 5.3 Displacement due to uniform acceleration

The average velocity during the time interval equals the sum of the initial velocity V^0 and final velocity V^f divided by 2. So

$$S = \left(\frac{V^0 + V^f}{2}\right)T$$

If we examine the graph in Figure 5.3(b), we note that the area beneath the velocity line is a trapezoid, whose area equals $(V^0 + V^f)/2 \times T$ (half the sum of the parallel sides multiplied by the distance between them). Since the displacement S above also equals $(V^0 + V^f)/2 \times T$, we conclude that, as before, the displacement equals the area under the velocity "curve."

This relationship is true whatever the nature of the velocity. Figure 5.4(a) shows a graph of velocity versus time in which the velocity varies irregularly for time T. The displacement in this case, during time T, also equals the area beneath the velocity curve. To determine this displacement, we have only to devise a technique for obtaining the area beneath the irregular curve.

A simple, accurate method involves the substitution of a series of rectangles for the original irregular area. In Figure 5.4(b), a vertical line ab is

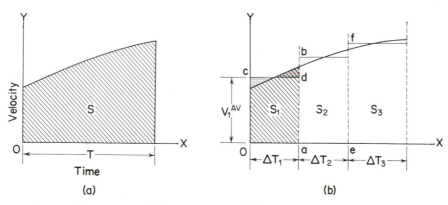

Figure 5.4(a) Displacement equals area under velocity curve

Figure 5.4(b) Graphical integration

drawn from the x axis across the curve. Next a horizontal line cd is drawn at such a level that the stippled triangular areas between cd and the curve are equal. It has been established that these areas can be satisfactorily matched by eye. The rectangle $Ocda$ can now be seen to equal the area beneath the curve between lines Oc and ab. The altitude Oc of this rectangle, at the assigned velocity scale, will equal the average velocity during the time interval ΔT_1. This time interval ΔT_1 equals Oa multiplied by the assigned time-scale factor.

The area of the rectangle equals $Oc \times Oa$, which equals S_1, or the displacement at the end of time ΔT_1. In the same manner, we draw a second vertical ef and construct another rectangle, the area of which (at the velocity and time scales) equals displacement S_2 during time interval ΔT_2. The total displacement at the end of these two intervals (from O to e on the time axis X) will equal the sum of displacements S_1 and S_2. Displacement S_3, obtained in the same way, must be added to S_1 and S_2 to obtain the total displacement during the time T described in Figure 5.4(a).

The question of how to determine a proper width ΔT of the rectangles is answered by the nature of the velocity curve. If the curvature is gradual and regular, the triangles are easily matched and the ΔT dimension may be quite large. If the curve is complex and irregular, the ΔT dimension must be made smaller, since the areas above and below the curve are less similar in shape and therefore more difficult to match by eye. While it is convenient to make the bases of all rectangles identical, the nature of the curve may demand variation, as shown in Figure 5.5.

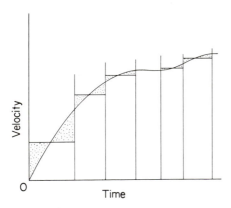

Figure 5.5 Varying ΔT to suit velocity curve

In the interest of efficiency, it is recommended that all the drafting on this type of problem be done first. All measurements can then be taken and tabulated. Suitable column headings for the table are shown in Figure 5.6.

The dimensions of ΔT and V^{av} are measured directly in inches and recorded. The calculations of displacements (S) can then be made. Both ΔT and V^{av} must be multiplied by scale factors to change them into time and velocity values in order that their product yields displacement in inches. It is most efficient to "package" these factors in a single constant K. The value of K is determined as follows:

In the equation $S = \Delta T \times V^{av}$, all values of ΔT in inches must be multiplied by the number of seconds which 1 in. represents on the scale of the graph. For example, if the time scale is 1 in. $= \frac{1}{5}$ sec, each value of ΔT must be multiplied by $\frac{1}{5}$. All measurements of V^{av} in inches must similarly be multiplied by the number of inches per second which 1 in. represents on the

Interval	ΔT (inches)	V^{av} (inches)	S ($V^{av} \times \Delta T \times K$)	Total S

Figure 5.6 Tabular form for integration

velocity scale of the graph. So, if the velocity scale is 1 in. $= 10$ in./sec, each value of V^{av} must be multiplied by 10. Thus,

$$S = (\Delta T \times \tfrac{1}{5})(V^{av} \times 10)$$

or, if we combine the scale factors into one term called K (Figure 5.6), K equals $\frac{1}{5} \times 10$ or 2.

The final column in the table (Total S) would be the sum of the individual displacements from the start to the end of the interval in question. The individual displacement values are useful when a graph of displacement versus time is required. If only the total displacement for the entire period of motion is sought, it is only necessary to add the displacements for each interval together to get the grand total.

This operation of adding up a series of products of varying velocity and time increments, which we have performed geometrically here, has its counterpart in analytical calculus, where it is called *integration*. Since we have employed a drawing to replace numbers, our method is known as *graphical integration*.

In terms of calculus, displacement is defined as the integral of velocity with respect to time, which is symbolized $\int V dt$. Graphically, an integral is an area beneath a curve. This method can be used to integrate any value which has been plotted with respect to another. For example, if angular velocity of a body were plotted against time, the angular displacement during a given period would equal the average angular velocity multiplied by the time interval, or $\theta = \int \omega dt$, which equals the area under the angular velocity curve. If acceleration is plotted against time, the area beneath the curve represents velocity (average acceleration multiplied by time or $\int a dt$). Precision drawing is required, but both the drafting and the calculations are simple.

5.4 Graphical Differentiation—Tangent Method

A velocity-time graph for a point which moves with constant acceleration is shown in Figure 5.7. As we noted before, the "curve" is a straight line, since the ratio of the change in velocity (ΔV) to the time interval required to make that change (ΔT) is the same at all times. This ratio is the acceleration of the point since $a = \Delta V/\Delta T$. This acceleration can be obtained by con-

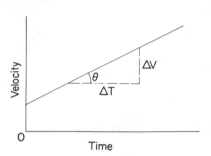

Figure 5.7 Determining uniform
acceleration graphically

structing a right triangle with a portion of the velocity line as hypotenuse and having a base ΔT and an altitude ΔV. The changes, ΔV and ΔT, can be measured, and ΔV can be divided by ΔT to yield the acceleration. The ratio $\Delta V/\Delta T$ defines the angle θ and is called the *slope* of the "curve." (In terms of trigonometry, this ratio of the side opposite an angle of a right triangle to the side adjacent is called the *tangent of the angle*.) Graphically, the acceleration is the slope of the velocity curve.

If a point moves with a velocity which changes at an irregular rate, it has variable acceleration, which is different at each instant. A velocity-time curve of such motion is shown in Figure 5.8. Suppose we wish to determine the acceleration after time t has elapsed. The velocity at this instant is defined by point P on the curve. The direction of the velocity at P is defined by a straight line TT, tangent to the curve at that point. If we build a right triangle with this tangent as its hypotenuse, the altitude will be ΔV and the base the corresponding ΔT. The acceleration then, at point P, is the change in velocity ΔV divided by the time ΔT in which this change took place:

$$a = \frac{\Delta V}{\Delta T}$$

Both ΔV and ΔT can be measured from the drawing, so the only problem is that of drawing an accurate tangent line TT. This tangent can be drawn with reasonable accuracy by eye, but if we must investigate many points along the curve, each tangent partially obscures the curve and distracts the eye, making it difficult to draw the next. This method does not yield as accurate results as

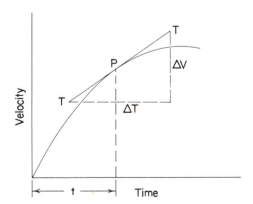

Figure 5.8 Determining instantaneous
acceleration graphically

graphical integration, so more points along the curve must be examined in order to average out the small errors. The inevitable overlapping tangents add to the confusion, as shown in Figure 5.9.

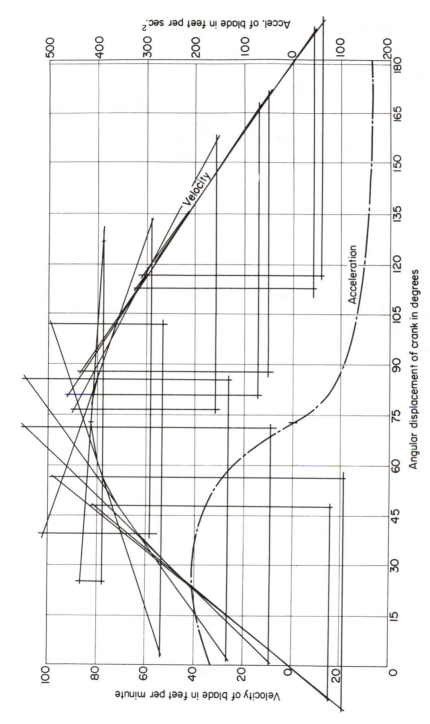

Figure 5.9 Graphical differentiation using tangents

Accel. of blade in feet per sec.²

Velocity

Acceleration

Angular displacement of crank in degrees

Velocity of blade in feet per minute

5.5 Graphical Differentiation—Step Method

A more satisfactory graphical method eliminates drawing tangents by eye and simplifies the drafting process. Figure 5.10 shows a point P on a velocity-time curve at which the acceleration is desired. Two vertical lines are drawn, equidistant from point P, intersecting the curve at a and c. A horizontal line

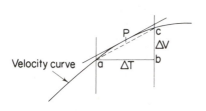

drawn from a, meeting the second vertical at b, defines ΔT, and the vertical segment bc defines ΔV. The acceleration at point P then becomes $\Delta V/\Delta T$. The equivalent distances bc and ab can be measured as before.

Figure 5.10 Tangent versus "step" method

If a tangent is drawn to the curve at P, it will prove to be very nearly parallel to ac. If ΔT is made smaller, the tangent and ac will be exactly parallel for all practical purposes. Therefore, the slope of ac will be the slope of the curve at P. This method is much faster and far more simple to draw than the actual tangents, so that, when a complete study is required, the accelerations of many more points can be obtained and small errors averaged out. It has been found by experiment that this method yields as high a degree of accuracy as when tangents are used.

When a complete survey is required, the process is organized for efficiency as follows: Figure 5.11 shows a velocity-time plot of a moving point. Let us assume that the acceleration-time curve for that point is to be defined.

First we select a ΔT measurement which is consistent with the curve, and we lay out a series of these distances, such as Oa, along the X axis. (Usually, we use a uniform ΔT throughout the entire study). Perpendiculars to the X axis are next drawn from the ΔT divisions to meet the curve at b, d, f, h, etc.

The ΔT's and ΔV's for each interval can all be laid out at one time. The first ΔT is Oa (along the X axis), and the vertical ab is the corresponding ΔV. We proceed, drawing the horizontal bc, the vertical cd, the horizontal de, the vertical ef, etc., throughout the curve, as shown in Figure 5.11.

A scale factor is next devised to interpret measured distances in velocity and time units. At the scales shown, each inch of ΔV must be multiplied by 5 to obtain velocity in feet per second, and each inch of ΔT must be multiplied by $\frac{1}{10}$ to obtain time in seconds. Thus we have

$$a = \frac{\Delta V \times 5}{\Delta T \times \frac{1}{10}} = \frac{\Delta V}{\Delta T} \times 50$$

Since all ΔT's are equal, we can include ΔT in the constant factor. If the ΔT measurements are all 1 in.:

$$a = \frac{\Delta V}{1} \times 50 \quad \text{or simply} \quad \Delta V \times 50$$

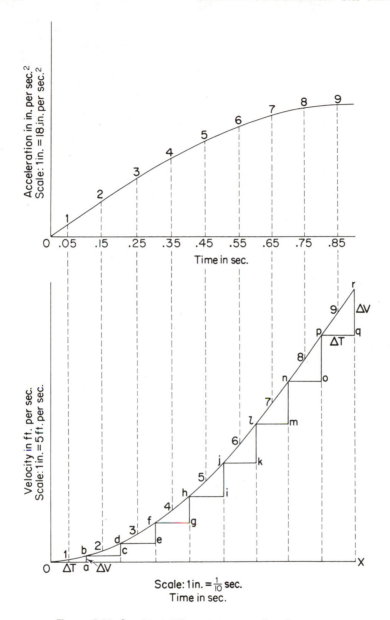

Figure 5.11 Graphical differentiation using "step" method

The constant factor K in this instance equals 50, and only ΔV need be measured.

All measurements of ΔV may now be made at one time and recorded in a table with headings, as indicated in Figure 5.12.

Point	ΔV (inches)	$a = \Delta V \times K$
1		
2		
3		

Figure 5.12 Table of values for differentiation

Calculations of acceleration are next performed, and the results recorded in the table. The first acceleration, for example, equals distance $ab \times 50$. This is the slope of the curve at point 1, midway between the Y axis and line ab, as shown in Figure 5.11. For convenience, the acceleration curve is plotted directly above the velocity graph in this illustration, with the same time scales. The first acceleration value ($ab \times 50$) is plotted on a vertical line through point 1 ($\frac{1}{2}$ in. from O) at a suitable scale. The second acceleration ($cd \times 50$) is plotted on a vertical through point 2 ($1\frac{1}{2}$ in. from O), etc.

We check results by attempting to draw a smooth, continuous curve through the plotted acceleration values. If this proves impossible, yet the variations from an apparent smooth curve are very small, an average curve, drawn between the offending points, should be very nearly correct. If the variations from an apparent smooth curve are appreciable, extra values may be obtained at points b, d, f, etc., by the dotted construction shown in Figure 5.13. A

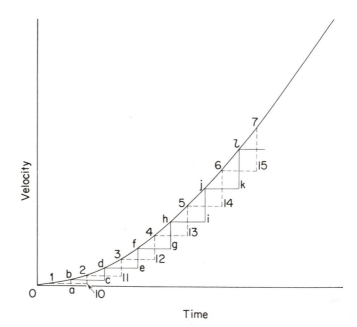

Figure 5.13 Adding steps to better define acceleration

horizontal line is drawn from point 1 to meet a vertical through point 2 at point 10. Distance 2-10 equals ΔV and distance 1-10 the corresponding ΔT (which measures 1 in. as before). The length 2-10 multiplied by 50 yields the acceleration at point b on the curve. Length 3-11 multiplied by 50 yields the acceleration at d, etc. These extra points should aid in defining the acceleration curve or justify drawing an average smooth curve between plotted points.

In terms of calculus, acceleration equals dv/dt, the derivative of velocity with respect to time, and the mathematical process of finding this derivative is called *differentiation*.* The graphical counterpart of this analytical operation is logically known as *graphical differentiation*, which is simply the art of determining the slope of a curve at a point by use of a drawing.

As is the case with integration, graphical differentiation is not limited to finding accelerations from velocity-time relationships, but can be applied to find the rate of change of any values which can be represented on a graph. For example, if we can plot angular velocity of a body versus time, we can differentiate and determine angular acceleration, using the same process. If we can plot a displacement-time curve, we can differentiate this and obtain velocity. Since velocity is the rate of change of displacement with respect to time, the slope of a displacement curve at any point is the instantaneous velocity at that point.

5.6 Graphical Differentiation—"Displaced-Curve" Method

In the exploratory stages of the design of mechanisms, we find graphical methods of differentiation effective tools, since they afford overall pictures of velocity or acceleration throughout the motion cycle. The "tangent" method is the most elaborate method and perhaps the most accurate, but it is rather time-consuming for preliminary design evaluation. The "step" method is definitely faster and involves less drafting and measurement but relies for its authority upon averaging and is less reliable on other than "well-behaved" curves. When exploring the motion properties of proposed mechanisms, we usually compromise somewhat on accuracy in favor of speed but we prefer to retain the pictorial lucidity which graphical methods provide. It is essential not only to obtain data efficiently but to be able to comprehend it readily for rapid evaluation.

The "displaced-curve" method, while very similar in concept to the "step" method, eliminates considerable drafting effort and retains the valuable asset of graphical display. This method presumes the availability of Ozalid printing facilities, since printing replaces much of the drafting. These services are in-

---

*The analytical calculus demands the ability to express the variables in an equation before integration or differentiation can be performed. As stated previously, devising this equation is frequently very difficult for even an accomplished mathematician.

expensive and readily accessible today in most schools and engineering environments.

Let us assume that we wish to investigate the acceleration of a significant point on a mechanism and have obtained sufficient data on the velocities of this point to plot the velocity-time curve shown in Figure 5.14. This curve is precisely plotted at the scales indicated and carefully drawn in a sharp, fine, continuous line on vellum (tracing paper) or some other translucent film. Next, we make an Ozalid print (black or blue lines on a white background) of this

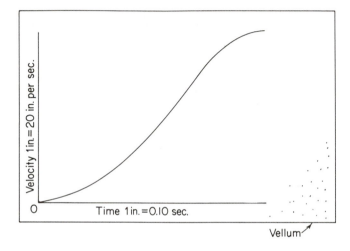

Figure 5.14 Velocity-time curve

original plotted curve. Since this is a contact printing process using dry developing, the print will be an exact, full-scale reproduction of the original.

We now attach this print to the drafting board and place the original vellum drawing on top of it, so that the horizontal axes are coincident. By noting the region of most intricate curvature, we next judge the largest value of ΔT which would be reasonable if a uniform value of this time increment were to be used over the entire curve. This decision is also required in the "step" method. We then displace the original tracing to the right of the vertical axis of the print beneath by a distance equal to the ΔT decided upon. The horizontal axes of the tracing and print will still be aligned. Figure 5.15 shows the tracing secured over the print in this manner.

Now, any vertical dimension measured between these two curves would be the ΔV corresponding to a ΔT which is equal to the displaced distance chosen above. Following the principle of the "step" method, this ΔV (in velocity units) divided by the ΔT (in time units) equals the acceleration at a point on the curve. The curve itself is located a distance to the left of the vertical ΔV line equal to one half the distance that the curves are displaced

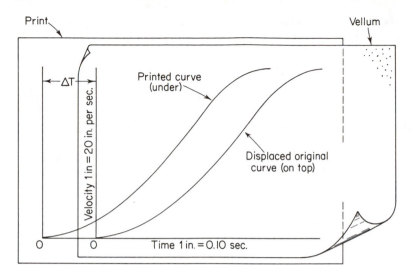

Figure 5.15 Displaced velocity-time curves

from one another or equal to one half the ΔT dimension. Figure 5.16 shows the "step triangle" drawn in completely, from which we can compute the equivalent acceleration at point P when the "step" method is used. The displacement here is 1.2 in.

Since the displacement of the two curves is uniform throughout, no horizontal ΔT lines need be drawn. (No consecutive "steps" are needed.) Also, since the ΔT dimension is constant, we observe that the accelerations vary directly with ΔV throughout the curve. The acceleration at any point P can be found by simply drawing a vertical line between the curves (ΔV), displaced to the right of point P by a distance equal to one half the horizontal curve displacement, ΔT. We measure this ΔV and multiply it by a constant composed

Figure 5.16 "Step-triangle" on displaced curves

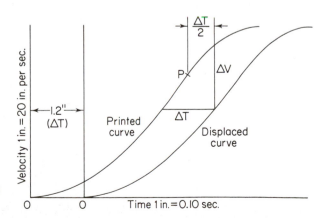

of the velocity-scale factor divided by the product of ΔT and its time-scale factor. In this example:

$$a_p = \Delta V \text{ (in inches)} \times \frac{20}{1.2 \times 0.1} = \Delta V \times \left(\frac{500}{3}\right)$$

A complete graphical study of acceleration for this example is shown in Figure 5.17, where the solution is plotted as an acceleration-time curve on the same axes, using the same time scale. Points 1, 2, 3, etc. on the acceleration curve are the plotted values of accelerations at the corresponding points 1, 2, 3 on the printed velocity curve having the same origin at O. In this instance, the acceleration scale has been selected so that the measured acceleration ordinates will be just twice as long as the corresponding ΔV ordinates. This greatly simplifies the layout of the acceleration curve, since no measuring or computation of acceleration values ($\Delta V \times \frac{500}{3}$) is required before plotting. Each acceleration is plotted by simply transferring the ΔV dimension to the accelerationt plot with the dividers. (Each ΔV is stepped off twice when plotting acceleration). The acceleration scale may of course be chosen so that the acceleration ordinates are some other multiple of the ΔV distances in order that the accleration curve will be well-proportioned.

Of course, it is still necessary to measure the ΔV lines in order to compute the actual magnitudes of accelerations at significant points along the curve. The graphical plot of acceleration only reveals relative values, but this "picture" is of primary interest to the designer, since it indicates the critical regions where specific values are needed.

Properly executed, the displaced-curve method is just as accurate as the step method. The use of a print greatly reduces the drafting effort, and the method of plotting the acceleration curve, described above, eliminates tedious measurement. This technique would not be well-suited to complex velocity

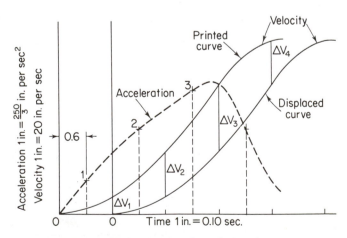

Figure 5.17 Accelerations from displaced curves

curves of widely varying curvature. The tangent method is the most reliable tool of graphical analysis in such cases. There is no reason why the tangent method might not be applied at any complex regions in a curve and some other method used for the portion in which the curvature is more uniform.

5.7 Techniques of Graphical Calculus

The graphical form of calculus has the advantage of simplicity and is applied with equal facility to any variables which can be plotted on a graph. However, the advantage of graphical methods may only be enjoyed by those who possess the skill of precise drafting techniques and exercise sound judgment in the choice of scales. If these requirements are not met, the graphical solution becomes only a rank approximation and is not respected as a professional technique.

The selection of scales is most important to accuracy. In the acceleration problem of Figure 5.11, a poor choice of velocity scale might result in a very "flat" curve, as shown in Figure 5.18. The ΔV distances all become very small and therefore demand an accuracy of measurement far beyond the ability of the most precise draftsman. We can measure any distance only to the nearest hundredth part of an inch, so if the measurement is small the percentage of error becomes very high.

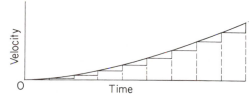

Figure 5.18 A poor choice of scale for differentiation

An example is offered in Figure 5.19, where a ΔT interval must be selected which is consistent with the curve. As in the case of graphical integration, a curve with rapidly changing curvature demands a shorter ΔT. A relatively large ΔT is permissible in that portion of the curve where curvature is fairly uniform. The tangent at point P is parallel to the line ac using the ΔT shown. At point Q, however, the tangent is visibly not parallel to the line df if the same large ΔT is used. The ΔT length must be shortened to obtain an accurate slope at Q because the curvature is rapidly changing within the large interval. If uniform ΔT's are to be used, they must be short enough to enclose a curve segment of fairly uniform curvature at the most irregular portion of the curve.

The selection of a scale with regard to simplifying layout work can usually save a great deal of time. For example, in the problem of Figure 5.11,

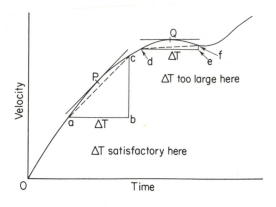

Figure 5.19 Adapting ΔT to the velocity curve

where acceleration equals $\Delta V \times 50$, if the acceleration scale is made 1 in. = 50 ft/sec/sec, the distance representing acceleration would be exactly equal to the length of ΔV in each case. (The number of inches representing each acceleration value equals the acceleration divided by 50. Since $a = \Delta V \times 50$,

$$\frac{a}{50} = \frac{\Delta V \times 50}{50} = \Delta V \; . \biggr)$$

The acceleration graph could then be laid out by transferring the ΔV values with the dividers, eliminating scaling.

5.8 Tabular Integration

When graphical facilities are not available, or when a preliminary approximate evaluation of motion must suffice, the principles of graphical integration and differentiation may be applied by computation. In such instances, we sacrifice accuracy and the advantage of graphical display but we can still avoid the need for mathematical competence at the level of calculus. This purely numerical study would determine the general adaptability of a certain mechanism and would indicate critical areas where more accurate investigation must be made.

If we have a velocity-time table we can calculate approximate displacements without any drawing. Figure 5.20(a) shows such a table, giving the velocity of a point at a series of equal time intervals. If these intervals are relatively short, we can assume that the velocity varies uniformly in each interval, so displacement equals average velocity multiplied by time. Thus, in the first 0.01-sec interval,

$$V^{\mathrm{av}} = \frac{V^f + V^0}{2} = \frac{5 + 0}{2} = 2.5 \text{ in./sec}$$

At the end of this interval:

$$S = V^{av} \times \Delta T = 2.5 \times 0.01 = 0.025 \text{ in.}$$

In the second 0.01-sec interval:

$$V^{av} = \frac{V^f + V^0}{2} = \frac{8.5 + 5}{2} = 6.75 \text{ in./sec}$$

At the end of this interval:

$$S = V^{av} \times \Delta T = 6.75 \times 0.01 = 0.0675 \text{ in.}$$

The approximation can be seen in the graph of the given velocities in Figure 5.20(b). In graphical integration, displacement equals the shaded area under the velocity curve. In this tabular method, we have assumed the curve to be a straight line from O to a, so we are actually obtaining the area of the triangle Oab.

Time (sec)	Velocity (in./sec)
0	0
.01	5
.02	8.5
.03	10

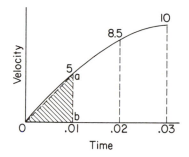

Figure 5.20 Approximations of tabular integration

The process described above can be repeated until the entire survey is complete. The table shown in Figure 5.20(a) can be expanded, as shown in Figure 5.21, to record all significant values and results. No drawing is required.

Time (sec)	Velocity (in./sec)	$V^{av} = \dfrac{V^f + V^0}{2}$	ΔT	$S = V^{av} \times \Delta T$	Total S (in.)
0	0	0	0	0	0
.01	5	2.5	.01	.025	.025
.02	8.5	6.75	.01	.0675	.0925
.03	10	9.25	.01	.0925	.185
.04	14	12	.01	.12	.305

Figure 5.21 Record of values for tabular integration

5.9 Tabular Differentiation

Using the velocity-time table in Figure 5.22, we can obtain approximate accelerations without making any drawing.

Time (sec)	Velocity (in./sec)
0	0
.1	3
.2	5
.3	6.5
.4	7.5

Figure 5.22 Velocity-time table

For a relatively short time interval, acceleration equals the change in velocity during the interval divided by the elapsed time:

$$a = \frac{\Delta V}{\Delta T} = \frac{V^f - V^0}{\Delta T}$$

In the first 0.1-sec interval:

$$a = \frac{V^f - V^0}{\Delta T} = \frac{3 - 0}{0.1} = 30 \text{ in./sec/sec}$$

This is an instantaneous acceleration, and our experience with graphical differentiation shows that it more nearly approximates the acceleration at the midpoint of the interval than at the start or end. Thus the value above will be the acceleration 0.05 sec from the start.

In the second interval:

$$a = \frac{V^f - V^0}{\Delta T} = \frac{5 - 3}{0.1} = \frac{2}{0.1} = 20 \text{ in./sec/sec}$$

This is the acceleration 0.15 sec from the start.

The table in Figure 5.22 is shown in expanded form in Figure 5.23, in which all pertinent values and the calculated acceleration are recorded in orderly fashion.

While the process used in this tabular method is similar to that of graphical differentiation, we are entirely dependent upon the given table of velocity-time data in the tabular study. The larger the tabulated interval, the less accurate are the results. When we work from a plotted curve, we can visualize critical areas and find intermediate acceleration values where required. Though the method is the same, graphical differentiation permits a much more thorough study and therefore greater accuracy. The use of the tabular method should be limited to rough experimental surveys.

Time (sec)	Velocity (in./sec)	$\Delta V = V_f - V_0$	ΔT	$a = \dfrac{\Delta V}{\Delta T}$
0	0	0	0	
.05				30
.1	3	3	.1	
.15				20
.2	5	2	.1	
.25				15
.3	6.5	1.5	.1	
.35				10
.4	7.5	1	.1	
.45				

Figure 5.23 Record of values for tabular differentiation

PROBLEMS

5.1. A point moves along a straight path, starting from rest and continuing for 1 sec, when it again comes to rest. Its velocity at the end of each 0.1 sec is given in the accompanying table. Plot a velocity-time graph and integrate graphically to determine the total displacement of the point. (Suggested scales: time, 1 in. = 0.1 sec; velocity, 1 in. = 2 in./sec.)

Time (sec)	Velocity (in./sec)
0	0
0.1	1
0.2	3
0.3	6
0.4	9
0.5	11
0.6	12
0.7	11
0.8	8
0.9	4
1.0	0

5.2. In the Scotch yoke shown in Figure P5.2, crank AB turns with a constant angular velocity of 1 radian/sec. The acceleration of yoke Y is tabulated below for each $18°$ of angular displacement of AB during $90°$ of crank rotation, starting in position AB_0.

(a) Determine and tabulate the displacement of *AB* in radians for each 18°
position of *AB*. (θ in radians = θ in deg $\times$ $\pi/180$.)
(b) Determine and tabulate the elapsed time in sec for each position.
(c) Draw an acceleration-time graph for yoke *Y* during the 90° displacement of
AB (see scales below).
(d) Integrate this curve graphically to obtain the velocity of yoke *Y* at each interval and tabulate ($V = \int a\,dt$ = area under acceleration-time curve).
(e) Plot the velocity of *Y* on the graph of Part (c) (see scales below).

θ_{AB} (deg)	θ_{AB} (radians)	Time (sec)	Accel. of *Y* (in./sec/sec)
0	0	0	2.00
18			1.90
36			1.62
54			1.18
72			0.62
90			0.00

Suggested scales for graph:

Time: 1 in. = $\pi/10$ sec

Acceleration: 1 in. = $\frac{1}{3}$ in./sec/sec

Velocity: 1 in. = 0.4 in./sec

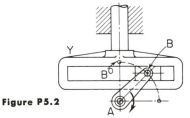

Figure P5.2

5.3. (a) Using the velocity-time table in Problem 5.2, plot a velocity-time curve at
the scales suggested in that problem.
(b) Differentiate the curve graphically; obtain acceleration values ($a = dv/dt$).
Tabulate these values.
(c) Plot an acceleration-time curve. [Use the same graph as in Part (a) and an
acceleration scale of 1 in. = $\frac{1}{3}$ in./sec/sec.]

5.4. A crank and slider linkage is shown in Figure P5.4. Crank *OP* is 2 in. long and
turns counterclockwise with a constant angular velocity of 10 radians/sec. Connecting rod *PN* is 4 in. long.
(a) Draw the mechanism (skeletonized) in at least nine positions (starting with
crank at OP_1 and proceeding for 180° rotation to OP_2), measuring and tabulat-
ing displacements of *N*.
(b) Plot a curve of the displacement of *N* versus time for the 180° rotation of
OP. (Devise suitable scales.)
(c) Differentiate this curve graphically and determine the velocity of *N* in at
least nine positions. Tabulate all values required ($V = ds/dt$).
(d) Plot a curve of velocity of *N* versus time. [Use same graph used in Part(b)
with a suitable velocity scale.]

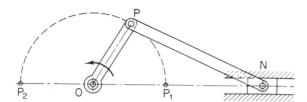

Figure P5.4

5.5. Cam C in Figure P5.5 turns counterclockwise about fixed shaft A at a constant speed of $\pi/18$ radians/sec driving follower F.

(a) Starting with cam C in the position shown, determine velocities of F as C turns 60°. Tabulate at least six velocities.

(b) Plot a velocity-time curve for F at suitable scales.

(c) Differentiate this velocity curve graphically to determine accelerations of F for the 60° rotation of C.

(d) Plot an acceleration-time curve on same graph as Part (b), using suitable acceleration scales.

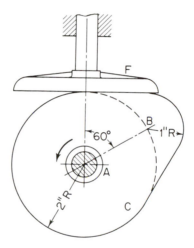

Figure P5.5

5.6. Using the velocity-time table in Problem 5.1:

(a) Integrate using the tabular method to determine total displacement in the 1-sec interval. Make a neat table recording all values computed.

(b) Differentiate by the tabular method to obtain acceleration values for the entire 1-sec interval. Tabulate neatly.

NOTE: If the student has completed Problem 5.1 by the graphical method, he may compare results with those obtained by the tabular method in Problem 5.6.

5.7. A four-bar linkage has a 2.6 in. crank AB, a 5.4 in. coupler BC, and a 3 in. crank CD. The fixed centers A and D are located as shown in Figure P5.7. Crank AB turns clockwise at a constant speed of 60 rpm. It is required to study

the angular speed of crank CD while AB turns $180°$ from position AB_0 to AB_1.
(a) Lay out the linkage full size at each $15°$ position of AB starting at AB_0 and measure the angular displacements of CD in each position.
(b) Plot a curve of angular displacement versus time for CD at the following scales:

Time: 1 in. $= \frac{1}{24}$ sec (abscissae)

Displacement: 1 in. $= 15°$ (ordinates)

(c) Differentiate the displacement-time curve graphically to obtain angular speed of CD.
(d) Plot this angular speed of CD versus time at the following scales:

Time: 1 in. $= \frac{1}{24}$ sec

Speed of CD: 1 in. $= 10$ rpm

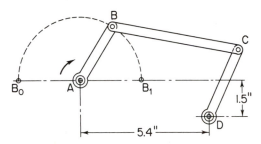

Figure P5.7

6

Gearing

The transmission of rotary motion from one shaft to another is a problem to be dealt with in the design of nearly every machine. The shafts may be required to turn at the same or at different speeds, but the ratio of their angular velocities, whatever its value, must remain unchanged as the shafts turn. If the speed ratio of follower to drive shaft ($\omega_F : \omega_D$) is 3 : 1, this ratio must remain constant for any amount of rotation. If the drive shaft makes one revolution, the follower shaft must turn exactly three. If the drive shaft turns 1°, the follower shaft must turn exactly 3°, so that for any number of turns or any minute fraction of a turn, the speed ratio remains always the same.

The simplest mechanism which will geometrically fulfill this exacting requirement is of course a pair of cylinders in pure rolling contact (Figure 6.1). If there is no slip and the diameter of D is three times the diameter of F, the speed ratio* will at all times remain constant, as we observed in velocity study:

$$\frac{\omega_F}{\omega_D} = \frac{D_D}{D_F}$$

In many cases the speeds, loads and friction forces are such that pure rolling contact drives are satisfactory, but more frequently, conditions demand that an absolutely uniform speed ratio must be guaranteed and a positive type of contact is demanded rather than dependence upon friction alone.

*To avoid confusion, the term *speed ratio* wherever used in this text shall mean the ratio of the speed of the follower to the speed of the driver in that order.

173

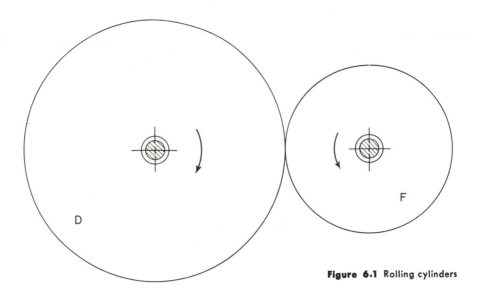

Figure 6.1 Rolling cylinders

This problem was encountered many years ago and was solved by substituting wheels with interlocking pins or cogs for the smooth cylinders. This produced a positive drive, tooth for tooth, barring fracture of one of the teeth. In the days when speeds were slow, such crude "gears" were adequate, but our present high-speed drives demand an absolutely uniform speed ratio during even a small fraction of a revolution, or vibrations and dynamic forces would develop, causing immediate failure. In the modern gear it is not sufficient that

Table 6.1 Symbols and Abbreviations Used in Gearing

P,	Pitch point
D^P,	Pitch diameter
R_E^P,	Pitch radius of gear E*
D^A,	Addendum diameter
R_F^A,	Addendum radius of gear F
D^B,	Base circle diameter
R_G^B,	Base circle radius of gear G
P^C,	Circular pitch
P^D,	Diametral pitch
P^N,	Normal pitch
T,	Number of teeth
ω_F/ω_D,	Angular speed ratio of gears F and D
θ	Pressure angle
sin	Sine
cos	Cosine

*In general, subscripts identify gear to which symbol pertains, (P_M^C, circular pitch of gear M).

174

the teeth simply intermesh. These teeth must be of such form that, as one drives another, they compel the gears to turn with a constant speed ratio.

Modern gears are highly standardized and usually purchased from gear manufacturers' stock, and not designed to order for each application. Since standard gears cannot be combined indiscriminately, a basic knowledge of gear design and limitations is necessary in order to devise proper gear drives. It is not sufficient to acquire a catalog and select at random.

6.1 Gearing Terminology

In order to study gearing intelligently, we must first learn some terms and definitions. Since gears must have the same motion properties as rolling cylinders, let us consider modifying a pair of cylinders into the form of gears.* Figure 6.2 shows that the teeth might be formed by cutting a series of notches into each cylinder and building out projections between these notches to form interlocking teeth. The physical surface of the cylinder disappears when teeth are formed in this way, but the cylinder circle remains an important reference

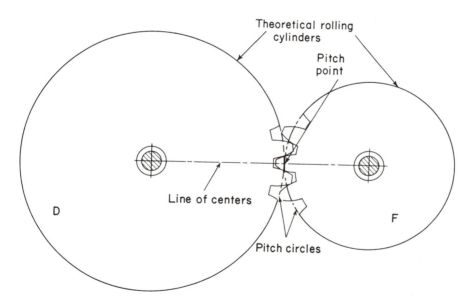

Figure 6.2 Teeth built upon rolling cylinders

*For practical reasons gears are not actually made this way. This assumption is only a convenient way of showing the relationship between gears and rolling cylinders.

line for gear dimensions. These theoretical circles are called the *pitch circles* of the gears, and their diameters are called *pitch diameters* (abbreviated D^P). When two gears are run together the pitch circles (like the rolling cylinders they represent) are tangent to one another at a point called the *pitch point* (*P*). This point will lie on the line joining the centers of the gears, which is called the *line of centers*. The speed ratio of the gears may be stated in terms of their pitch diameters:

$$\frac{\omega_F}{\omega_D} = \frac{D_D^P}{D_F^P}$$

This is the same speed ratio as that of the rolling cylinders from which the gears were assumed to be built.

Figure 6.3 describes some more gear terms and dimensions, many of which are measured on or from the pitch circle as a reference.

Addendum: Radial distance from the pitch circle to the outside circle of the gear.

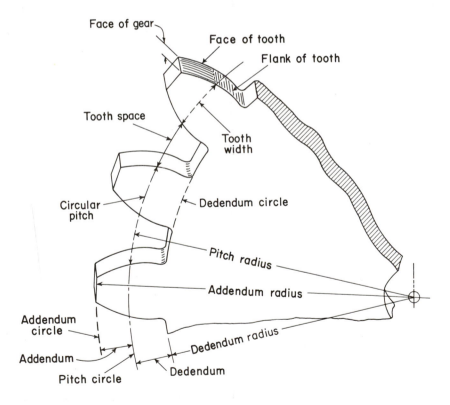

Figure 6.3 Gear terminology

Addendum radius R^A: The maximum radius of the gear (pitch radius plus addendum).

Dedendum: Radial distance from pitch circle to bottom of space between teeth.

Dedendum radius: Radius of circle defining bottom of space between teeth (pitch radius minus dedendum).

Tooth width: Arc of pitch circle subtending width of a tooth.

Space width: Arc of pitch circle subtending space between teeth.

Circular pitch (P^C): Segment of arc of pitch circle assigned to one tooth and one space. (Circular pitch equals tooth width plus space width or circumference of pitch circle divided by number of teeth in the gear: $P^C = \pi D^P/T$.) Since the tooth width and space width on a gear are nearly equal, the P^C is a measure of the size of a tooth, although its shape will vary with different pitch circle radii.

Diametral pitch (P^D): Number of teeth on a gear divided by its pitch diameter:

$$P^D = \frac{T}{D^P}$$

The diametral pitch is a convenient number in gear calculations rather than a physical dimension. A gear of a given pitch diameter may have 20 teeth ($P^D = 2$) or 60 teeth ($P^D = 6$). In order to fit around the same pitch circle, the teeth on the 60-tooth gear ($P^D = 6$) must be much smaller than those on the 20-tooth gear ($P^D = 2$). Thus the smaller the P^D number, the larger the teeth. P^D therefore denotes relative size of teeth. Figure 6.4 shows the approximate size of teeth of different P^D numbers. Since both involve pitch diameter and number of teeth, circular pitch and diametral pitch are related to one another:

$$P^C = \frac{\pi D^P}{T} \quad \text{or} \quad \frac{\pi}{P^C} = \frac{T}{D^P} \quad \text{but} \quad \frac{T}{D^P} = P^D \quad \therefore \quad P^D = \frac{\pi}{P^C}$$

Face of gear: Thickness of the gear measured parallel to the axis of rotation.

Face of tooth: Contacting surface of tooth from pitch circle to addendum circle.

Flank of tooth: Contacting surface of tooth from pitch circle to dedendum circle.

Fillet: Rounded corner between flank and dedendum circle.

When two gears run together, their pitch circles turn in rolling contact, like the rolling cylinders which they represent. Equal portions of their pitch

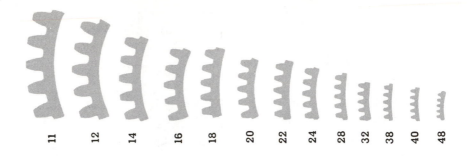

11 12 14 16 18 20 22 24 28 32 38 40 48

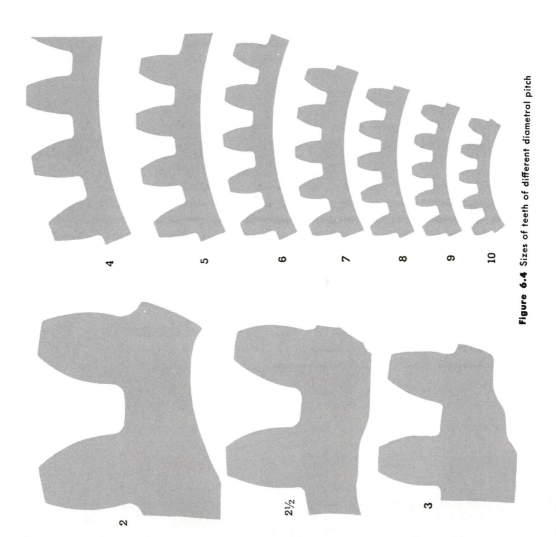

4 5 6 7 8 9 10

2 2½ 3

Figure 6.4 Sizes of teeth of different diametral pitch

circles are turned past a reference point in any given time interval. Figure 6.5 shows teeth in contact on two gears, D and F. When a tooth and a space on D pass the line of centers, a space and a tooth on F must pass also. Since one tooth and one space comprise circular pitch, it then follows that *mating gears have the same circular pitch.* Since $P_D^C = P_F^C$ and $P^C = \pi/P^D$, then

$$\frac{\pi}{P_D^D} = \frac{\pi}{P_F^D} \quad \text{or} \quad P_D^D = P_F^D$$

Mating gears have the same diametral pitch.

One gear drives another through contact at the face and flank of each tooth, so contact between the addendum surface of one tooth and the dedendum surface of the other is undesirable, The space measured on the line of centers between the addendum circle of one gear and the dedendum circle of the other is called the *clearance.*

If the tooth width of one gear is slightly less than the space width on the mating gear, the drive will still operate properly, but if the sense of rotation is suddenly reversed, the driving tooth will break contact with one of the follower teeth and lash back, striking the adjoining tooth. The difference between the driver tooth width and follower space width (measured on their pitch circles) is called *backlash.* Any appreciable amount of backlash is detrimental if the drive involves frequent reversals.

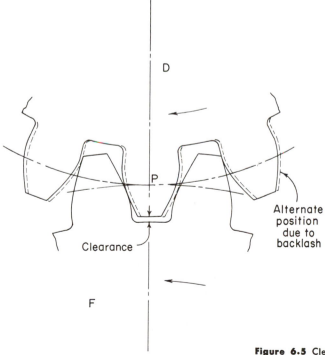

Figure 6.5 Clearance and backlash

6.2 The Nature of Tooth Contact

Since all teeth on a gear are identical, the behavior of one pair during contact is typical of all other pairs. This is true whether the motion is transmitted by a single pair or by several pairs simultaneously and applies in either sense of rotation.

To preserve a constant speed ratio, the contact between a pair of teeth must be continuous from the time they first meet until they separate. At all times there must be at least one pair in contact. Intermittent contact would cause the speed ratio to fluctuate.

Figure 6.6 shows a single pair of teeth in a series of consecutive positions during contact. With the driver turning clockwise, the first point of contact occurs at the right of the line of centers on the flank of the driver tooth and at the outermost tip of the face of the follower tooth. Thence it proceeds diagonally downward to the left until, in the final position of contact, a point at the addendum circle of the driver touches the flank of the follower tooth.* Note that the pitch point is one of the contact points.

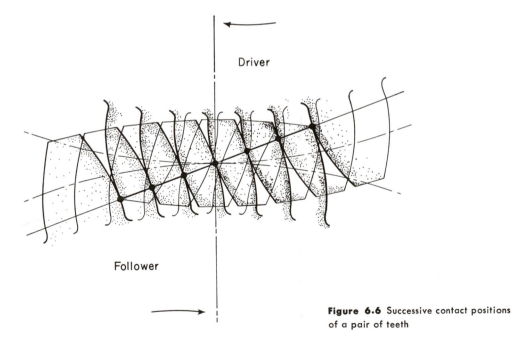

Driver

Follower

Figure 6.6 Successive contact positions of a pair of teeth

---

*Gear teeth of standard shape are shown here. With standard gear teeth, the successive contact points all lie in one straight line. This might not be true if other tooth forms are used.

A pair of teeth are shown in contact at point C in Figure 6.7. The straight line drawn from the contact point through the pitch point is called the *pressure line*. The angle θ between the pressure line and the common tangent to the pitch circles is designated as the *pressure angle*.

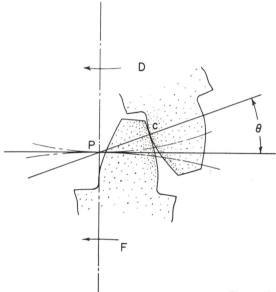

Figure 6.7 Pressure line and pressure angle

Figure 6.8 shows a tooth on the driving gear D in the position where it first comes in contact (at point a) with a tooth on the follower F. This same pair of teeth is shown dotted in the final position of contact (at point b). Assuming that these are standard gears, the pressure angle is constant and the pressure line therefore is one straight line ab through the pitch point. The locus of all points of contact will then be the line ab which is called the *path of contact*.

The arc of the pitch circle through which a gear turns, from that position where contact starts to the position where contact ceases (arc ec on the drawing), is called the *arc of contact*. Since pitch circles move with pure rolling contact, this arc of contact will be of the same length on the pitch circle of either gear.

The *angle of action* is that angle turned by a gear during contact between any pair of teeth. Therefore, it is the angle subtended by the arc of contact. The angle of action for the driving gear is labeled in Figure 6.8. Since arcs of contact are equal on both gears whereas the pitch radii are not necessarily equal, angles of action are of different size on driver and follower unless the gears happen to be identical.

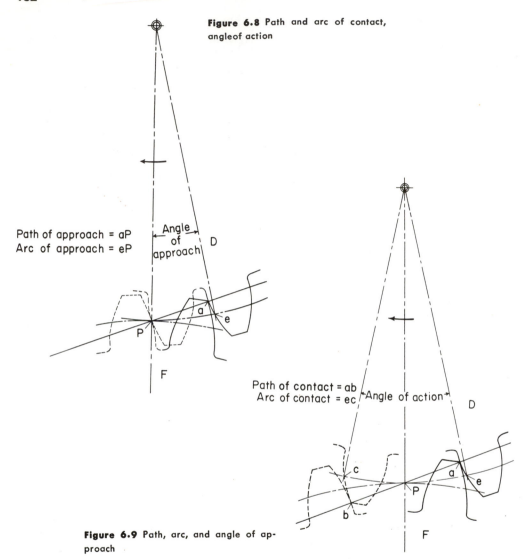

Figure 6.8 Path and arc of contact, angleof action

Path of approach = aP
Arc of approach = eP

Figure 6.9 Path, arc, and angle of approach

Paths, arcs, and angles are divided for study into two parts. The period during which the teeth are approaching the position of pitch point contact is designated as *approach*. Figure 6.9 shows a pair of teeth starting contact at *a* and later in the position of pitch point contact (dotted). The path during approach (line *aP*) is called the *path of approach*. The arc of the pitch circle turned during approach (arc *eP*) is the *arc of approach*. The angle subtended by arc *eP* is the *angle of approach*, as labeled in Figure 6.9. *Paths and arcs of approach are equal on driver and follower*, but the angles are only equal when the gears are identical.

The period during which the teeth are receding from the pitch point contact position until they cease contact is designated as *recess*. Figure 6.10 shows the same pair of teeth in pitch point contact and later at the final position of contact (dotted). The path during this interval (line *Pb*) is the *path of recess*, and the arc of the pitch circle turned (*Pc*) is the *arc of recess*. These are equal for both gears. The angle subtended by arc *Pc* is the *angle of recess*, which again differs for each gear unless they are identical.

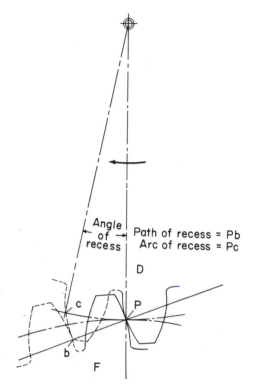

Figure 6.10 Path, arc, and angle of recess

6.3 Determining the Length of the Path of Contact

If the path of contact is a straight line, as is the case with standard gears, the length of the path of contact is easily determined. Contact is only possible within the area defined by the overlapping addendum circles of the gears, shown shaded in Figure 6.11. If the path of contact is a straight line, it coincides with the pressure line. Therefore, the path of contact is that portion of the pressure line lying between the overlapping addendum circles within the shaded area shown. Only the pitch circles, pressure–angle, and addendum cir-

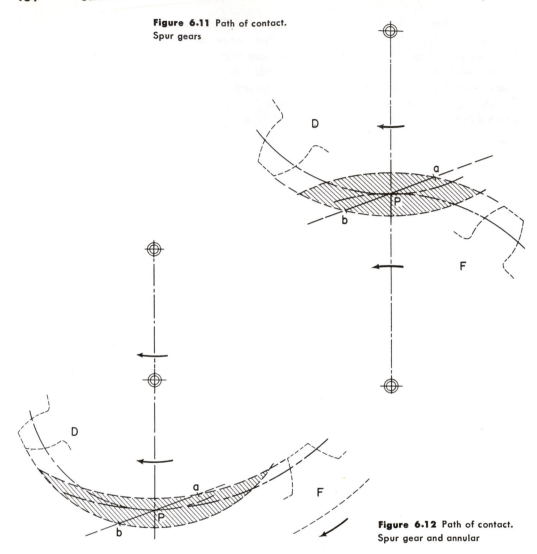

Figure 6.11 Path of contact. Spur gears

Figure 6.12 Path of contact. Spur gear and annular

cles are required to determine the path of contact. Figure 6.11 shows the layout for *spur gears*, while Figure 6.12 illustrates the path of contact for a spur gear running with an *annular gear*,* and Figure 6.13 depicts the path when a spur gear drives a rack.† Line *ab* is the path of contact in each case. The tooth outlines are sketched dotted only to orient the reader. No teeth need be drawn to find the path of contact.

*An *annular gear* is a ring with teeth cut on its inner surface. The minimum diameter is the addendum diameter on this type of gear.

†A *rack* is a gear of infinite radius with teeth cut on a straight bar. It translates only, and its addendum and pitch "circles" become straight lines.

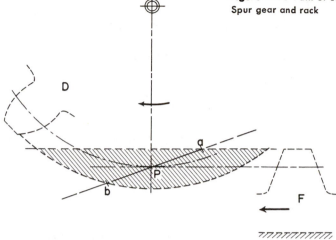

Figure 6.13 Path of contact. Spur gear and rack

A graphical layout is the simplest means of determining paths of contact. If large scales are employed, the paths can be measured very accurately.

If extreme accuracy is required, paths can be calculated. This process requires finding the path of approach and recess separately and adding them together, so it is much more laborious. Figure 6.14 shows the relationship of pitch radius, addendum radius, pressure angle, and path when two spur gears run together as in Figure 6.11. In triangle fPa, two sides (pitch and addendum radii) are known and one angle (90° plus the pressure angle). The other side (the path of approach aP) may be found by using the law of sines from trigonometry, which states, in terms of this triangle:

$$\frac{fP}{\sin \beta} = \frac{fa}{\sin (90° + \theta)}$$

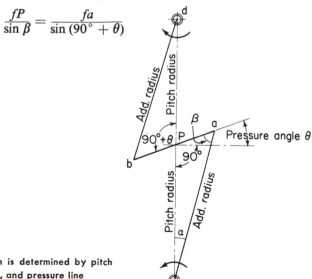

Figure 6.14 Path is determined by pitch radius, addendum, and pressure line

Since $\sin(90° + \theta) = \cos\theta$

$$\sin\beta = \frac{fP\cos\theta}{fa}, \quad \text{so we solve for angle } \beta*$$

$$\text{Angle } \alpha = 180° - (90° + \theta) - \beta†$$

Now, using the law of sines again:

$$\frac{aP}{\sin\alpha} = \frac{fP}{\sin\beta} \quad \text{we solve for } aP.$$

Similarly, we may solve for Pb (path of recess) in triangle dPb. The sum of the paths of approach and recess equals the path of contact.

6.4 The Speed Ratio of a Pair of Gears

We have observed that, since pitch circles of gears behave like rolling cylinders, the speed ratio of a pair of gears is:

$$\frac{\omega_F}{\omega_D} = \frac{D_D^P}{D_F^P}$$

Since pitch circles are only imaginary circles and not physical surfaces which can be measured, it is much more practical to define speed ratio in terms of the numbers of teeth on the gears. Diametral pitch relates pitch diameters and tooth numbers:

$$P^D = \frac{T}{D^P} \quad \text{and rearranging:} \quad D^P = \frac{T}{P^D}$$

So

$$D_D^P = \frac{T_D}{P_D^D} \quad \text{and} \quad D_F^P = \frac{T_F}{P_F^D}$$

Substituting in the speed ratio equation:

$$\frac{\omega_F}{\omega_D} = \frac{T_D/P_D^D}{T_F/P_F^D}$$

Since D and F are mating gears, they have the same diametral pitch. So $P_D^D = P_F^D$, and these terms will cancel:

$$\frac{\omega_F}{\omega_D} = \frac{P_F^D \times T_D}{P_D^D \times T_F} = \frac{T_D}{T_F}$$

The angular speeds therefore vary inversely as the numbers of teeth. This equation holds true for spur, annular, and bevel gears when mounted on fixed axes.

---

*Values of sines of angles are tabulated in the appendix. In trigonometry the sine of an angle equals the side opposite the angle divided by the hypotenuse.
†The sum of the interior angles of a triangle equals 180°.

6.5 Tooth Forms for Constant Speed Ratio

If we are to replace rolling cylinders with gears for a positive drive, the gears must provide a constant angular velocity ratio or they will not be kinematically equivalent. Since one gear drives another through contact of the teeth, the shape of the contacting surfaces must be governed by the uniform speed–ratio requirement. There are a variety of matched curves which will satisfy this requirement, but all share a basic geometric relationship which should first be defined before we select the form best adapted to gear teeth.

6.6 Conjugate Curves

When the contacting surfaces on two rotating bodies are so shaped that one drives the other with a constant speed ratio, these surfaces are said to be *conjugate* to one another and are called *conjugate curves*. The fundamental geometry governing the design of these surfaces will be revealed by a study of familiar velocity relationships.

A given body D (Figure 6.15), turning about a fixed axis O, is to drive another body F, turning about fixed axis L. The contacting surfaces of D and F are to produce a constant angular–velocity ratio as they turn. Let us assume that point A on this surface of D is the present contact point. The velocity of A on D equals $\omega_D \times OA$ and is perpendicular to OA. The velocity of A on F must be perpendicular to LA. Since these two velocities are in different directions, we know that sliding contact takes place at A. Therefore V_A^D and V_A^F have a common effective component along line NN perpendicular to the direction of sliding, i.e., perpendicular to the surface of D at point A. This *ec* defines the magnitude of V_A^F, as shown in Figure 6.15.

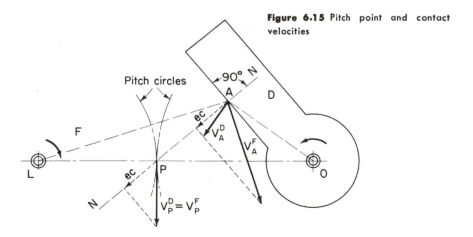

Figure 6.15 Pitch point and contact velocities

If D and F have a constant speed ratio, they turn as if driven by two cylinders rolling without slip. In gearing terminology we call these theoretical cylinders *pitch circles* and their point of tangency (on line OL in this instance) the *pitch point*. We can establish that this pitch point P lies at the intersection of the line NN (the common normal to D and F at contact point A) and the line of centers OL. If we expand body D to include its pitch circle, the velocity of P on body D will be perpendicular to OL and will have the same *ec* along NN as point A, as dictated by the rigid–body principle. Point P on the pitch circle of body F will likewise have a velocity perpendicular to OL and will also have the same *ec* along NN as point A. Then V_P^D, equals V_P^F, since they have the same direction and the same *ec* along NN. When two cylinders roll without slip, they have identical velocities at the contact point. Point P satisfies these conditions and is therefore the point of tangency of the pitch circles or the pitch point. P is the only point on OL at which D and F will have equal velocities, when the contact is at A.

If another point, such as Q in Figure 6.16, is examined, the velocities of Q on D and F will not be equal if the contact remains at A. The effective components of the velocities of A on D and F are only equal along NN. They are not equal along AQ, so V_Q^D will not equal V_Q^F.

Figure 6.16

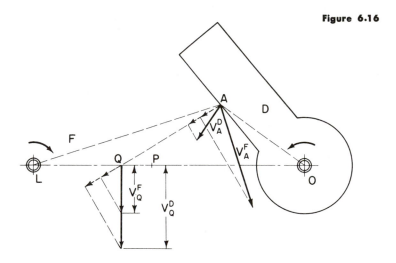

Pitch point P establishes the speed ratio $\omega_F/\omega_D = OP/PL$, and this ratio can only be attained when the pitch point remains at P. It follows then that, if D turns to a new position, the new contact point B must lie on a normal to the surface of D through P, as shown in Figure 6.17. Conversely, we can state that, *wherever D and F make contact, their common normal must pass through the pitch point*, if the speed ratio is to remain constant. This is called the *conjugate law*. It is the fundamental geometric rule by which conjugate curves are designed.

Figure 6.17 The conjugate law

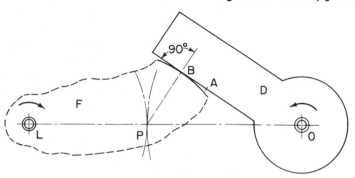

6.7 Designing Conjugate Curves

Applications of conjugate curves are not limited to gear teeth. Theoretically, it is possible to design a surface which will be conjugate to any given body, and these matched surfaces have many uses.

Let us consider this design process in a typical example. In Figure 6.18 a circular body D with center at E turns counterclockwise about fixed axis A at constant speed. It is required to design another body F, conjugate to D, which will turn clockwise about fixed axis B with the same constant speed as D.

Since the speed ratio is 1 : 1, D and F will have the same motion as two cylinders of equal diameter, rolling without slip. These hypothetical cylinders are represented as *pitch circles*, tangent at a pitch point P, located at the midpoint of line AB. These circles offer a convenient reference for this layout.

In order to define the curved surface of F, we must plot a series of points and connect them with a smooth curve. Let us consider the layout of some typical points, as the process is repetitive.

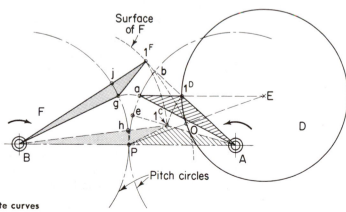

Figure 6.18 Design of conjugate curves

When D is in the position shown in Figure 6.18, its surface is in contact with F at some point. To locate this present point of contact, we are guided by the conjugate law, which states that, whenever D and F make contact, the common normal to their surfaces at the contact point must pass through pitch point P. When two curves touch one another, they have the same straight line tangent. The common normal is a perpendicular to this tangent at the contact point. It is normal to *both* D and F. We know little of F, but D is completely defined, so a line through P normal to D will also be normal to F and will determine the contact point. A radial line is normal to a circle, so line PE is normal to D, and point 0, on the surface of D, will be the contact point when D is in the position shown. Point 0 will then be one point on the surface of F.

Now let us consider another point on D, such as point 1^D, and locate the corresponding point on F. We can draw a normal to D through 1^D. When D has turned about A so that this normal $E1$ passes through P, point 1 will be in the contact position 1^C. It is usually impractical to draw the entire body D in the rotated position, so we extend the normal to meet the pitch circle of D at a and rotate only the segment $a1$.

It is easier to visualize the motion of this segment if we consider it a side of the triangle $Aa1^D$, shown shaded in Figure 6.18. As we swing this triangle about A, point a travels along the pitch circle to P, so that side Aa becomes AP. Point 1 also travels along a circular arc about A to position 1^C, so that the original triangle $Aa1^D$ becomes triangle $AP1^C$ in the rotated position. Point 1^C may be accurately located by extending side $A1^D$ to the pitch circle at b, making arc Pe equal to arc ab and locating 1^C on line eA. The normal $a1^D$ is now located at $P1^C$, passing through the pitch point. Point 1^C is therefore a contact point—a point on F as well as on D.

Since D rotated to bring point 1 to the contact position, F must also have turned in order to bring the point on its surface into contact. To determine the original position of this contact point on F, we must retrace its path. Since F is a rotating body, this path will be a circular arc about B through point 1^C.

We can more easily visualize the motion of the normal $P1^C$, as it swings back around B, if we picture it as the side of a triangle $BP1^C$, shown stippled in Figure 6.18. The point at P on this triangle will move up along the pitch circle of F to g. Pitch circles, like rolling cylinders, turn equal arc lengths in any given time, so while a was moving down the pitch circle of D to P, g was moving an equal distance down the pitch circle of F to P. Therefore g is so located that arc Pg equals arc Pa. We now can reproduce the triangle $BP1^C$ in the position $Bg1^F$, as shown, by making arc gj equal to arc Ph. This locates point 1^F, the original position of the contact point on F.

We now have two points (0 and 1^F) on the surface of body F. Other points must be determined in order to define the curve of the surface of F, and they may be found by the same process outlined for point 1^F. Since the construction for additional points would be obscure if added to Figure 6.18, the

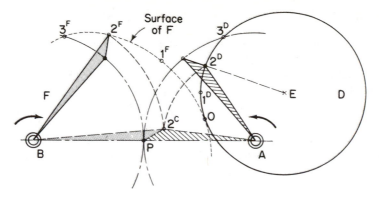

Figure 6.19 Design of conjugate curves

construction for another point (2^F) is shown on Figure 6.19 and a portion of the contour of F, as defined by further points, has been added.*

6.8 Applications of Conjugate Surfaces

In addition to gear tooth contours, conjugate curves are used in the design of rotors for blowers of the type shown in Figure 6.20. Two rotors, having two, or sometimes three, lobes are mounted on fixed parallel shafts and en-

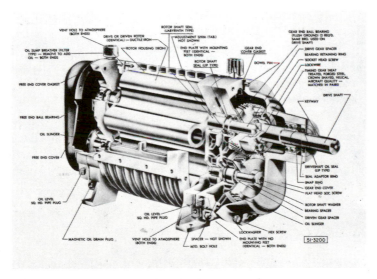

Figure 6.20 Conjugate rotor blower (Courtesy of MGD Pneumatics, Inc., Racine, Wis.)

*When the student understands the process of construction, the device of rotating triangles will not be necessary and the drafting will be thus simplified.

closed in a surrounding case. The rotors are driven by a pair of gears so that they turn in opposite directions at the same angular speed. The lobes are conjugate curves of such shape that they are identical. These rotors do not contact one another but are designed to have a very minute clearance maintained at all times by the constant speed ratio of the gears which drive them. There is likewise a small clearance between each rotor and the walls of the case.

Figure 6.21 shows a cross section of a blower having two-lobed rotors. As the rotors turn, air is drawn in at the opening in the bottom and entrapped between the lobes and the case. It is carried around the outside as shown by the arrows and forced out at the top. The leakage through the small clearance between rotors and case is negligible, and the lack of actual contact prevents noise and wear. The conjugate design is required to ensure the required clearance between the lobes in all positions even though they do not actually touch one another.

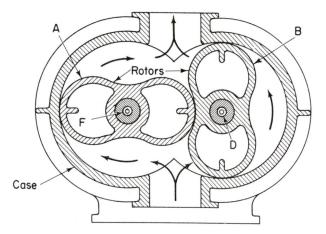

Figure 6.21 Cross section showing conjugate rotors

Figure 6.22 shows the layout work in designing the conjugate lobes, indicating the number of design points needed to provide sufficient accuracy. If the lobes are designed for contact with one another, then the contours are ground slightly undersize to establish the necessary clearance.

6.9 Involutes of Circles Are Conjugate

The end of a taut cord unwinding from a fixed cylinder traces a path called an *involute*. Involutes are especially adapted for use as conjugate curves, since the conjugate of an involute is another involute. Figure 6.23 shows an involute curve traced by a point P. The circle from which the cord is unwound is called the *base circle*. In any position the cord is always tangent to the base circle, and the radius of curvature of the involute at any point (such as P_1) is

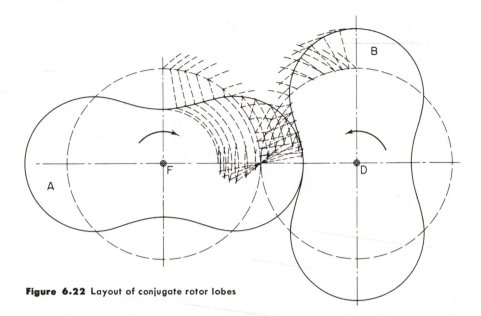

Figure 6.22 Layout of conjugate rotor lobes

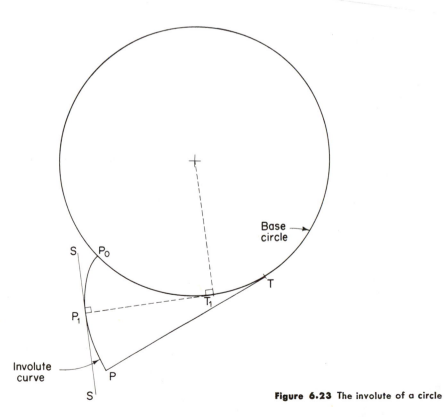

Figure 6.23 The involute of a circle

the length of this tangent (P_1T_1). A straight line tangent to the involute curve at any point will therefore be perpendicular to the tangent to the base circle at that point. For example, SS is perpendicular to P_1T_1. If the cord starts to unwrap at P_0, the arc of the base circle P_0T_1 will equal the tangent P_1T_1 in length.

In Figure 6.24 two base circles, A and B, are so placed that their involutes are in contact at point c. The tangent to the involute of A at c is line ss. This line ss is also the tangent to the involute of B at c, since when two curves touch one another they have a common tangent. The tangents to the base circles, cT and cQ, are both perpendicular to ss, so QT is one straight line and becomes the common normal to the involutes at the contact point. This common normal crosses the line of centers ab at point P.

Now let us rotate the base circles about their centers a and b, so that their involutes come into contact at some other point, such as e. The involutes will have a new common tangent tt. If we draw tangents to the base circles A and B from e, these tangents will both be perpendicular to tt and therefore will form a continuous straight line. This line can only be QT, as only one line can be drawn tangent to two circles in this manner. The new contact point e must then lie upon QT, and QT becomes the common normal to the involutes in this new position.

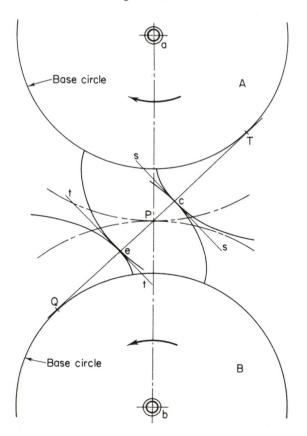

Figure 6.24 Contact between involutes

In fact, we can rotate the base circles about these centers a and b, bringing their involutes into contact at many different points, and all of these contact points will fall upon the common tangent to the base circles QT. In each case QT will be a common normal to the involutes and will cross the line of centers at point P. If we designate P as the pitch point, these involutes will conform to the conjugate law, for in each position of contact their common normal will pass through the pitch point.

6.10 Involute Gear Teeth

Since involute curves are conjugate to one another, they can be used as profiles for the contacting surfaces of gear teeth, this insuring a constant speed ratio. This application is worthy of our attention, as the involute has been adopted as the tooth form for nearly all standard gears.

Figure 6.25 shows a pair of circular disks mounted on fixed axes at d and f. One tooth has been cut on each disk. Each tooth profile is designed as the involute of a base circle having the same diameter as the disk upon which it is formed. Since gears are designed to run in either direction, each tooth is sym-

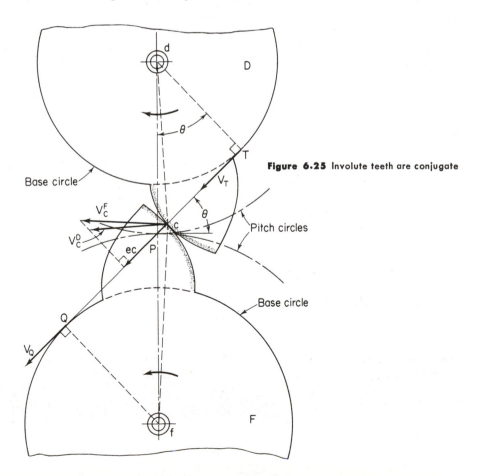

Figure 6.25 Involute teeth are conjugate

metrical, but we are only concerned here with the active surfaces, shown shaded. This pair of teeth is shown in contact at a typical point c. We know from the previous investigation that point c, like all contact points, lies upon the straight line QT, tangent to the base circles. In gearing terminology line QT is the *pressure line* and its intersection with the line of centers (df) is the *pitch point P*. Furthermore, we have established that line QT is always normal to both contacting surfaces at the contact point. This, coupled with the fact that QT passes through P, makes these surfaces conjugate.

The speed ratio is governed by contact between the teeth through which the drive is transmitted. Since the teeth turn with the base circles, the speed ratio must be established by determining the angular velocities of the base circles of F and D. Velocity vectors offer a familiar approach. The velocity of point T on the base circle of D equals $\omega_D \times dT$ and is directed along the tangent to the base circle QT. Contact point c on D has an effective component along QT equal to V_T by the rigid body principle. The velocity of point c on D is directed perpendicular to line dc and thus can be determined from the effective component, as shown in Figure 6.25. The velocity of point c on F is directed perpendicular to line fc, which has a different slope than dc. The contact at c is thereby defined as sliding contact. Since QT is the common normal, the effective component along QT is directed perpendicular to the common tangent of the tooth surfaces, along which sliding takes place. This effective component is thus perpendicular to sliding and therefore is the effective component of c on F as well as the effective component of c on D. Since we know the direction of the velocity of c on F, it can be determined as shown. The velocity of the tangent point Q on the base circle of F has an effective component, along QT, equal to that of point c. The velocity of the tangent point Q is also directed along QT and is therefore equal to the effective component along QT.

We note now that, since the velocity of T equals the effective component of the contact point c and the velocity of Q equals that same component, $V_T = V_Q$, but since $V_T = \omega_D \times dT$ and $V_Q = \omega_F \times fQ$, it follows that $\omega_D \times dT = \omega_F \times fQ$ and the speed ratio

$$\frac{\omega_F}{\omega_D} = \frac{dT}{fQ} \quad \text{or} \quad \frac{\omega_F}{\omega_D} = \frac{\text{base radius of } D}{\text{base radius of } F}$$

If P is a proper pitch point, the speed ratio in terms of pitch radii ($\omega_F/\omega_D = dP/fP$) must be the same as the speed ratio in terms of base circle radii ($\omega_F/\omega_D = dT/fQ$) or, in other words,

$$\frac{dP}{fP} = \frac{dT}{fQ}$$

This can be easily established by geometry. Right triangle dTP is similar to right triangle fQP, since their sides are respectively parallel. Therefore

$$\frac{dP}{fP} = \frac{dT}{fQ}$$

as corresponding sides are proportional. The speed ratio, expressed in terms of pitch radii, is then:

$$\frac{\omega_F}{\omega_D} = \frac{\text{pitch radius of } D}{\text{pitch radius of } F}$$

This is the same ratio as that of the base circles established by tooth contact. Since the involute teeth are conjugate, this ratio will remain constant at all times.

6.11 Relationship of Pitch and Base Circles

We have noted in Figure 6.25 that QT is the pressure line, that it crosses the line of centers at pitch point P, and that it is tangent to the base circles of the involute teeth. For a given speed ratio and center distance (df), the pitch point has a specific location and the pitch circles a specific size. The base circles may vary in size, however, as long as the ratio of their radii is the same as that of the pitch radii. The pressure angle is the angle θ, shown in Figure 6.25, between QT and the common tangent to the pitch circles. If the pressure angle is made smaller, the base circles will be proportionately larger in order to remain tangent to the pressure line. Thus there is a choice of pressure angle when designing for a given speed ratio.

The ratio of the size of the base circle radius to the pitch radius on any involute gear is determined by the size of the pressure angle. In the right triangle dTP, angle PdT is equal to the pressure angle θ, since dT is perpendicular to PT and dP is perpendicular to the tangent to the pitch circles. In this triangle the ratio of side dT to side dP is the cosine* of the pressure angle θ. Therefore

$$\frac{\text{radius of base circle}}{\text{radius of pitch circle}} = \cos \theta$$

Since the size of the pressure angle is usually specified and the pitch radii are determined by the speed ratio, this relationship affords an easy way to calculate this radius of the base circles.

6.12 Standard Involute Gears

There are many requirements which must be satisfied in the selection of tooth contours. The involute curve fulfills these essential conditions so well that it is used on nearly all standard gears. Among the virtues of the involute tooth form are the following:

*In a right triangle the ratio of the side adjacent to an angle to the hypotenuse is called the cosine of that angle. Values of cosines for various angles are tabulated in the Appendix, for reference.

1. Contacting surfaces of involute form are always conjugate.
2. The involute tooth is so shaped as to be strong and rigid.
3. The shape of the space between involute teeth affords sufficient clearance for the entry and departure of meshing teeth on mating gears,
4. Several different methods of cutting involute teeth are available and economical.
5. Since the path of contact is a straight line, the pressure angle is constant. A small pressure angle may be used so as to reduce bearing friction and thus increase efficiency.
6. The center distance at which mating gears are mounted is not critical. The geometry of the involute is such that the same pair of teeth will transmit the same constant speed ratio even though the mounting distance of the gears is somewhat oversize. As the shafts and teeth deflect (bend) under loads, this adaptability of the involute is of special advantage.

Pressure angles and tooth proportions are standardized so that any one gear will run with a wide variety of others and is not limited to one mating gear especially designed for it. Experience has shown that pressure angles of either $20°$ or $14\frac{1}{2}°$ are most desirable in the majority of cases.

The addendum distance, dedendum distance, clearance, fillet radius, and tooth width are tabulated in Figure 6.26 for the three systems of tooth proportions generally used: $14\frac{1}{2}°$ full depth, $20°$ full depth, and $20°$ stub teeth. The applications of each system will be explained later. With the exception of tooth width, these dimensions are given in terms of the diametral pitch, as this is a sort of scale factor for tooth sizes. Since mating gears have the same diametral pitch, these dimensions will be the same on any mating pair. The tooth width is given in terms of circular pitch, as it is measured on the pitch circle. Under normal conditions gears are designed to run without backlash, so tooth width is equal to space width in each case. All gears are designed to run in either direction, so the teeth are symmetrical with respect to their radial center lines.

	$14\frac{1}{2}°$ Full depth	$20°$ Stub	$20°$ Full depth
Pressure angle	$14\frac{1}{2}°$	$20°$	$20°$
Addendum	$1/P_D$	$0.8/P_D$	$1/P_D$
Dedendum	$1.157/P_D$	$1/P_D$	$1.157/P_D$
Clearance and fillet radius	$0.157/P_D$	$0.2/P_D$	$0.157/P_D$
Tooth width	$P_C/2$	$P_C/2$	$P_C/2$

Figure 6.26 Standard involute gear tooth dimensions

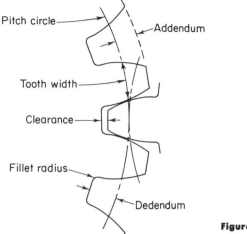

Pitch circle

Addendum

Tooth width

Clearance

Fillet radius

Dedendum

Figure 6.26 (Cont.)

Flanks are straight radial lines extending from the base to fillets. Contacting surfaces from base circle to addendum circle are involutes of the base circle. The outer tips of the teeth follow the curve of the addendum circle, and the inner surfaces between teeth follow the dedendum circle.

6.13 Layout of Standard Involute Gears

The established dimensions, forms, and production tools for standard gear teeth make it unnecessary for a designer to make an actual layout of tooth profiles for any standard gears. These gears can be purchased ready-made from gear manufacturers or cut by tools which generate the correct form without the use of a template or contour layout.

In order to become acquainted with the geometry and application of standards, there is no better method than to "build" some gears on paper. This is interesting and instructive for the student of mechanisms (if not necessary in industry) and is indirectly very valuable preparation for the design of non-standard gears.

Let it be required to draw a standard spur gear D driving a rack F, with the following specifications: diametral pitch = 1, pressure angle = $14\frac{1}{2}°$, number of teeth on gear = 24, system = full-depth involute. Three teeth are to be shown on both the gear and rack with one pair in contact at the pitch point.

First, we draw a line of centers and lay out the pitch radius of the gear, locating the pitch point P in relation to the center A. Since $P^D = T/D^P$, $D^P = \frac{24}{1}$ and the pitch radius = 12 in. (all dimensions of length on gears are assumed to be in inches).

Next, we draw the pressure line at the correct pressure angle. This line passes through P and is tangent to the base circle. Since the base circle will be required in this layout, it is efficient to draw that first and locate the pressure line tangent to it, rather than to lay out the pressure angle. The relation between the pitch diameter and base circle diameter has been established:

$$\frac{\text{Base diam}}{\text{Pitch diam}} = \text{cosine of pressure angle } \theta$$

$$\text{Base diam} = D^P \times \cos 14\tfrac{1}{2}° = 24 \times 0.968*$$

$$\text{Base diam} = 23.23$$

$$\text{Base rad} = 11.62 \text{ in.}$$

Figure 6.27 shows the layout at this stage including the line of centers, pitch and base circles, and pressure line. Note that the pressure line slopes outward from the center of the driver in the direction of rotation.

The addendum and dedendum distances are obtained from the table in Figure 6.26. Addendum = 1 in., and dedendum = 1.157 in. These may be laid out from the pitch point and the addendum and dedendum circles drawn. The tooth width = space width = $P^c/2$. These distances must be measured along the pitch circle. This can be done by first laying them out on a straight

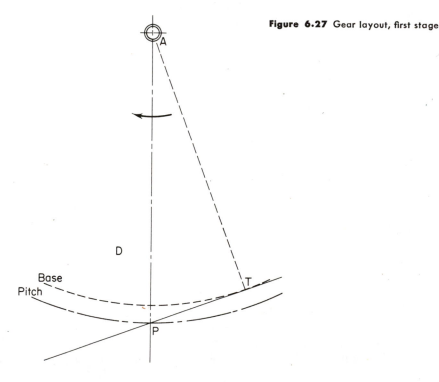

Figure 6.27 Gear layout, first stage

*See Appendix for table of trigonometric functions, sine, cosine, tangent, etc.

line and then stepping the distances off along the circle in several steps with the dividers. However, it is often easier to find the angle subtended by P^c and construct the angle. P^c is that portion of the pitch circle circumference devoted to one tooth and one space. Here there are 24 teeth, so 360° divided by 24 equals 15°, the angle subtended by P^C. A 15° angle may be constructed by using the table of chords in the Appendix or by simply bisecting a 30° angle. The 15° angle may in turn be bisected to obtain a tooth or space width, and these may be laid out from the pitch point P (Pb, Pc, ab, etc, as shown in Figure 6.28).

Now we must construct an involute of the base circle for the contacting tooth surface. This construction is shown in Figure 6.28, and, since it is an approximate method, *requires very precise drafting*. With the small dividers set to about $\frac{1}{20}$ of the base radius ($\frac{1}{2}$ in. in this case is satisfactory), we step off about 15 of these divisions on the base circle, preferably off to one side of the drawing. These are numbered 0, 1, 2, 3, etc. The involute originates at point 0. At point 1 a tangent to the base circle is constructed, and one of the $\frac{1}{2}$-in. steps is laid out on the tangent. At point 2 another tangent is constructed, and two steps are laid off from point 2 on this tangent. We lay out three steps at

Figure 6.28 Gear tooth layout, second stage

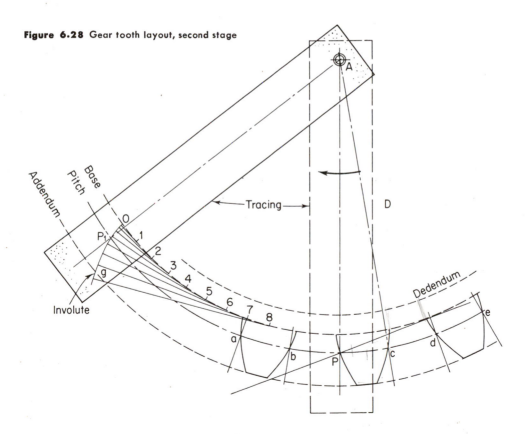

point 3, etc. All of these distances are measured in the general direction of point 0, and the points at the end of each distance define an involute. A smooth curve is drawn through these terminal points from the base circle to the addendum circle, as shown in Figure 6.28.

To transfer this involute profile to the position of the gear teeth, we draw a radial reference line AP_1 from the center of the gear through the intersection of the pitch circle and the involute profile. Corresponding radial lines (*Aa*, *Ab*, *Ac*, *Ad*, and *Ae*) are drawn to the segments of the pitch circle defining tooth positions. A tracing of the involute profile is now made including line AP_1 and the intersections of the involute with the base circle (0) and addendum circle (*g*). This tracing may be positioned so that AP_1 coincides with *AP*, *Aa*, and *Ad* and the profile traced in these positions. We then turn the reference tracing over, so as to reverse the curve, and trace the opposite sides of the teeth, aligning AP_1 with *Ab*, *Ac*, and *Ae*, as shown in Figure 6.28,

The flanks of the teeth from the base to the dedendum circle are drawn radially towards *A*. Fillet curves having the radius specified in Figure 6.26 and tangent to these flanks and the dedendum circle complete the tooth profiles on the gear (Figure 6.29).

Figure 6.29 Completed layout of gear and rack.

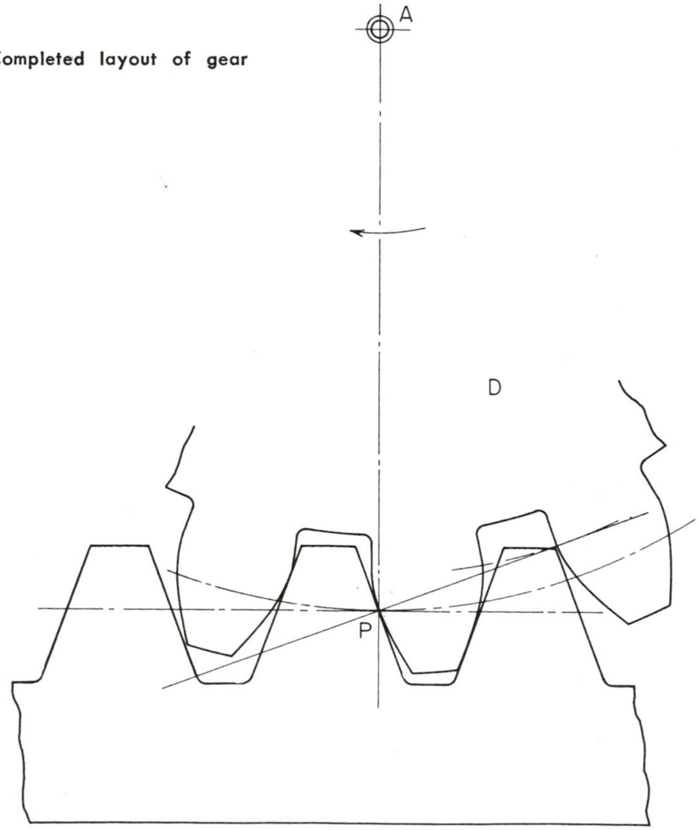

Since a rack has an infinite radius, the pitch "circle" of the rack is a straight line through P, perpendicular to AP. The addendum and dedendum distances of the rack are the same as those of the gear, obtained from the table in Figure 6.26. The addendum distance is measured from P toward A and the dedendum in the opposite direction. Like the pitch "circle" the addendum and dedendum "circles" are straight lines, perpendicular to AP, as shown in Figure 6.29.

Tooth and space widths are both equal to 1.57 in. and are measured along the pitch line of the rack from P. The involute of a straight line is another straight line, so the flank and face of the rack teeth are defined by single straight lines from addendum to fillet. Since the rack and gear teeth must be normal to the pressure line when they are in contact (to obey the conjugate law), the contacting surfaces of the rack teeth are drawn perpendicular to the pressure line. The rack teeth have the same slope on each side so as to make the drive reversible. Rack fillets have the same radius as gear fillets.

The three rack teeth may now be drawn, completing the layout, as shown in Figure 6.29.

6.14 Geometrical Relationships in Terms of Pressure Angle

We have observed that a graphical layout is often the easiest method of determining significant measurements, such as paths of contact. Wherever calculations are simple, however, the analytical approach is preferred.

The size of the pressure angle establishes the relationship of a number of quantities which are frequently used in gearing study. In Article 6.11 we found that the ratio of the radii of the base circles to the pitch circles was equal to the cosine of the pressure angle:

$$\frac{R^B}{R^P} = \cos \theta$$

It so happens that the ratio of other significant quantities is also equal to $\cos \theta$. Since these relationships are easy to remember and to use, they are worthy of recognition.

The distance along the pitch circle from one tooth to the next has been defined as *circular pitch*. A corresponding distance between adjacent teeth measured along the base circle is also useful and is designated *normal pitch* (P^N). The involute contours of the teeth are in theory formed by unwrapping a taut cord from the base circle. Since the straight unwrapped portion of this cord may coincide with the pressure line, it follows that the distance between a pair of teeth measured along the base circle (normal pitch) equals the distance between these teeth measured along the pressure line (Figure 6.30).

Just as circular pitch equals the circumference of the pitch circle divided

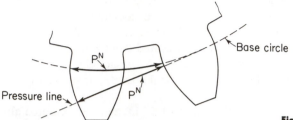

Figure 6.30 Normal pitch (P^N)

by the number of teeth on a gear, so is the normal pitch equal to the circumference of the base circle divided by the number of teeth:

$$P^N = \frac{\pi D^B}{T}$$

We have observed the ratio of the diameters of the base and pitch circles:

$$\frac{D^B}{D^P} = \cos \theta$$

If we multiply both numerator and denominator by π/T, this ratio is not altered:

$$\frac{\pi D^B/T}{\pi D^P/T} = \cos \theta$$

But $\pi D^B/T =$ normal pitch (P^N) and $\pi D^P/T =$ circular pitch (P^C)

$$\therefore \frac{P^N}{P^C} = \cos \theta$$

In Figure 6.31, a single tooth on gear D is shown in that position where it first makes contact and later (dotted) when it is in contact at the pitch point. The path of approach is aP (along the pressure line), and the arc of approach is eP (along the pitch circle). Again let us consider the pressure line a taut cord unwound from the base circle. If we wind this cord back around the base circle, point a will follow the involute to b and P will follow the involute to c. The arc bc of the base circle will equal the straight line path of approach aP. The segment cc' of the base circle equals the segment bb', since it is the same measurement on the tooth in each position. Therefore the arc $c'b'$ equals the arc cb, which in turn equals the path of approach aP.

It is clear from Figure 6.31 that the arc of approach eP and the arc $c'b'$ both subtend angle α (the angle of approach). Since an arc equals its radius multiplied by the subtended angle (in radians):

$$eP = OP \times \alpha \quad \text{and} \quad c'b' = Ob' \times \alpha$$

$$\text{or} \quad \frac{eP}{OP} = \alpha \quad \text{and} \quad \frac{c'b'}{Ob'} = \alpha$$

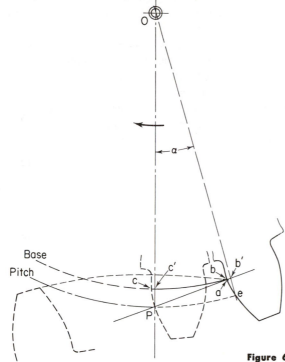

Figure 6.31 Path and arc relationships

$$\therefore \frac{eP}{OP} = \frac{c'b'}{Ob'} \quad \text{or} \quad \frac{Ob'}{OP} = \frac{c'b'}{eP} \qquad \text{(multiplying both sides by } Ob'/eP)$$

But Ob' equals the radius of the base circle, and OP equals the radius of the pitch circle. Also $c'b'$ equals the path of approach, and eP equals the arc of approach, so

$$\frac{R^B}{R^P} = \frac{\text{path of approach}}{\text{arc of approach}} \quad \text{but} \quad \frac{R^B}{R^P} = \cos \theta$$

$$\therefore \frac{\text{Path of approach}}{\text{Arc of approach}} = \cos \theta$$

Similarly:

$$\frac{\text{Path of recess}}{\text{Arc of recess}} = \cos \theta$$

$$\frac{\text{Path of app} + \text{path of rec}}{\text{Arc of app} + \text{arc of rec}} = \frac{\text{path of contact}}{\text{arc of contact}} = \cos \theta$$

Since all these ratios equal the cosine of the pressure angle, the only problem in recalling them arises in setting the quantities in correct order. This is simplified if we note that *all of the denominators of the ratios (diameter of pitch circle, circular pitch. and arcs of contact) are measurements of the pitch circle,* whereas the numerators pertain to the base circle.

These relationships between paths and arcs are very useful in determining arc lengths. We have noted that paths may be found by a simple drawing. In Figure 6.32 *aP* is the path of approach and *Pg* the path of recess. These are straight line distances and can be measured readily from the layout which involved only the pitch radii, addendum radii, and pressure angle. Having determined these paths, the cos θ relationships permit us to calculate arcs without out drawing the involute tooth profiles, which is a laborious process.

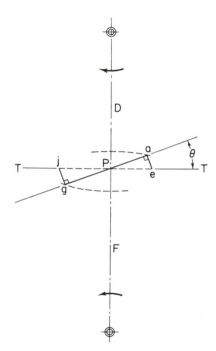

Figure 6.32 Graphical relationship of path to arc

Arcs of action, approach, and recess can easily be found graphically, if desired. If we draw perpendiculars to the pressure line at *a* and *g* to meet the common tangent *TT* at *e* and *j*, we observe that, in the right triangle *Pae*, $aP/eP = \cos\theta$. The ratio of the path of approach to the arc of approach is cos θ. Therefore, since *aP* is the path of approach, *eP* must equal the arc of approach. By similar reasoning, *Pj* equals the arc of recess. The arc of contact will then be *ej*, the sum of *eP* and *Pj*. By this device the arc lengths are rectified to straight-line measurements for direct scaling.

6.15 Limitations in the Use of Standard Gears

Since many manufacturers specialize in the production of standard gears, it is more economical to purchase them ready-made, whenever possible, than to design and produce special gears for each application. Once the speed ratio is determined, we select standard gears of proper size from a wide variety offered in the manufacturers' catalogues.

It has been established that a pair of gears, to run together, must have the same diametral pitch, the same pressure angle, and the same system of tooth proportions (stub or full depth involute). It might appear that if they correspond in the above respects, one might combine gears of any available tooth numbers into a satisfactory drive. This is not the case. There are limitations upon the tooth numbers of gears which will run together correctly even though they have the same standard specifications of tooth form. In general the restrictions apply to the use of gears having small tooth numbers. This is why a knowledge of gear-tooth design is important even in the use of standard gears.

6.16 The Minimum Permissible Path of Contact

In order to maintain a constant speed ratio when a pair of gears run together, it is essential that at least one pair of teeth be in contact at all times. If there is any position of the gears in which no teeth are in contact, the drive will be intermittent, causing variations in the speed ratio, noise, and vibration at high speeds.

A pair of adjacent teeth on each of two meshing gears are shown in Figure 6.33. The path of contact, determined by the intersections of the addendum circles and pressure line, is *ab*. The left-hand pair of teeth are shown in the final position of contact at *b* and are just about to separate. The right-hand pair first came into contact at *a* (in the position shown dotted) and have maintained contact down the pressure line to point *k*, the present position. Thus, with these gears, there always will be at least one pair of teeth in contact, as required in a correct drive. We note that the arc of contact (*ec*) is greater than the circular pitch (*jc*) and the path of contact (*ab*) is greater than the normal pitch (*kb*, measured on the pressure line).

Figure 6.34 shows a second pair of adjacent teeth on each of two meshing gears. The path of contact is *ab*, and the left-hand pair are shown in the final position of contact, just about to separate. The right-hand pair have not yet come into contact, as their active surfaces have not turned as far as point *a*, where contact starts. In this case, if the gears are turned slightly in the direction of the arrows, the left-hand pair of teeth will have separated; yet the right-hand pair will not have reached the first contact position at *a*. Thus, for

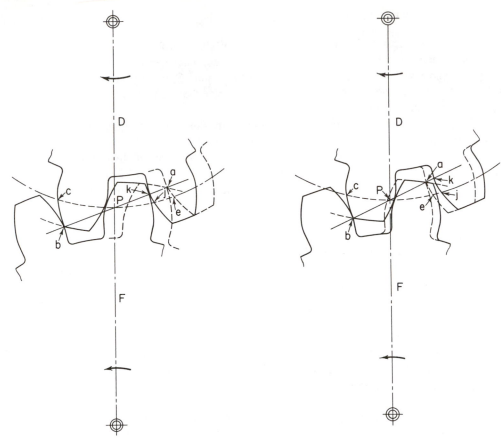

Figure 6.33 Arc of contact greater than circular pitch

Figure 6.34 Arc of contact less than circular pitch; intermittent contact

an instant, there is no positive tooth contact between the two gears, and the follower gear *F* will slow down until the right-hand driver tooth has turned far enough to establish contact again. This *intermittent contact* causes impact and variable speed ratio, both of which are undesirable. In Figure 6.34, note that the arc of contact (*ec*) is less than the circular pitch (*jc*) and that the path of contact (*ab*) is also less than the normal pitch (*kb*).

In Figure 6.35 we show a third gear drive in which the right-hand tooth pair have just made contact at *a*, while the left-hand pair are at the final contact position at *b*—just about to separate. In this case we have the limiting condition for a correct drive where the arc of contact exactly equals the circular pitch and the path of contact equals the normal pitch.

We may deduce from the above examples that the ratio of the arc of contact to the circular pitch or the ratio of the path of contact to the normal pitch is the critical value to be controlled if we are to prevent intermittent

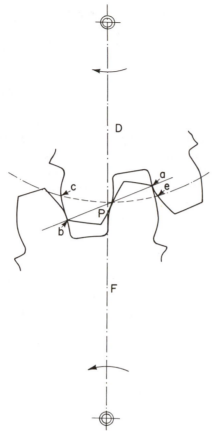

Figure 6.35 Arc of contact equal to circular pitch

contact. We can state the limiting conditions as follows: *the minimum arc of contact equals the circular pitch;* or the equivalent statement, *the minimum path of contact equals the normal pitch.* These two statements are equivalent, since the path equals the arc multiplied by the cosine of the pressure angle. If one is satisfied, the other will follow automatically.

6.17 The Contact Ratio

Since the path of contact is easier to obtain than the arc of contact, the minimum path of contact is the criterion generally used when checking for intermittent contact. *The ratio of the path of contact to the normal pitch is called the contact ratio.* Since the path can never be less than the normal pitch, it follows that, in order to prevent intermittent contact, the contact ratio must not be less than 1.

6.18 Checking Gears for Intermittent Contact

Let us assume that we wish to use two standard, 11-tooth, stub involute gears of 20° pressure angle to connect two parallel shafts which are to turn at the same speed. Before purchasing these gears, it is advisable to check this design for contact ratio*. It is necessary to determine the path of contact and the normal pitch.

Since the graphical method of obtaining the path is easy, let us use that first and avoid the more lengthy calculations until we determine if greater accuracy is required. Diametral pitch is not specified in this example, since it is only a scale factor. The ratio of the path to the normal pitch is the same for all gear pairs of specified tooth numbers, whatever the diametral pitch. As the layout should be reasonably large for accuracy, let us assume the gear to have a diametral pitch of 1. Since $P^D = T/D^P$,

$$D_P = \frac{T}{P^D} = \frac{11}{1} = 11$$

The line of centers is therefore 11 in. long, with the pitch point at the midpoint, as shown in Figure 6.36. The pressure line is drawn through P at 20° with the common tangent TT. The addendum is found (in Figure 6.26) to be $0.8/P^D = 0.8$ in., so the addendum radii will be $5.5 + 0.8 = 6.3$ in. The addendum circles are now drawn intersecting the pressure line at a and b. The

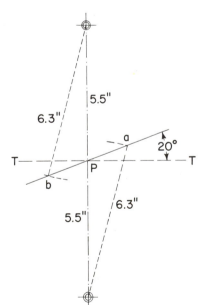

Figure 6.36 Layout to check for intermittent contact

*Experience shows that when standard gears are used and mounted at the correct center distance, intermittent contact does not occur. The contact ratio is a significant design factor.

path of contact *ab* is next measured from the layout and found to be 3.42 in. The normal pitch is easily calculated:

$$P^N = \frac{\pi D^B}{T}$$

$$\frac{D^B}{D^P} = \text{Cos } \theta \quad \text{So} \quad D^B = D^P \text{ Cos } \theta$$

$$\therefore P^N = \frac{\pi D^P \text{ Cos } \theta}{T} = \frac{\pi \times 11 \times 0.9397}{11} = 2.95 \text{ in.}$$

The path is much greater than the normal pitch, so these gears will operate without intermittent contact. The contact ratio is:

$$\frac{\text{Path}}{P^N} = \frac{3.42}{2.95} = 1.16$$

6.19 Remedies for Intermittent Contact

If intermittent contact is found to take place when two gears are run together, it can be eliminated by increasing the path of contact until it exceeds the normal pitch.* This can be accomplished in several ways:

1. *Increasing the addendum:*
   The layout of the line of centers, pressure line, and addendum circles is shown in Figure 6.37 for a gear pair in which the path of contact (*ab*)

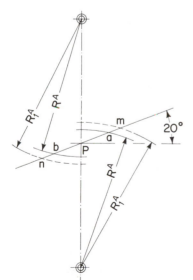

Figure 6.37 Increasing the addendum lengthens the path

*There is a limit to the amount in which the path of contact may be increased. This maximum limit will be discussed in Article 6.20.

is less than the normal pitch. If the addendum distance is increased (other specifications remaining the same), the addendum radii will be larger (R_1^A), intersecting the pressure line at m and n. If the new path (mn) is equal to, or greater than, the normal pitch, the contact will no longer be intermittent.

2. *Increasing the Pitch Diameters:*

   Figure 6.38 shows the layout of the same gear pair as in Figure 6.37, in which the path ab is less than the normal pitch. The resulting intermittent contact can be remedied by increasing the pitch diameters. If the other specifications are not altered, only the tooth numbers will be affected. If the diametral pitch is not changed, this will also increase the numbers of teeth. The relationship of the pitch diameters and tooth numbers will remain in the same ratio and the pressure angle is unchanged, so the normal pitch ($\pi D^P \cos \theta / T$) is the same.

The change in pitch radii increases the addendum radii (R_1^A), and the new path of contact becomes mn—a little larger than ab.

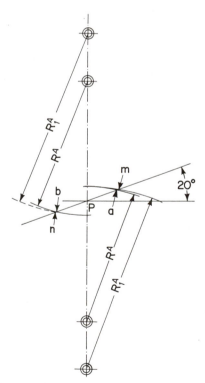

Figure 6.38 Increasing pitch radii lengthens the path

3. *Decreasing the Pressure Angle:*

If we hold all values constant except the pressure angle and make that smaller, the path of contact will become larger, as shown in Figure 6.39. The normal pitch ($\pi D^P \cos \theta / T$) will also increase, since the cosine becomes larger as the angle grows smaller. This increase in the normal pitch is much less, however, than the increase in path, so that ratio of path to normal pitch increases as the pressure angle is made smaller. The original path of contact *ab* (with a 20° pressure angle) is increased to *mn* when the pressure angle is reduced to $14\frac{1}{2}°$ as in Figure 6.39.

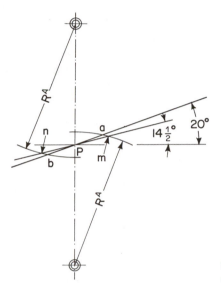

Figure 6.39 Decreasing the pressure angle lengthens the path

Intermittent contact may be overcome by increasing the addendum, by using gears with larger numbers of teeth, or by decreasing the pressure angle. From this we may generalize that this difficulty can be expected when stub-tooth gears of low tooth numbers and larger pressure angles are run together.

The implication is that all gear combinations having these characteristics should be checked for contact ratio before the drive is designed. Experience shows, however, that, while this fault may occur when two standard, 20°, stub-involute gears of low tooth numbers are run together, other difficulties arise to impose more drastic limitations on the size of the gears. Therefore intermittent contact is not the critical factor in the case of standard gears. This limiting condition is only important in the design of special gears having short addenda and large pressure angles and especially when drives are composed of gears having few teeth.

6.20 The Maximum Permissible Path of Contact

A second restriction in the use of involute gears is important in the selection of standard gears as well as those of unique design. The active surfaces of gear teeth are carefully designed so as to maintain a constant speed ratio as contact proceeds along each tooth. Only these specially designed active surfaces can be brought into contact, for if other surfaces are permitted to touch one another, the speed ratio may be caused to fluctuate or motion may be prevented altogether. It is not only necessary that the teeth be conjugate to one another, but it is equally important that they clear one another as they approach and depart from the contact positions. This undesirable contact between nonconjugate surfaces is called *interference*, and further limitations must be imposed upon the design of gear drives in order to avoid it.

Since the involute surfaces of the teeth extend only from the base circle to the addendum circle, any contact with the radial flank, fillet, or dedendum surface must be prevented, as only involute surface contact will ensure constant speed ratio.

In a gear drive, it is advantageous to have as long a path of contact as possible. The long path of contact brings more pairs of teeth into contact at one time, distributing the load rather than imposing it all upon one pair. A long path of contact requires a long addendum on the gear tooth. There are limitations to the length of this addendum which in effect establish the maximum permissible path of contact.

6.21 Maximum Path of Contact Due to Pointed Teeth

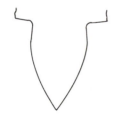

Figure 6.40 Pointed involute tooth

The first limitation on the addendum is an obvious one arising directly from the geometry of the tooth shape. The converging involute faces of a tooth intersect when extended, as shown in Figure 6.40. This point of intersection, barring other restrictions, definitely establishes the maximum addendum radius of a gear, and the intersection of this addendum circle with the pressure line frequently determines the maximum path of contact.

6.22 Checking for Pointed Teeth

There is no simple calculation to determine the exact addendum radius of the pointed tooth. A simple graphical process is shown here which does not require a layout of the complete involute but which is sufficiently accurate for checking purposes.

The radial centerline of a tooth is shown as a vertical line through C in Figure 6.41. The pitch point P is located so that the pitch radius equals CP, and a portion of the pitch circle is drawn. The radius of the base circle (CB) is calculated, and a larger portion of the base circle is drawn. Next an arc of the pitch circle equal in length to one half of the tooth width is laid out (see arc PE). From E, a line tangent to the base circle at G is drawn, and an arc of the base circle (GJ) is stepped off with the dividers equal to the tangent GE.

Points J and E will be two points on the involute curve of the tooth, but we need one or two more to sufficiently define the curvature for an accurate intersection with the tooth centerline. For example, the tangent $9L$ is laid out with dividers equal to base circle arc $9J$ and the tangent $12M$ is made equal to arc $12J$, using the usual procedure for involute plotting. Point 9 is nine steps

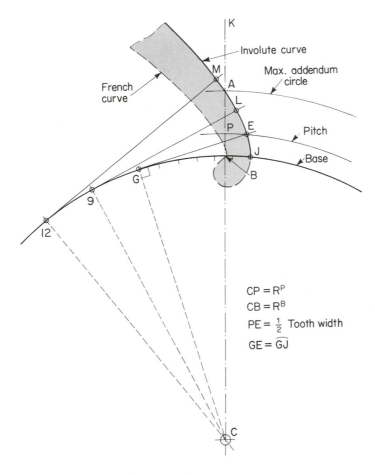

Figure 6.41 Finding addendum radius of pointed tooth

on the arc from J and point L is nine steps on the tangent from point 9. Similarly, point 12 is twelve steps from J and point M is twelve steps from point 12.

We now have four points on the involute tooth curve (J, E, L, and M), and by applying the irregular curve in the usual fashion, we can draw a portion of the involute crossing the tooth centerline at A. The point of the tooth will be at point A, and the required addendum circle will be drawn as shown with radius CA. Like all graphical solutions, this layout must be made at reasonably large scale to obtain valid results. One may guess at the position of tangents like $9L$ and $12M$ so as to bring points like L and M near the centerline of the tooth. A good choice of the first tangent may yield a point so near the centerline that only one more will be needed. Obviously, there is no reason to draw the other side of the tooth form.

The standard addendum values always terminate the tooth before it becomes pointed, so this limitation is not significant when standard gears are to be used. Pointed teeth are used in specially designed gears to provide a longer path of contact on small pinions* and to facilitate meshing of teeth in variable-speed transmissions.

6.23 Maximum Path of Contact Due to Interference

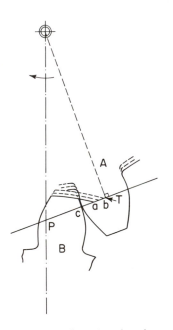

A pair of teeth A and B are in contact at c in Figure 6.42. With the addendum on gear B shown, the path of contact would start at point a. This path could be extended to point b, if the addendum of tooth B were increased and the dedendum of A adjusted to maintain clearance, as shown dotted.

The same teeth are shown in Figure 6.43, in which the addendum of tooth B is extended until the tooth is pointed. This extends the path of contact to point e, which appears to be the maximum permissible extension of the path at this end. Actually this is not true, since interference between the point of tooth B and the radial flank of tooth A takes place when the tooth is made so long.

This difficulty can be seen in Figure 6.44, where a series of successive positions of the contacting teeth are shown, indicating the overlap of the point of B with the flank of A.

The path of contact (in Figure 6.43) is shown to extend beyond point T, the point of tangency of the base circle and pressure line. This point T has

Figure 6.42 Extending the addendum

*Small spur gears are often called *pinions*.

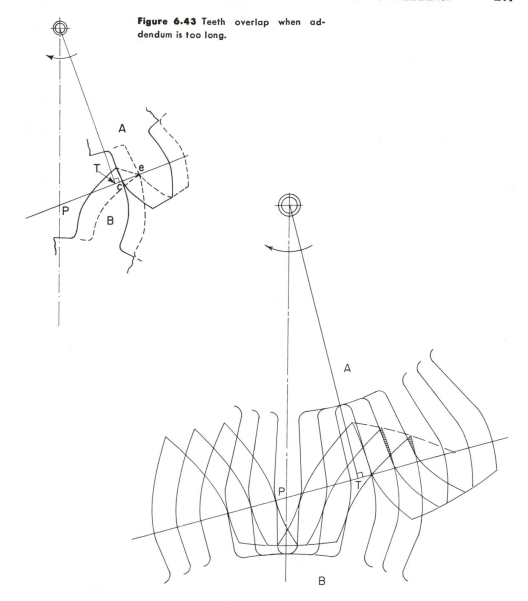

Figure 6.43 Teeth overlap when addendum is too long.

Figure 6.44 Successive contact positions showing interferences (shaded)

been found experimentally to be the limit of the path of contact if interference is to be prevented.

It is of course possible to undercut the flank of tooth *A* so as to provide clearance, but this usually weakens the tooth, since the section between the flank and fillet is already quite narrow, as shown in Figure 6.45.

Figure 6.45 Tooth with undercut flank

Contact of the teeth beyond the point of tangency of the base circle is furthermore improper, since it may be observed that (in Figure 6.43) the involutes of *A* and *B* are not tangent at point *e*. Therefore they can have no common tangent and cannot be conjugate to one another. The conclusion is that the path of contact may not extend in either direction beyond the points of tangency of the base circles and pressure line. Also, if the teeth become pointed before they reach the points of tangency, this condition will establish the end of the path of contact. These two possibilities must be taken into account. Whichever situation first limits the path establishes maximum length permissible.

6.24 Checking for Interference

Since a graphical check is faster than calculations, it is recommended that a layout should first be made when checking gears for interference. If borderline conditions are indicated by the layout, then calculations may be used for greater accuracy. To check a gear pair for interference, we need only know the pressure angle, addendum, and tooth numbers. The diametral pitch only determines the relative size of the layout, so a P^D number may be selected to provide a scale sufficiently large for accuracy.

A specimen layout is shown in Figure 6.46. The pitch radii are calculated from the tooth numbers and diametral pitch ($P^D = T/D^P$), and the line of centers *AB* and pitch point *P* is determined. The pressure line is drawn using the specified pressure angle. Perpendiculars from *A* and *B* to the pressure line locate the points of tangency of base circles. The addendum distances are added to the pitch radii to obtain the addendum radii *Ae* and *Bc*. The ends of the path of contact *a* and *b* are at the intersections of the addendum circles with the pressure line. In this case point *a* lies between *P* and *T*, which is a proper condition, but *b* lies outside of the line *PQ*, indicating that interference will occur. This combination of gears cannot be used, as interference at either end of the path renders the drive improper.

Since the distance *Qb* is significantly large in this instance, no calculations are required to substantiate the graphical results. If distance *Qb* were very small, it would be safer to calculate distances *Pb* and *PQ* and compare these figures.

If such an analytical check is required, it is a little easier to compare distances *Tb* and *TQ*, which achieves the same result as comparing *Pb* and *PQ*, since $Tb = Pb + PT$ and $TQ = PQ + PT$. The method is outlined below, using Figure 6.47 as a guide.

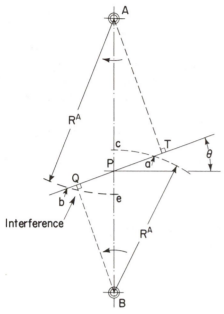

Figure 6.46 Graphical check for interference   **Figure 6.47**

1. Calculate Tb, using the triangle AbT in which

$$(Ab)^2 = (Tb)^2 + (AT)^2.*$$

Then

$$Tb = \sqrt{(Ab)^2 - (AT)^2}.$$

2. Calculate TQ as follows:
   In triangle BQP, angle QBP = pressure angle θ.

$$\sin\theta = \frac{PQ\dagger}{BP} \quad \text{so} \quad PQ = BP\sin\theta$$

   In triangle APT, angle $PAT = \theta$.

$$\sin\theta = \frac{PT}{AP} \quad \text{so} \quad PT = AP\sin\theta$$

$$\therefore TQ = PQ + PT$$

3. In the example shown, Tb is greater than TQ, which shows that point b lies to the left of the point of tangency Q, indicating that interference takes place.

*The Pythagorean theorem states that in a right triangle the square of the hypotenuse equals the sum of the squares of the other two sides.

†In trigonometry, the sine of an angle of a right triangle equals the side opposite the angle divided by the hypotenuse. (Values of the sines of angles are given in the Appendix.)

6.25 Remedies for Interference

Unlike intermittent contact, interference occurs with standard gears as well as with those of special design, so it is important to check gear pairs for this fault when designing a drive composed of standard gears. As stated before, standard gears do not have pointed teeth, so that is not an issue. When standard gears are used, interference is the only check necessary.

If a pair of special gears is being designed, they must be checked for intermittent contact as well as for interference and pointed teeth in order to establish proper addendum distances. Experiment indicates that interference is a much more common fault than intermittent contact, so the *tests for interference and pointed teeth should be made first*, before checking for intermittent contact.

Several remedies for interference are suggested below, should it be found to exist.

1. *Decrease the Addendum:*
   Interference occurs when the path of contact is so long as to extend beyond the points of tangency of the base circles and pressure line. The most obvious method of shortening the path of contact is to decrease the addendum which defines the path. If standard full depth involute gears are found to interfere, a change to stub teeth may shorten the path sufficiently. (Standard addendum for full depth teeth is $1/P^D$, for stub teeth it is $0.8/P^D$).

2. *Increase the Pitch Radius:*
   Increasing the pitch radius of a gear results in increasing the number of teeth, if the diametral pitch is held constant. Figure 6.48 shows the addendum circles of a pair of gears with centers at A and B. The resulting path of contact (ab) exceeds the limits established by the points of tangency (T and Q), and there is interference. If larger gears are used with pitch radii in the same proportion, so as to have the same speed ratio, their centers might be at C and D. Although the addendum is unchanged, the addendum radii are larger than the first pair whose addendum circles are shown dotted. Although these new addendum circles actually increase the path of contact slightly, the base circle tangent points (V and W) are much farther apart, with the result that they fall outside the new path of contact (cd) and interference no longer exists. If the pitch diameters are increased while the diametral pitch remains the same, the larger gears will have more teeth than the original pair.

3. *Increase the Pressure Angle:*
   A layout of addendum circles and pressure line is shown in Figure 6.49 for a pair of gears in which interference occurs. The path of con-

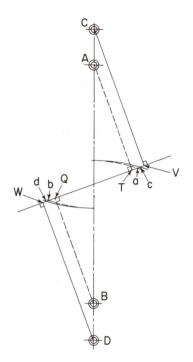

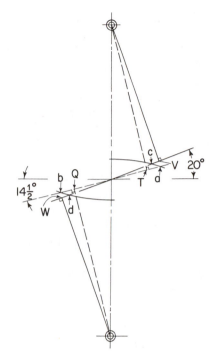

Figure 6·48 Increasing the pitch radius to avoid interference

Figure 6·49 Increasing the pressure angle to avoid interference

tact is *ab*, and the base circle tangent points are *T* and *Q*. If the pressure angle were made larger, as shown above, the new path of contact would be *cd* and the new tangent points would be at *V* and *W*. This increase in pressure angle shortens the path of contact and lengthens the distance between the tangent points, with the result that interference no longer exists.

If interference is found to take place in a drive composed of standard gears of $14\frac{1}{2}°$ pressure angle, this difficulty might be removed if similar standard gears of 20° pressure angle were substituted.

From the discussion above we can conclude that interference is likely to occur when small gears (low tooth numbers) having long addenda and small pressure angles are run together. Whether they be standard or special gears, these are the cases which should be checked for interference. It should be noted that, in the case of intermittent contact, gears having low tooth numbers were also subject to faulty operation, so in general it is true that drives involving gears with few teeth are suspect in each case, while those having larger tooth numbers need not be examined for either fault.

6.26 Limits of Speed Ratio with a Given Gear

Since there are limitations upon the size of gears which can be combined to form a correctly operating pair, it follows logically that there are limits to the speed ratios obtainable using a given gear as driver. Interference is usually the governing limitation.

A specific design problem will provide the clearest illustration. Suppose it is required to determine the limits of speed ratio which may be attained using a standard 14-tooth, full-depth involute gear with 20° pressure angle as a driver. The follower gear is also to be a standard full-depth involute gear. Since speed ratio is dependent upon tooth numbers, this problem reduces, first to finding the gear with the least number of teeth which will run with the given driver and, second, to finding the gear with the greatest number of teeth which will run correctly.

A graphical investigation should first be made, since it is simple and fast. This can be later supported by calculations, if the solution is sensitive to small errors. Large scale is important in this layout. The diametral pitch is not specified in the problem, since it is a scale factor, determining the size of the pitch diameter and not affecting the ratio of tooth numbers of the pair. In order to attain a reasonably large scale, let us assume a diametral pitch of 2.

First, we calculate the pitch radius of the given gear D ($\frac{1}{2} \times T/P^D =$ 3.5 in.) and locate the pitch point P from center O (Figure 6.50). Next the radius of the base circle is determined ($R^B = R^P \times \cos \theta = 3.29$ in.) and the arc of the base circle added. The pressure line is now drawn through P, tangent to the base circle. A perpendicular to the pressure line through O locates the point of tangency T. The addendum distances cP and eP are equal to 0.5 in. (see table in Figure 6.26). The addendum radius of D ($R^A = R^P +$ addendum) equals 4 in., and the addendum circle of D cuts the pressure line at b, defining one end of the path of contact.

Now let us determine the smallest follower gear F which will run with the given gear D. We know, from our study of limitations, that the path of contact is a critical factor. In this instance we know only the path of recess Pb. To avoid interference, the point of tangency of the base circle and the pressure line of any follower gear must not fall within the limits of the path of recess (Art 6.23). So F must be of such size that its base-circle tangent point Q falls to the left of point b. Furthermore, the center M of the follower gear lies at the intersection of the line of centers and a perpendicular to the pressure line, drawn from the point of tangency Q. Such a line determines the pitch radius of follower F. For example, a pitch radius PM_1 defined by perpendicular Q_1M_1 would be acceptable (since Q_1 lies to the left of b), while a follower with a pitch radius PM_2 would cause interference (since Q_2 lies to the right of b). From these experiments we conclude that the closer point Q approaches b (without falling on path Pb), the smaller the gear F.

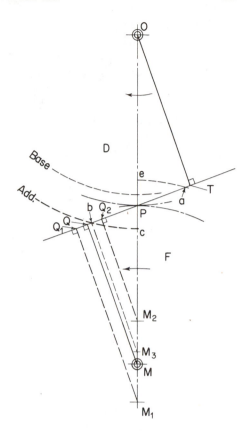

Figure 6.50 Smallest gear to run with a given gear

With this criterion, the theoretical minimum gear F would have a tangent point at point b and therefore a pitch radius of PM_3. If we scale PM_3, it will measure 3.05 in., giving F a pitch diameter of 6.1 in. With a diametral pitch of 2, this gear F would have ($T = D^P \times P^D = 6.1 \times 2$) 12.2 teeth. A gear must have an integral number of teeth, so the smallest actual gear F must have 13 teeth (the next integer greater than 12.2), with a pitch radius PM of 3.25 in. and base circle tangent point at Q, which falls safely outside of the path of recess Pb.

The addendum radius of F is Me, and its addendum circle cuts the pressure line at a, defining the path of contact ab. Just as the path of recess (Pb) must not exceed PQ, so must the path of approach (Pa) be less than PT, or we have interference. In this case we note that point a falls safely between P and T.

If D and F had the same number of teeth, the distance PT would equal PQ, so a would fall on PT just as b falls on PQ. However, if F were larger than D, point a would fall beyond point T and interference would ensue. If this were the case, the only solution is to reduce the addendum distance on *F*, since the pitch radius of *F* cannot be further reduced without interference at *Q*.

From this we conclude that if two identical standard gears fail to run together, owing to interference, there is no gear, larger or smaller, having the same addendum distance, which may be used with this size of gear. Either larger standard gears or special addendum distances are required.

Since diminishing the size of gears tends to shorten the path of contact, it might appear that this minimum gear *F* should be checked for intermittent contact. Experience shows that this difficulty rarely occurs with combinations of standard gears. This conclusion is valid here, since path *ab* measures 2.12, where the normal pitch $(\pi D^B/T)$ equals 1.475. However, with gears of special design, the contact ratio may be a critical factor.

One limiting value of the speed ratio is established by this minimum gear *F*:

$$\frac{\omega^F}{\omega^D} = \frac{T_D}{T_F} = \frac{14}{13} = 1.076$$

Now let us determine the largest standard gear *G* which will run with the driver *D*. The layout of the base and addendum circles, pressure line, and pitch point of the given gear *D* (Figure 6.50) is reproduced in Figure 6.51. Addendum *Pe* of gear *G* equals the addendum of gear *D*. The point of tangency of the base circle of *D* is at *T*, which marks the end of the longest permissible path of approach. The addendum circle of the follower gear must pass through *e* yet cut the pressure line to the left of point *T* or interference occurs. As the addendum radius increases, the pitch radius increases. The larger the gear *G*, the "flatter" is the addendum arc. Thus, we may increase the addendum radius until the addendum circle passes through *T*, but this is the limit. If the center of gear *G* is called *N*, the lengths *eN* and *TN* will then be equal. This suggests that *N* must lie on the perpendicular bisector of the line *eT*. *N* also lies upon the line of centers, so it must fall at the intersection of the perpendicular bisector with the line of centers.*

The theoretical maximum pitch radius of *G* will be N_1P, which measures 6.56 in., and the number of teeth on *G* will be 26.24 ($T = P^D \times D^P = 2 \times 13.12$). The largest integral number not exceeding 26.24 is 26. So the maximum gear *G*, that will run correctly with *D*, has 26 teeth and a pitch radius *NP* of 6.5 in. Its addendum radius will equal 7 in., making the path of contact 2.28 in., which is well above the normal pitch (1.475 in.). As in the previous example, a check of the contact ratio for intermittent contact is unnecessary, since the gears are standard.

*In many instances the angle between the perpendicular bisector and the line of centers will be very small, with the result that the intersection *N* cannot be accurately located graphically. In such instances calculations should be substituted for graphical methods.

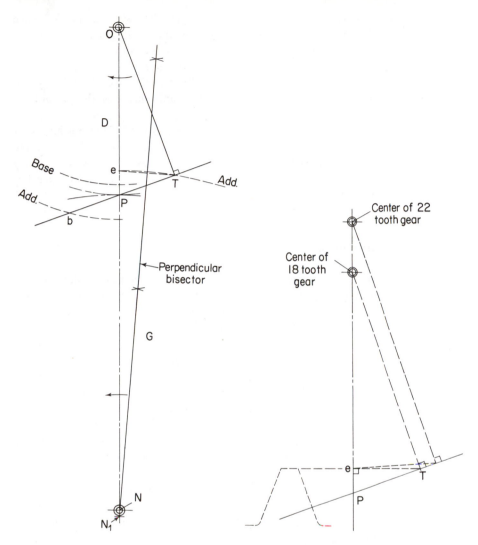

Figure 6.51 Largest gear to run with a given gear

Figure 6.52 Gears without maximum limit to size of their mates

The limit of the speed ratio established by the maximum gear G is

$$\frac{\omega_G}{\omega_D} = \frac{T_D}{T_G} = \frac{14}{26} = 0.538$$

It is interesting to note that, if gear D were somewhat larger (about 18 teeth), the tangent point T would fall on a perpendicular to the line of centers through the addendum point e, as shown in Figure 6.52. In this case, the perpendicular bisector of eT would be parallel to the line of centers, so that they would never intersect. This means that the largest follower gear would have an infinite pitch radius. Even a rack (a gear having a straight line pitch "circle") will run with this driver. There is no maximum limit to the size of a follower gear which will run with this 18-tooth gear D. If a still larger driver were used, there would also be no maximum limit to the size of mating gears. Therefore only the smaller gears impose a maximum limit upon the size of their mates.

Any given gear can easily be checked to determine if there is any limit to the maximum size of mating gears. The addendum Pe is laid off on the line of centers from the pitch point. A line eT is drawn to the point of tangency of the given gear and angle TeP examined. If this angle is 90° or greater, there is no limit to the maximum size of the mating gear.

6.27 Separation of Involute Gears

It was stated in Article 6.12 that one of the virtues of involute gears is that they may be run on shafts whose center to center dimension is not correct

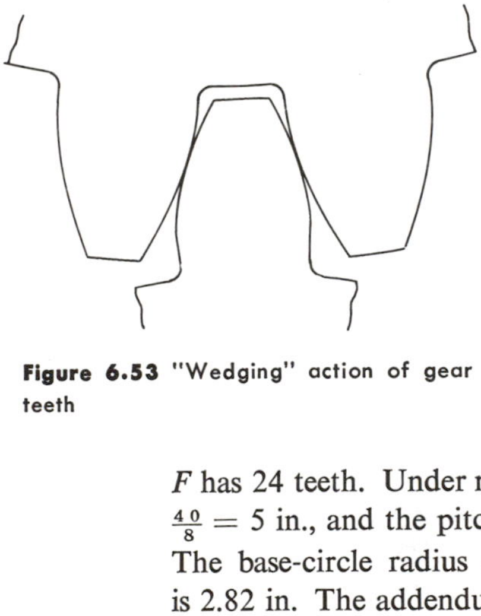

Figure 6.53 "Wedging" action of gear teeth

without destroying conjugate action. Small errors in the intended mounting distance often occur as a result of inaccurate location of bearings, wear, or deflection of the shafts under heavy loads. Sometimes a pair of gears are intentionally mounted at abnormal center distances so as to simplify the design of a gear train. These variations of mounting are always elongations of the center distance, as the "wedging" action of the teeth (Figure 6.53) prevents mounting the gears any nearer together than the designed center-line spacing.

Let us consider a pair of full-depth involute gears designed with a 20° pressure angle and diametral pitch of 4. Driver D has 40 teeth, and follower F has 24 teeth. Under normal conditions, the pitch radius of D is $\frac{1}{2}(T/P^D) = \frac{40}{8} = 5$ in., and the pitch radius of F is 3 in., making the center distance 8 in. The base-circle radius of D is $R^P \cos \theta = 4.70$ in., and the base radius of F is 2.82 in. The addendum is $1/P^D = \frac{1}{4}$ in., so the addendum radius of D is 5.25 in. and the addendum radius of F is 3.25 in. The layout shown in Figure 6.54 shows the path of contact ab equal to 1.22 in. The speed ratio is $\omega_F/\omega_D = \frac{5}{3}$.

Now let us mount these gears on a center distance AB' of 8.16 in. and observe what takes place.

The base circles are physically indicated on the gears, since they define where the radial flanks meet the involute surfaces of the teeth. Since the change in mounting distance does not alter the tooth form, the base circle radii re-

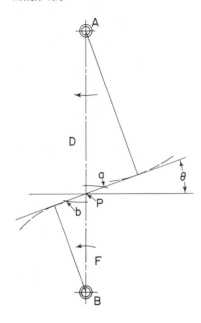

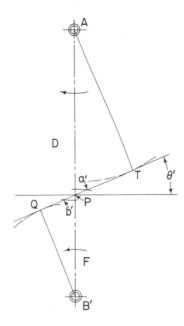

Figure 6.54 Standard gears mounted at correct center distance

Figure 6.55 Standard gears mounted at extended center distance

main as originally calculated: $R_D^B = 4.70$ and $R_F^B = 2.82$. These are shown, about centers A and B', in Figure 6.55.

The pressure line can now be drawn, tangent to the base circles at T and Q. The intersection of the pressure line and line of centers AB' locates the pitch point P. The pitch radii can be measured, but the calculations are simple and provide exact values.

Since the speed ratio is determined by the tooth numbers, it remains the same:

$$\frac{\omega_F}{\omega_D} = \frac{T_D}{T_F} = \frac{40}{24} = \frac{5}{3}$$

This also establishes the ratio of the pitch radii

$$\frac{\omega_F}{\omega_D} = \frac{R_D}{R_F} = \frac{5}{3}$$

The sum of the pitch radii equals the center distance $= R_D + R_F = 8.16$, so the new pitch radii can be calculated from these two equations:

$$R_D = \frac{5R_F}{3} = 8.16 - R_F \quad \text{and} \quad 5R_F = 24.48 - 3R_F$$

$$\therefore R_F = 3.06 \quad \text{and} \quad R_D = 5.10$$

We note that these pitch radii differ from those of the original design. This is permissible since pitch circles are only imaginary circles not actually cut on the gears. Their ratio remains the same since it must be consistent with the tooth ratio, which is unchanged.

Diametral pitch relates tooth numbers and pitch diameters, so this has assumed a new value. Using gear D, for example,

$$P^D = \frac{T}{D^P} = \frac{40}{10.2} = 3.92$$

The pressure angle has changed because of the new center distance, but the ratio of the pitch and base radii is still equal to the cosine of the pressure angle. In triangle APT, angle PAT is equal to the new pressure angle θ'; so base radius AT divided by the pitch radius AP equals the cosine of the pressure angle:

$$\frac{AT}{AP} = \frac{R^B}{R^P} = \frac{4.7}{5.1} = 0.9216 = \cos \theta'$$

If we find a cosine value of 0.9216 in the tables of trigonometric functions,* it will designate a pressure angle of $22°50'$.

The path of contact is defined by the intersections a' and b' of the addendum circles and pressure line. The addendum radii are cut physically on the gears and so remain unchanged, but the separation of the gears and attendant increased pressure angle both tend to decrease the path of contact, which now measures only 0.8 in.

6.28 Checking Performance of Separated Gears

It has been stated that involute gear teeth remain conjugate even though mounted at extended center distances. Figure 6.56 shows two involutes of different base circles. If these are brought into contact at any random point, such as C, they will have a common straight–line tangent TT. A line from C tangent to one base circle at N will be perpendicular to TT, since it represents the radius of curvature of the involute at point C. Similarly, a line from C tangent to the other base circle at N' will also be perpendicular to TT. Thus, we see that NN' is one straight line normal to both involute curves or, in other words, a common normal to the involutes through their contact point. Owing to the nature of the involute curve, it will always be possible, whenever two involutes are brought into contact, to draw one common normal to the curves through the contact point, no matter where contact takes place. These common normals will always be tangent to the two base circles, whether they are placed close together or far apart.

If we refer back to the layout of the separated gears in Figure 6.55, we

*See Appendix.

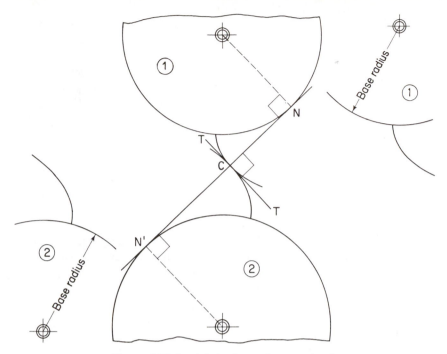

Figure 6.56 Involute teeth are always conjugate

may note that the pressure line TQ is tangent to both base circles and contains the pitch point P. Therefore, TQ is a common normal to the involute tooth surfaces in every position of contact which passes through the pitch point. Thus the teeth are conjugate regardless of the mounting distance of the gears.

We have noted that, as the center distance AB is extended, the path of contact grows shorter. There is a minimum limit on this path, if intermittent contact is to be prevented, which equals the normal pitch. We must check these separated gears for intermittent contact. The normal pitch is determined by the base circle diameter and number of teeth—both of which remain unchanged by variations of mounting. Calculated on gear D:

$$P^N = \frac{\pi D^B}{T} = \frac{\pi \times 9.4}{40} = 0.74$$

Comparing this value with the path of contact, we see that the new path $a'b'$ (0.8 in.) is still greater than the normal pitch, so intermittent contact does not take place.

The limit to the separation of a pair of gears is thus defined. The center distance can only be extended until the path of contact equals the normal pitch.

When a pair of gears is mounted at greater than normal center distances,

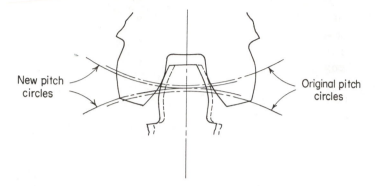

Figure 6.57 Backlash due to extended mounting

a considerable amount of backlash develops, since the larger pitch circles make the tooth width narrower and the tooth space wider, as shown in Figure 6.57. If this backlash is objectionable in a certain application, it becomes a limitation upon the amount of separation of the gears.

6.29 Application of Separation

A selective-speed transmission is shown in Figure 6.58, comprised of five involute gears designed with a pressure angle of $14\frac{1}{2}°$ and diametral pitch of 3. These are all standard gears except that they have an addendum of 0.5 in. Gear D is a driver having 54 teeth. It is keyed to turn with shaft A but may be positioned to drive any one of the four follower gears: F (30 teeth), G (29 teeth), H (28 teeth), or J (27 teeth), thus providing a choice of four speeds of

Figure 6.58 Selective speed transmission with separation

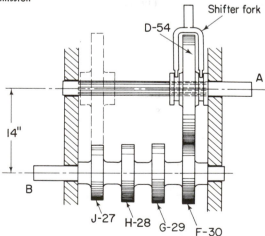

shaft B. Shafts A and B are mounted parallel to one another 14 in. apart. The possibility of running several different follower gears with the same driver and at the same center distance is due to the ability of involute gears to operate properly even though their centers are spaced farther apart than usual practice dictates.

Let us consider gears D and F. Using the given specifications, the pitch radius of D is

$$\frac{1}{2}\left(\frac{T}{P^D}\right) = \frac{1}{2} \times \frac{54}{3} = 9 \text{ in.}$$

The pitch radius of F is 5 in. The center distance of 14 in. is correct for D and F, so there is no separation. The original pressure angle of $14\frac{1}{2}°$ is unchanged, and the diametral pitch is 3. The radius of the base circle of D is $R^P \cos 14\frac{1}{2}° = 8.71$ in., and the base radius of F is 4.84 in. The speed ratio of this entirely normal pair is

$$\frac{\omega^F}{\omega^D} = \frac{T_D}{T_F} \times \frac{54}{30} = 1.8$$

Next let us examine the drive from D through J, with gear D in the dotted position shown in Figure 6.58. The center distance is still 14 in.: D has 54 teeth, and J has 27. Under normal conditions the pitch radius of D would be 9 in. (as noted above), and the pitch radius of J would be

$$\frac{1}{2}\left(\frac{T}{P^D}\right) = \frac{27}{(2 \times 3)} = 4.5 \text{ in.}$$

Therefore the usual mounting distance would be 13.5 in. instead of 14 in. Gears D and J in this design are separated 0.5 in. beyond their normal center to center measurement. This will alter their pitch radii, since the ratio of the pitch radii must be equal to the ratio of the tooth numbers.

$$\frac{T_D}{T_J} = \frac{R_D}{R_J} = \frac{54}{27} = \frac{2}{1}$$

$$R_D + R_J = 14 \quad \text{but} \quad R_D = 2R_J$$

and substituting:

$$R_J = 14 - 2R_J$$

$$\text{So} \quad R_J = \frac{14}{3} = 4\frac{2}{3} \text{ in.} \quad \text{and} \quad R_D = 9\frac{1}{3} \text{ in.}$$

The base circle radii are not changed, so

$$R_D^B = 8.71 \text{ in.}$$

The base radius of J is determined by the original specifications ($14\frac{1}{2}°$ pressure angle and diametral pitch $= 3$) to which J was designed. The original or intended pitch radius of J was

$$\frac{1}{2}\left(\frac{T}{P^D}\right) = \frac{27}{(2 \times 3)} = 4.5 \text{ in.}$$

and the original pressure angle was $14\frac{1}{2}°$. So

$$R^B = R^P \cos\theta = 4.5 \times 0.968 = 4.36 \text{ in.}$$

The addendum radii are not affected by the extended center distance, so the addendum radius of D is the original pitch radius plus the specified addendum:

$$R_D^A = 9 + 0.5 = 9.5 \text{ in.}$$

Similarly, the addendum radius of J is:

$$R_J^A = 4.5 + 0.5 = 5 \text{ in.}$$

A layout of the base circles with the tangent pressure line is shown in Figure 6.59. The addendum circles define the path of contact ab, which measures 1.3 in.

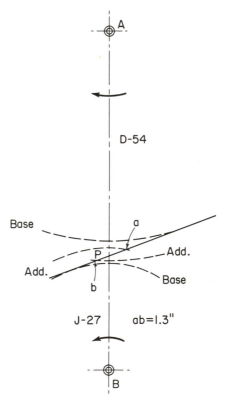

Figure 6.59 Checking for intermittent contact

Since considerable separation is evident here, we must check for intermittent contact. The normal pitch, as calculated on the driver D, equals:

$$P^N = \frac{\pi D^B}{T} = \frac{\pi \times 17.42}{54} = 1.014 \text{ in.}$$

The path of contact of 1.3 is safely larger than 1.014, so there is no intermittent contact.

    Gears H and G also are separated in this design, but since they have more teeth than J, the separation is not as drastic and there is no need to check them for intermittent contact because it does not occur with gear J.

    Four different speed ratios are possible with this transmission:

$$\frac{\omega_F}{\omega_D} = \frac{54}{30} = 1.8 \qquad \frac{\omega_G}{\omega_D} = \frac{54}{29} = 1.86$$

$$\frac{\omega_H}{\omega_D} = \frac{54}{28} = 1.93 \qquad \frac{\omega_J}{\omega_D} = \frac{54}{27} = 2$$

PROBLEMS

6.1. Two standard full-depth involute gears are shown in contact position in Figure P6.1. Assume that gear D turns clockwise about center d, driving gear F. The drawing is at reduced scale, but the actual pitch diameter of D is 12 in. and the diametral pitch is 1.5. Gear F has 28 teeth. The pressure angle is 20° (cos 20° = 0.9397).

Identify the following, using letters given on the drawing.*

1. Pitch radius of D
2. Pressure line
3. Dedendum radius of F
4. Addendum distance of D
5. Tooth width of D
6. Tooth space of F
7. Clearance cut on F
8. Pressure angle
9. Circuit pitch
10. Path of contact
11. Path of recess
12. Arc of approach
13. Angle of approach on F
14. Angle of action on D
15. Arc of contact

6.2. Calculate the following values for gears D and F in Problem 6.1. Use the data given above and known relationships. Only the angles of the drawing (Figure P6.1) may be measured, as linear measurements are not to scale.

1. Number of teeth on D
2. Line of centers (df)
3. Speed ratio (ω_F/ω_D)
4. Circular pitch
5. Addendum distance
6. Pitch diameter of F
7. Base diameter of D
8. Normal pitch
9. Angle of action of D (measure)
10. Arc of contact
11. Path of contact
12. Angle of approach of F (measure)
13. Path of approach
14. Contact ratio

6.3. A drive is composed of two gears, A and B. A has 50 teeth, and B has 35 teeth. Either gear may be driving at 1000 rpm.
(a) What is the fastest output speed attainable with these gears?
(b) What is the slowest output speed?

---

    *Two letters define a straight line. Two letters marked $\overset{\frown}{AB}$ define an arc. Three letters define an angle—the middle letter denoting the vertex.

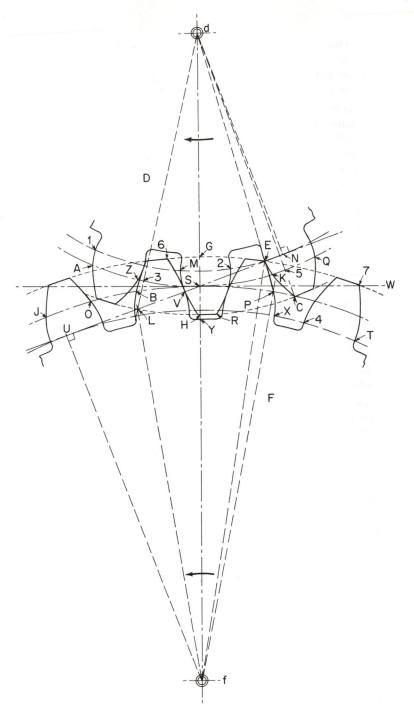

Figure P6.1

6.4. Gear D turns 200 rpm, driving gear F at 500 rpm. They are mounted on shafts $5\frac{1}{4}$ in. apart and have a diametral pitch of 6.

(a) How many teeth are there on each gear if they turn in opposite directions?

(b) How many teeth are there on each gear if they turn in the same direction?

6.5. Gear D has 20 teeth and a diametral pitch of 16. Gear F is driven by D in the same sense of rotation. The speed ratio (ω_F/ω_D) is $1:8$. What is the center to center distance of their shafts?

6.6. The gear box shown in Figure P6.6 supports two shafts, A and B, which carry a pair of gears whose speed ratio is $2:1$.

(a) What is the smallest diametral pitch number (largest tooth size) which may be used if the gears are to run inside the box? Use only standard P^D numbers (6, 8, 10, 12, 14, 16, 20) and an addendum equal to $1/P^D$.

(b) Using the gears selected, what is the minimum inside dimension (h) of the box if a clearance of $\frac{1}{16}$ in. is maintained between gears and box.

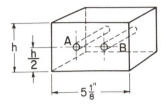

Figure P6.6

6.7. Design a two-gear drive for a speed ratio of $1:4$ with the relative sense of rotation not specified. The minimum number of teeth is 24. Use the largest tooth size permissible. Select standard P^D numbers: 4, 5, 6, 8, 10, 12. No P^D number larger than 12 is permitted, and the center distance is not to exceed 4 in. State the diametral pitch, numbers of teeth on each gear, and actual center distance.

6.8. A pair of standard, full-depth involute gears have a 20° pressure angle, a diametral pitch of 2, and a speed ratio of $5:4$, with the follower turning opposite to the driver. The center-to-center distance is 15.75 in. What is the length of the path of contact?

6.9. A 12-tooth pinion* drives a 24-tooth annular gear. The pressure angle is 20°, and the diametral pitch is 2. The pinion has an addendum of 0.4 in. and the annular an addendum of 0.25 in. Determine the length of the path of contact.

6.10. An 18-tooth pinion drives a rack, the diametral pitch is 1, and the pressure angle 20°. All teeth are full-depth involute with the standard addendum. Determine the length of the path of contact.

6.11. A two-lobed rotor (D) for a blower is shown in Figure P6.11 which turns about fixed shaft A. It is required to design the mate (F) to rotor D. F is to turn about fixed shaft B at the same angular speed as D and must be conjugate to D at all times. Design rotor F using the conjugate law. Plot and draw the path of contact.

6.12. Construct an involute of the 6-in. diameter base circle shown in Figure P6.12. Start at point O, and use 11 equally spaced points about $\frac{5}{8}$ in. apart around the circle.

*A pinion is the smaller of two gears in a pair.

Considering this involute to be one side of a 1-tooth gear D, which turns counterclockwise about fixed axis A, design the mating tooth F which is to rotate clockwise about fixed axis B at the same speed as D. F must be conjugate to D from point 1 to point 11. Plot and draw the path of contact. Identify the pressure angle θ.

The base circle of F is tangent to the pressure line. Check several points to show that F is also an involute.

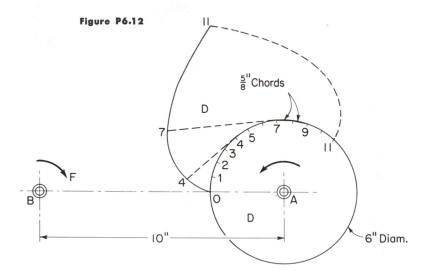

Figure P6.12

6.13. Make a full-sized layout of a pair of standard full-depth involute gears having a pressure angle of 20° and a diametral pitch of 1.5. The driver D has 18 teeth, and the follower F has 28 teeth. Show three consecutive teeth on the driver with the middle tooth centered on the line of centers. Draw two consecutive teeth on the follower in contact with the driver.
(a) Label the path of contact ab. Measure and report its length.
(b) Label the arc of contact ce. Calculate and report its length.
(c) Measure the normal pitch and check against the calculated value.
(d) Report the contact ratio (path $\div P^N$).

6.14. Make a full-sized layout of a standard full-depth 18-tooth involute gear driving a standard rack. Pressure angle, 20°; diametral pitch, 1. Show three consecutive teeth on the gear and four on the rack with one pair of teeth in contact at the pitch point.
(a) Label the path of contact ab. Measure and report its length.
(b) Draw and label the angle of approach (α) and the angle of recess (β).
(c) Report the contact ratio (path $\div P^N$).

6.15. It is proposed that a 12-tooth gear D be used to drive a 14-tooth gear F. Both gears are to be standard $14\frac{1}{2}°$ involute gears with a diametral pitch of 1. Check these gears graphically for interference and intermittent contact. Do not draw teeth. Will these gears operate correctly as purchased? Describe faulty performance if any and recommend any remedial alterations necessary.

6.16. Will a standard 15-tooth gear run properly with a standard rack if the pressure angle is $14\frac{1}{2}°$? Check graphically first, without drawing teeth. Any diametral pitch may be used, but $P^D = 1$ is suggested for large scale. If there is difficulty, describe it and show any required alterations to the rack.

6.17. The design of a machine requires a gear pair with a speed ratio of 26 : 21 and diametral pitch of 2. A pair of standard $14\frac{1}{2}°$ gears having 26 and 21 teeth provide an acceptable center-to-center dimension. Check this pair of gears graphically for proper performance. If any difficulty is discovered, recommend (and check) a satisfactory pair of standard gears. The speed ratio, diametral pitch, and center distance must not be changed. The path of contact should be kept as long as possible.

6.18. A pair of standard gears must be selected to drive two parallel shafts, located 9.75 in. apart, with a speed ratio of 2 : 1. The load and wear calculations indicate that a diametral pinch of 2 must be used. The path of contact is to be made as long as possible. Determine the tooth numbers of the gears.
Check the gears for proper performance, using $14\frac{1}{2}°$ pressure angle first, as this will provide the longest path of contact. Next try 20° with full-depth teeth and finally 20° with stub teeth. The center distance, diametral pitch, and speed ratio cannot be altered.

6.19. A belt drive connecting two shafts on a machine has proven to be unsatisfactory, and it is proposed that it be replaced by a gear drive. This problem requires the specification of suitable standard gears and a layout of their assembly on the existing machine. The present belt drive is shown in Figure P6.19. The drive pulley D is 3.5 in. in diameter, and the follower F is 7.5 in. in diameter.

Figure P6.19

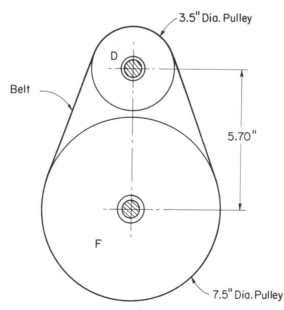

The center-to-center distance of 5.70 in. cannot be altered, and the sense of rotation of the input and output shafts must remain unchanged. Only standard spur gears are to be used (annulars are too costly and large), selected from the excerpts of a typical gear manufacturer's catalog on pages 416 and 417 of the Appendix. In order to avoid interference, it is suggested that no gear of 20° pressure angle having less than 13 teeth and no gear of $14\frac{1}{2}$° pressure angle with less than 24 teeth be used. As to design factors, aside from function, the first consideration is that the largest tooth size (lowest diametral pitch number) should be used. Second, the total cost of the gears specified is to be as low as possible. (Disregard expenses of other alterations to the machine which might be required.)

(a) When your scheme is complete, fill out a specification table, like that shown below, for each gear to be used. Identify gears by assigning a part number.

Gear no.	Pressure angle	Diametral pitch	Number of teeth	Face width	List price
			Total		

(b) Draw a full-size layout showing the assembly of your drive. Include pitch, addendum, and dedendum circles of the gears, carefully labeled, but do not draw any gear teeth. Determine and report the contact ratio of gear pairs.

6.20. A pair of gears has a pressure angle of 20° and a diametral pitch of 1.5. The driver D has 18 teeth, and the follower F has 28 teeth. If the addendum and dedendum of both gears are increased as required, what is the length of the maximum path of contact which can be attained by this special design? Describe the conditions which limit the path of contact at each end.

6.21. Determine the number of teeth on the smallest standard gear which will run properly with the following standard gears:
(a) Gear A, 27 teeth, pressure angle $14\frac{1}{2}$°.
(b) Gear B, 27 teeth, pressure angle 20°, full-depth teeth.
(c) Gear C, 27 teeth, pressure angle 20°, stub teeth.
(Assume diametral pitch of 2 for a suitable scale for graphical investigations.)

6.22. Determine the number of teeth on the smallest pair of identical standard gears which will run together correctly. State pressure angle and addendum (full-depth or stub).

6.23. What is the slowest linear speed at which a standard rack may be driven by a standard gear turning 100 radians/min, if their diametral pitch is 2 and their pressure angles are both 20°? State the number of teeth on the gear used and the addendum (full-depth or stub) on gear and rack.

6.24. A standard gear *D* has 13 teeth, diametral pitch of 2, 20° pressure angle, and full-depth teeth. If *D* turns 1000 rpm, driving standard follower gears of the same specifications, what are the fastest and slowest angular speeds of the followers which may be attained? State number of teeth on follower in each case.

6.25. A gear drive is to be designed to turn two parallel shafts in the same direction with a speed ratio (follower to driver) of exactly 5 to 3. Stress and wear conditions demand a tooth size at least as large as 10 diametral pitch. The driver and follower gears must be mounted $2\frac{5}{8}$ (2.625) inches from center to center. The drive is to be composed entirely of standard spur gears selected from the typical gear manufacturer's catalog excerpts on pages 416 and 417 in the Appendix. The gears selected must run together properly as purchased without undercutting of flanks, modified tips, or other alterations. The total cost of the gears is to be kept as low as possible. The arrangement and assembly of the gears is to be as compact as possible as the space is limited. Other things being equal, the contact ratio is to be as high as is consistent with other design factors, and in any case, this must be greater than 1.

(a) Select the gears to be used from the catalog excerpt, noting diametral pitch, pressure angle, tooth numbers, and cost. Check for interference. (A large-scale graphical check is suggested.)

(b) Lay out the assembly of the gears at least 4 times size. Show location of shafts and center-to-center dimensions. Draw pitch circles, addendum circles, base circles, and pressure lines, and label these clearly. Do not draw any teeth. Identify driver and follower gears. (Full circles need not be drawn—arcs in the general area of contact will be sufficient.)

(c) Obtain by computation or by graphical construction the dimensions and specifications of all gears as indicated in the table below:

	Driver	Follower
Pressure angle		
Diametral pitch		
Number of teeth		
Pitch diameter		
Addendum distance		
Path of contact		
Circular pitch		
Face width		
Normal pitch		
Contact ratio		
Price of gear		

Also specify the total cost of the gears to be used and the number of cubic inches in the smallest rectangular prism which will contain the assembly.

6.26. An 18-tooth gear A drives a 28-tooth gear B. They are both standard, full-depth involute gears of 20° pressure angle and 1.5 diametral pitch. The correct mounting distance is $15\frac{1}{3}$ in. from center to center. Make a layout of these gears (do not draw teeth), using an extended mounting distance of 15.57 in. and determine the following:
(a) New pressure angle
(b) New pitch diameter of A
(c) New path of contact
(d) New diametral pitch
(e) New contact ratio
(f) Will the gears perform properly?

6.27. If gears A and B in Problem 6.26 were mounted on shafts 15.9 in. center-to-center, would they run together satisfactorily? (State nature of faulty performance, if any.)

6.28. A machine design requires that a standard gear D, having a 20° pressure angle, diametral pitch of 1 and 12 stub involute teeth, be used to drive two racks. Gear D turns 100 radians/min. Rack A is to have a linear velocity of 600 in./min and rack B a velocity of 650 in./min. Make a layout showing a complete drawing of one tooth of each rack, reporting the following for each rack:
(a) Pressure angle
(b) Circular pitch
(c) Diametral pitch
(d) Addendum (use maximum addendum permissible without alterations to gear D)
(e) Tooth width

6.29. In the design of a gear train it is required to obtain a speed ratio of nearly 1. This is accomplished by the gear train shown in Figure P6.29. The center lines of shafts A and B are parallel to one another. Gear D turns freely on shaft A and drives gear E, which is attached to shaft B. Gear G is also attached to shaft B, so it turns the same speed as E. Gear G drives F, which turns shaft A. (If D turns 3721 rpm, F will turn 3720 rpm.) Tooth numbers are given on the diagram. All gears are standard, $14\frac{1}{2}°$, full-depth involute with a diametral pitch of 4. Since the sum of the pitch radii of D and E will obviously not

Figure P6.29

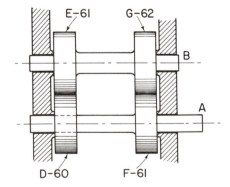

equal the sum of the pitch radii of G and F, the smaller pair (D and E) will have to be mounted at the same center distance as the larger pair (G and F) if the shafts are to be parallel. Calculate the center distance for G and F under normal conditions. Make a layout of D and E, using the same center distance as for G and F.

Draw the addendum circles (using standard addendum), base circles, and check for proper performance. If difficulty is encountered, describe it and recommend alterations to gear D which will correct the problem. All other gears must remain standard.

Gear Trains

Our study of gearing up to this point has been concerned with tooth design and performance of single pairs of gears. As we turn to applications, we discover practical limitations upon speed ratios when only two gears are used. Since speed ratios depend upon tooth numbers ($\omega_F/\omega_D = T_D/T_F$), large speed ratios demand that one gear be very large and the other very small if only one pair is employed. In practice the purchase of standard gears is far more economical than the design and manufacture of special gears. The range of sizes offered by manufacturers of standard gears does not include those of very large tooth numbers. Furthermore, in the interest of a compact design, the space required by large gears becomes quite as objectionable as their cost. For these reasons we are often forced to sacrifice the simplicity of a two-gear drive and use several pairs, connected in series, which are known as a *gear train*.

A simplified diagram of a three-pair gear train is shown in Figure 7.1. All gears are mounted in fixed bearings. The input is at the 100-tooth driver D, and the output is at the 18-tooth follower F. Gears A and B are attached to the same hub, and gears C and E also turn together. D meshes with A, forming the first pair in which the speed ratio* is

$$\frac{\omega_A}{\omega_D} = \frac{T_D}{T_A} = \frac{100}{24}$$

so that A turns faster than D. B and C form the second pair with a speed ratio of

$$\frac{\omega_C}{\omega_B} = \frac{T_B}{T_C} = \frac{90}{20}$$

*In this text, speed ratio always equals the follower divided by the speed of the driver.

243

in which C turns faster than B. In the third pair, E and F, there is a further increase in speed. Since

$$\frac{\omega_F}{\omega_E} = \frac{T_E}{T_F} = \frac{72}{18}$$

gear F turns faster than E.

7.1 Speed Ratio of a Gear Train (Tooth Ratio—t.r.)

Let us combine these speed ratios into a single over-all speed ratio for the entire train.

$$\frac{\omega_F}{\omega_E} = \frac{T_E}{T_F}$$

But $\omega_E = \omega_C$, so, substituting ω_C for ω_E,

$$\frac{\omega_F}{\omega_C} = \frac{T_E}{T_F}$$

Since $\omega_C/\omega_B = T_B/T_C$, then:

$$\omega_C = \omega_B \times \frac{T_B}{T_C}$$

Substituting this for ω_C:

$$\frac{\omega_F}{\omega_B \times T_B/T_C} = \frac{T_E}{T_F} \text{ or } \frac{\omega_F}{\omega_B} = \frac{T_B}{T_C} \times \frac{T_E}{T_F}$$

But $\omega_B = \omega_A$, so:

$$\frac{\omega_F}{\omega_A} = \frac{T_B}{T_C} \times \frac{T_E}{T_F}$$

Next,

$$\frac{\omega_A}{\omega_D} = \frac{T_D}{T_A} \text{ or } \omega_A = \omega_D \times \frac{T_D}{T_A}$$

Substituting again:

$$\frac{\omega_F}{\omega_D \times T_D/T_A} = \frac{T_B}{T_C} \times \frac{T_E}{T_F}$$

And the over-all speed ratio:

$$\frac{\omega_F}{\omega_D} = \frac{T_D \times T_B \times T_E}{T_A \times T_C \times T_F}$$

Gears D, B, and E are all drivers, and A, C, and F are all followers of their respective pairs. The right-hand side of the above equation could therefore be described as *the product of the tooth numbers of all driver gears divided by the product of the tooth numbers of all follower gears.* For simplicity this expression may be called the *tooth ratio* (*symbol* t.r.).

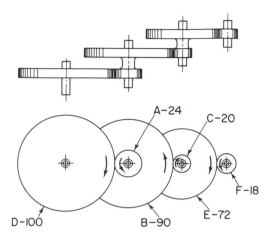

Figure 7.1 Gear train

The speed ratio of the train in Figure 7.1 is then

$$\frac{\omega_F}{\omega_D} = \text{t.r.} = \frac{100 \times 90 \times 72}{24 \times 20 \times 18} = \frac{75}{1}$$

If D turns 50 rpm,

$$\omega_F = \omega_D \times \text{t.r.} = 50 \times 75 = 3750 \text{ rpm}$$

The direction of rotation of the follower with respect to the driver of a train is often important. This can easily be determined by drawing arrows on a sketch of the gear layout, as shown in Figure 7.1. If the driver and follower turn in the same direction, a plus sign may be placed before the tooth ratio. If they turn opposite to one another, a minus sign may be used. In this example gear F turns opposite to gear D, so the t.r. is negative:

$$\omega_F = \omega_D \times (-\text{t.r.}) = -3750 \text{ rpm}$$

This is read: The angular velocity of F is 3750 rpm, opposite to the driver D. The correct sense of the arrows on adjacent meshing gears may be verified by the fact that the linear velocities at the pitch point must always be identical on both gears. Whether it be plus or minus, it is good practice to place this sign denoting relative sense before all tooth ratios, lest sense be overlooked.

7.2 Trains of Different Types of Gears

Our previous study of gearing has been limited to spur gears, annular gears and racks. A detailed study of all gear types would fill a book in itself. The spur gear offers a basic example of all types, and few curricula permit sufficient time for study of others. Gear trains often combine the different kinds of gears shown in Figure 7.2.

Spur gears have teeth on their outer circumference which run parallel to the axis of rotation. The smaller of a pair of gears is often called a *pinion*.

Figure 7·2 Types of gears

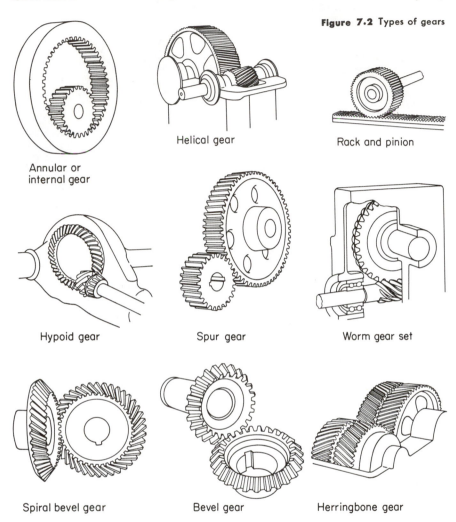

Annular or
internal gear

Helical gear

Rack and pinion

Hypoid gear

Spur gear

Worm gear set

Spiral bevel gear

Bevel gear

Herringbone gear

Annular or internal gears are rings with teeth on their inner surfaces.

Racks are straight gears of infinite radius, only capable of translation.

Bevel gears are basically truncated cones with teeth on their outer surfaces. An internal bevel is a ring with teeth on a conical inner surface. Bevel gears are used to drive nonparallel intersecting shafts. When a pair of identical bevel gears is mounted on shafts at 90° with one another, they are called *miter gears*.

Helical gears have teeth "wound around" cylindrical surfaces in the form of a helix. They may be used with parallel or nonintersecting shafts, perpendicular or otherwise, and may be external or internal. A variation of the helical gear is called the *herringbone gear*. This is composed of two bands of helical teeth, one right-hand and one left-hand so as to eliminate axial thrust. The advantage of helical gearing lies in the fact that a pair of teeth do not make contact across the entire face at one time, but contact progresses along each tooth as the gears turn. There are also more teeth in contact at a time to distribute the load.

Spiral gears are similar to helical gears except that the teeth are "wound around" conical surfaces, taking the form of a spiral. They are used as bevel gears. A variation of the spiral gear is the *hypoid gear*, which may be used between nonintersecting shafts at right angles.

Worm gears are composed of a screw called a *worm* and a matching gear. They are used to connect nonintersecting shafts at right angles for reduction of speed only.

7.3 Speed Ratios with Different Gear Types

The formula relating speed ratio and tooth ratio which we have developed for spur gears applies equally well to all types *except worm gears*. In order to obtain the required position of the follower with respect to the driver shaft, all types of gear pairs may be used in a gear train. Figure 7.3 shows a diagram of a train consisting of spur, internal, bevel, and helical gears. All gears are keyed to turn with the shafts upon which they are mounted. If shaft D turns 1800 rpm, let us determine the speed of shaft F.

$$\frac{\omega_F}{\omega_D} = \text{t.r.} \quad \text{so} \quad \omega_F = \omega_D \times \text{t.r.}$$

$$\therefore \omega_F = 1800 \times \frac{20 \times 18 \times 32 \times 22}{48 \times 86 \times 56 \times 40} = 49.34 \text{ rpm}$$

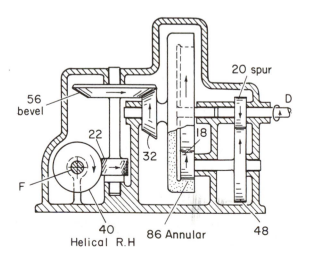

Figure 7.3 Speed changer

Shaft D turns as shown by the arrow in Figure 7.3. To determine the direction of rotation of F, we sketch arrows on the gears as shown and discover that F turns clockwise. Since F is at right angles with D, the sign convention on the t.r. is meaningless in this case.

7.4 Speed Ratio of Worm Gears

A worm gear set is similar to a helical gear pair in which the driver (worm) is small and has a very low helix angle. The follower (worm gear) has a concave face which wraps part way around the worm to provide a longer path of contact, as shown in Figure 7.4. The helix angle is so small that the worm has the characteristics of a screw. The helical teeth are called *threads* on this type of gear. Worms are made with right- or left-hand threads. They may be single, double, triple, or quadruple threaded.

A *single threaded worm* in one revolution advances the thread 1 pitch distance* along the worm shaft. This one turn of the worm will then rotate the meshing worm gear tooth through an amount equal to the circular pitch of

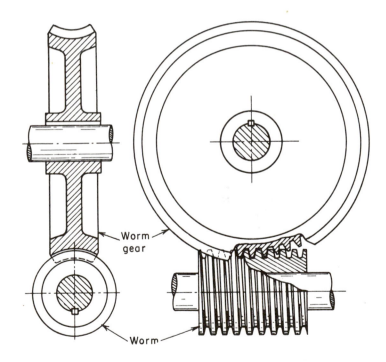

Figure 7.4 Worm and worm gear

*The pitch of a thread is the distance, measured parallel to the axis of the screw, from one thread to the next adjacent thread.

the worm gear. If the worm gear is to be driven at 1 rpm, the worm must turn as many revolutions as there are teeth on the worm gear in 1 min, or at the speed of T_F rpm.*

$$\frac{\omega \text{ of worm gear}}{\omega \text{ of worm}} = \frac{\omega_F}{\omega_D} = \frac{1}{T_F}$$

If the worm is double threaded it will advance the gear a distance equal to 2 teeth, making the speed ratio:

$$\frac{\omega_F}{\omega_D} = \frac{2}{T_F}$$

If the worm is triple threaded:

$$\frac{\omega_F}{\omega_D} = \frac{3}{T_F}$$

and if quadruple threaded:

$$\frac{\omega_F}{\omega_D} = \frac{4}{T_F}$$

7.5 Reverted Gear Trains

Simplicity and compactness are important in gear train design. The mounting arrangement shown in the commercial speed reducer in Figure 7.5 requires more space vertically than the *reverted train* in Figure 7.6, where driver and follower shafts are concentric.

The design of a reverted train imposes new restrictions on the tooth numbers of the gears beyond the requirement of producing the

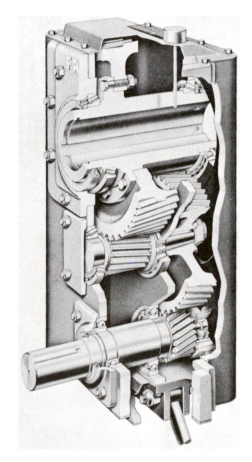

*T_F is the symbol for the number of teeth on the worm gear, which is always the follower. A worm gear cannot be used to drive a worm due to the low helix angle.

Figure 7.5   Shaft mounted drive (Courtesy of The Falk Corp., Milwaukee, Wis.)

necessary speed ratio. A simplified diagram of a reverted train similar to that in Figure 7.6 is shown in Figure 7.7. The 20-tooth gear is the driver which is connected to the motor. *D* drives the 60-tooth gear *A*, which is keyed to the same shaft as the 16-tooth gear *B*. Gear *B* in turn drives the 48-tooth follower *F*, which is on the output shaft. Gears *D* and *A* are 10 diametral pitch while *B* and *F* are 8 diametral pitch, since the slower pair carry more torque.

The speed ratio

$$\frac{\omega_F}{\omega_D} = \text{t.r.} = \frac{T_D}{T_A} \times \frac{T_B}{T_C} = \frac{20}{60} \times \frac{16}{48} = \frac{1}{9}$$

Figure 7.6 Speed reducer with reverted gear train (Courtesy of the Falk Corp., Milwaukee, Wis.)

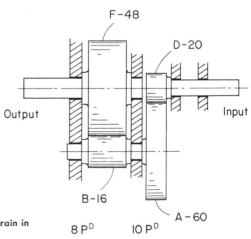

Figure 7.7 Diagram of gear train in Figure 7.6

The input and output shafts are on the same center line, which is parallel to the shaft carrying gears A and B. This requires that the sum of the pitch radii of D and A must equal the sum of the pitch radii of B and F, or the *sums of the pitch diameters of each pair must be equal.*

Using the relationship

$$D^P = \frac{T}{P^D}$$

$$D_D^P = \frac{20}{10} = 2$$

$$D_A^P = \frac{60}{10} = 6$$

$$D_B^P = \frac{16}{8} = 2$$

$$D_F^P = \frac{48}{8} = 6$$

So
$$D_D^P + D_A^P = 8 = D_B^P + D_F^P$$

as required for parallel shafts.

7.6 Selective Speed Drives

Many machines require a range of different speeds to meet varied operating conditions. Gear drives may be designed to afford a selection of output speeds from a constant input. Figure 7.8 shows a selective speed drive offering six speeds forward or reverse. The drive shaft D carries a *cluster* of six gears of tooth numbers shown. The follower shaft F carries the 16-tooth gear G mounted on splines so that it may be positioned as desired along the shaft and yet it is compelled to rotate with F at all times. Gear G is contained in a clevis of control lever C which turns freely around the splined shaft F. Gears A and B are carried on the control lever. A meshes with both B and G.

When control lever C is in the position shown in Figure 7.8, the 36-tooth gear on the cluster is driving A, while A in turn drives G, so the speed ratio is:

$$\frac{\omega_F}{\omega_D} = +\text{t.r.} = \frac{+36 \times T_A}{T_A \times 16} = \frac{36}{16} = \frac{9}{4}$$

Note that gear A is both a driver and a follower, so the tooth number of A appears in both the denominator and numerator of the t.r. and therefore cancels out. Such gears are called *idlers*, since they do not contribute any change of speed. They do change the relative sense of rotation of the 36-tooth driver and the follower G (see arrows on Figure 7.8) making the t.r. plus as shown.

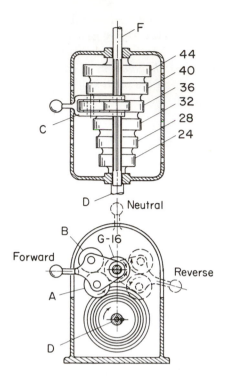

Figure 7·8 Selective speed drive

The number of teeth on an idler gear may be any convenient value, as it is not a factor in speed ratio.

The alternate dotted position of the control level brings gear B into contact with the 36-tooth gear. The speed ratio is:

$$\frac{\omega_F}{\omega_D} = -\text{t.r.} = -\frac{36 \times T_A \times T_B}{T_A \times T_B \times 16} = -\frac{9}{4}$$

While the speed ratio is the same as before, since gears A and B are both idlers, gear B makes the sense of rotation of F opposite to D.

The highest speed ratio (with A or B in contact with the 44-tooth gear is:

$$\frac{\omega_F}{\omega_D} = \frac{44}{16} = \frac{11}{4}$$

The lowest speed ratio (using the 24-tooth driver) is:

$$\frac{\omega_F}{\omega_D} = \frac{24}{16} = \frac{3}{2}$$

A second example of a selective speed transmission, shown in Figure 7.9, offers a direct 1:1 drive and three reduction ratios and also will permit changing speeds while the gears are in motion.

Figure 7·9 Selective speed transmission (Courtesy of Western Mfg. Co., Detroit, Mich.)

A diagram of the gear arrangement of such a transmission is shown in Figure 7.10. Gear *A* is attached to the *drive*, or *motor*, *shaft*. *B*, *C*, *E*, and *G* all turn together with the *jack shaft*. *H*, *J*, and *K* all turn with the *follower shaft F* and are mounted on splines to allow lateral positioning. Gear *A* meshes with *B* at all speeds. The gears are shown in neutral position in Figure 7.10 with the output shaft disconnected from the jack shaft.

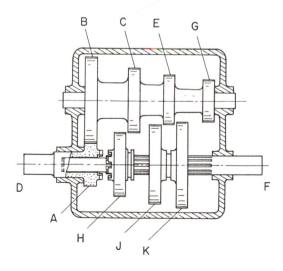

Figure 7·10 Neutral position

Figure 7.11 shows K in position to mesh with G, which gives F its lowest speed. The drive goes from A to B to G to K, giving a speed ratio:

$$\frac{\omega_F}{\omega_D} = \frac{14}{28} \times \frac{14}{28} = \frac{1}{4}$$

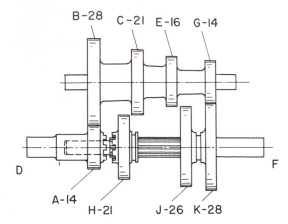

Figure 7.11 Lowest speed

Figure 7.12 shows the drive progressing from A to B to E to J, giving a speed ratio.

$$\frac{\omega_F}{\omega_D} = \frac{14}{28} \times \frac{16}{26} = \frac{1}{3.25}$$

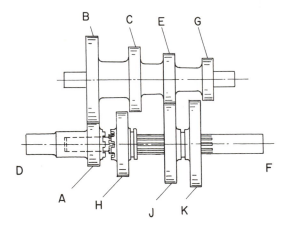

Figure 7.12 Second speed

In Figure 7.13, J and K are moved back into the neutral position and H is brought into mesh with C. The speed ratio is now:

$$\frac{\omega_F}{\omega_D} = \frac{14}{28} \times \frac{21}{21} = \frac{1}{2}$$

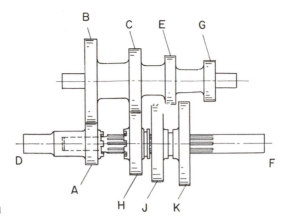

Figure 7.13 Third speed

If H is moved to the left (Figure 7.14), the clutch jaws on it engage those on gear A so that A and H turn together. This direct connection gives the follower shaft the same speed as the motor—the highest speed obtainable. (K and J of course remain in the disengaged neutral position.)

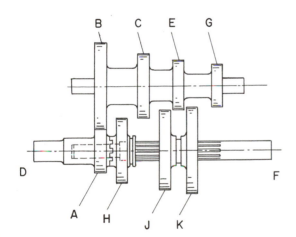

Figure 7.14 Direct drive for highest speed

7.7 Design of Regular Gear Trains

The design of gear trains, like most machine design problems, does not arise from the direct analytical solution of a given equation, but is a judicious compromise in an effort to satisfy a number of design requirements and practical limitations.

The usual factors governing a kinematic design are listed on the following page in order of importance.

1. *Speed Ratio and Sense of Rotation:*

   In some instances an exact speed ratio is essential, while others permit a considerable tolerance. Sense of rotation of the output shaft may or may not be important.

2. *Number of Gears:*

   In the interests of cost and compact design, the solution requiring the least number of gears is usually preferred. Often an additional small idler gear is more economical than the use of an annular, and in some cases an extra pair of inexpensive small gears offsets the cost of using several large gears.

3. *Use of Stock* (*Standard*) *Gears:*

   When possible it is always more economical to buy standard gears available from the stock of gear manufacturers than to employ special designs which must be made to order. These stock sizes vary somewhat with different manufacturers, so a wide choice is available within reasonable limits of size. When selecting stock gears of small size, checking for interference becomes mandatory, especially if these small gears are run together. The fact that a gear is standard does not guarantee its performance with all other standard gears.

4. *Diametral Pitch and Mounting Requirements:*

   The tooth size dictated by strength and wear analyses establishes the diametral pitch, which dictates the diameter of gears of given tooth numbers. These factors regulate the mounting distances, which must be reconciled with the available space, relative position of input and output shafts, and the location of bearings.

5. *Uniform Gear Ratios:*

   Just as limitations of space and stock size discourage the use of very large gears, so is it logical to divide the total speed change in a train equally between the several gear pairs of which it is composed. Exact equality of these component ratios is not essential, but reasonable uniformity is desirable. Figure 7.15 illustrates good and bad proportions.

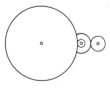

Uneven ratios

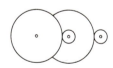

Uniform ratios   **Figure 7.15**

Let us design a typical spur gear train in which the required speed ratio (ω_F/ω_D) is exactly 1 : 300. The driver and follower shafts are to be parallel and are to turn in the same direction. Available stock sizes are all tooth numbers from 12 to 24, all even numbers from 24 to 100, and every fourth number from 100 to 120 teeth.

First we determine how many pairs of gears are needed. The largest gear has 120 teeth and the smallest has 12, so the highest possible ratio using a single pair is 12 : 120 or 1 : 10. The maximum ratio for two pairs of gears is then $\frac{1}{10} \times \frac{1}{10} = \frac{1}{100}$. Three pairs will allow a maximum of $\frac{1}{10} \times \frac{1}{10} \times \frac{1}{10} = \frac{1}{1000}$. Thus we see that there must be three pairs of gears in order to reach the required ratio of 1 : 300.

Next, we assign the value of each ratio, seeking to make them as nearly equal as possible. To obtain exact equality we would use the cube root of $\frac{1}{300}$ (since there are three pairs).

$$\sqrt[3]{\tfrac{1}{300}} = \frac{1}{6.7} \text{ approximately}$$

But since 300 is not a perfect cube, this is only useful as a guide. If we make the first two ratios 1 : 6.7, we can then calculate the required value of the third to produce the exact total speed ratio of 1 : 300. The numerator of the third may be made equal to the product of the denominators of the first two, thus canceling them. The denominator of the third may then be made 300:

$$\frac{1}{6.7} \times \frac{1}{6.7} \times \frac{44.89}{300} = \frac{1}{300}$$

Now we consider the possibility of replacing these numbers by the available tooth numbers without altering the ratios. The first two could be 10 : 67 or 20 : 134, for example, but these multiples fail, since there are no 10-tooth or 134-tooth gears available. As there are no other multiples of 6.7 between 10 and 20 which are integers and no factor of 44.89 which is an integer, we abandon these ratios.

Let us try 1 : 6 and 1 : 7 as the first two ratios and determine the third:

$$\frac{1}{6} \times \frac{1}{7} \times \frac{42}{300} = \frac{1}{300} \qquad \frac{42}{300} = \frac{1}{7.14} \text{ approximately,}$$

so these ratios are somewhat uniform. Now, to obtain available tooth numbers, we multiply the first two ratios by 14 and divide the third by 3:

$$\frac{14}{84} \times \frac{14}{98} \times \frac{14}{100} = \frac{1}{300}$$

This is an acceptable solution.

A third trial yields more uniform ratios. The ratio $1 : 6\frac{2}{3}$ is nearly equal to the ideal 1 : 6.7. Let us try ratios of $1 : 6\frac{2}{3}$ and determine the required value of the third (to simplify the figures $1 : 6\frac{2}{3} = 1 : \frac{20}{3} = \frac{3}{20}$)

$$\frac{3}{20} \times \frac{3}{20} \times \frac{400}{9 \times 300} = \frac{1}{300}$$

or (since the third ratio $= \frac{4}{27}$)

$$\frac{3}{20} \times \frac{3}{20} \times \frac{4}{27} = \frac{1}{300}$$

$\frac{4}{27} = \frac{1}{6.75}$, which is very nearly $1 : 6\frac{2}{3}$. To get available tooth numbers we multiply the first two by 5 and the third by 4:

$$\frac{15}{100} \times \frac{15}{100} \times \frac{16}{108} = \frac{1}{300}$$

The design of the gear train is not complete without a sketch to show the arrangement of the gears in addition to the tooth numbers. Since $\omega_F/\omega_D =$ product of teeth on drivers $\div$ product of teeth on follower gears, we know that the 15- and 16-tooth gears must be drivers, as shown in the sketch in Figure 7.16. The driver and follower of the train should be identified, using labels D and F.

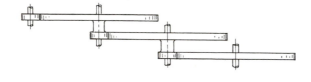

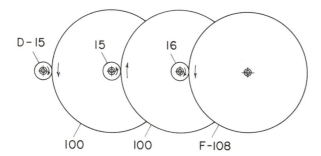

Figure 7.16 Sketch of gear train design

We now check the sense of rotation of D and F and find that F turns opposite to D. The specifications required that they turn the same way. We may make one of the large gears an annular or introduce an idler between any driver and follower (of such size as will run properly with the smaller driver). The addition of a 14-tooth idler is shown in Figure 7.17 and also an edge view of a more compact arrangement.

7.8 Design of Reverted Gear Trains

The factors which govern the design of regular gear trains all apply to reverted trains with added emphasis on mounting distances and diametral pitch, often at the sacrifice of uniform ratios.

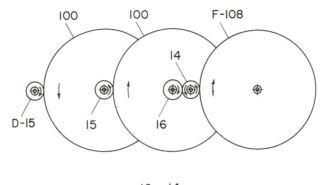

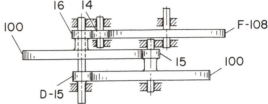

Figure 7.17 Use of an idler gear

Let us try a sample reverted train problem composed of spur gears. The speed ratio is 30 : 1 with driver and follower parallel and turning in the same sense. Available stock gears are all tooth numbers from 12 to 24, all even numbers from 24 to 100, every fifth number from 100 to 120 and also 128 and 144. All gears are to have the same diametral pitch.

The maximum ratio using one pair is 144 : 12 or 12 : 1. So two pairs will be sufficient.

An ideal value for each of the two ratios would be $\sqrt{30/1}$ which equals 5.48 : 1 approximately, but since there are other conditions which complicate this design, let us try ratios of integral numbers, such as 6 : 1 and 5 : 1.

The selection of tooth numbers must not only provide the correct tooth ratio (ω_F/ω_D = t.r. = 30/1) but also must insure parallel shafts. Figure 7.18 shows a typical reverted train with four gears D, A, B, and F. To keep the shafts parallel,

$$R_D^P + R_A^P = R_B^P + R_F^P \quad \text{or} \quad D_D^P + D_A^P = D_B^P + D_F^P$$

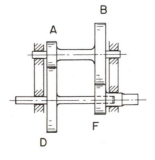

Figure 7.18 Reverted gear train

This equation can be expressed in tooth numbers and diametral pitch, since $D^P = T/P^D$:

$$\frac{T_D + T_A}{P^D} = \frac{T_B + T_F}{P^D}$$

(All gears have the same P^D here)

$$\therefore T_D + T_A = T_B + T_F \text{ (canceling } P^D)$$

This equation and the t.r. equation $\left(\dfrac{T_D \times T_B}{T_A \times T_F} = \dfrac{30}{1}\right)$ govern the selection of tooth numbers.

It is easier to visualize these two requirements if they are set down in the following pattern:

MULTIPLY
$\longrightarrow$

$$\text{ADD} \downarrow \quad \frac{T_D}{T_A} \times \frac{T_B}{T_F} = \frac{30}{1}$$
$$S = S$$

If we *multiply* the numbers from left to right they must equal $\frac{30}{1}$. If we *add* the numerator to the denominator of each ratio they must have equal sums, which we call S. Now, substituting the ratios for each pair in this arrangement, we have:

$$\frac{5}{1} \times \frac{6}{1} = \frac{30}{1}$$
$$\overline{6} \neq \overline{7}$$

The t.r. is satisfied, but the tooth sums are unequal. To correct this, we must find one number S, which is divisible into 6 parts (so that we can put 5 parts in the numerator and 1 in the denominator of the first ratio) and divisible into 7 parts (6 of which go in the numerator and 1 in the denominator of the second ratio). The smallest such number is 42, the product of 6 and 7, in this case. Now $42 \div 6 = 7$, so the first ratio becomes $\dfrac{7 \times 5}{7 \times 1} = \dfrac{35}{7}$ (keeping the $5 : 1$ ratio). In the second ratio, $42 \div 7 = 6$, making that ratio $\dfrac{6 \times 6}{6 \times 1} = \dfrac{36}{6}$ (preserving the $6 : 1$ ratio).

Substituting these numbers in the general expression

$$\frac{T_D}{T_A} \times \frac{T_B}{T_F} = \frac{30}{1}$$
$$S = S$$

We have

$$\frac{35}{7} \times \frac{36}{6} = \frac{30}{1}$$
$$\overline{42} = \overline{42}$$

which satisfies the numerical conditions, but we must use multiples of these figures in order to bring them within the range of available tooth numbers. Multiplying throughout by 2, we have

$$\frac{70}{14} \times \frac{72}{12} = \frac{30}{1}$$
$$\overline{84} = \overline{84}$$

Now, referring to the general form, $T_D = 70$, $T_A = 14$, $T_B = 72$, and $T_F = 12$. The solution of the problem is properly recorded in Figure 7.19.

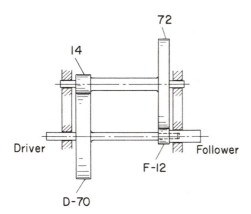

Figure 7.19 Sketch of design of reverted train

7.9 Design of Reverted Trains with Gears of Different Pitch

If each pair of gears in a reverted train has a different diametral pitch, the method of attack is similar, but the figures are slightly more complicated. For example, let us do the previous problem with all given data the same except that gears D and A are to have a diametral pitch of 6, while B and F have a diametral pitch of 8.

The equation for parallel shafts is the only change. The sum of the pitch diameters of D and A must still equal the sum of the pitch diameters of B and F. In terms of tooth numbers and the given diametral pitches ($D^P = T/P^D$)

$$\frac{T_D + T_A}{6} = \frac{T_B + T_F}{8}$$

And (multiplying by 24)

$$4T_D + 4T_A = 3T_B + 3T_F$$

The given speed ratio:

$$\frac{\omega_F}{\omega_D} = \frac{T_D \times T_B}{T_A \times T_F} = \frac{30}{1}$$

is not changed if we include the coefficient 4 in the numerator and denominator of the first pair and the coefficient 3 in the second. Following the arrangement of Article 7.8, the pattern looks like this:

$$\text{MULTIPLY} \longrightarrow$$

$$\text{ADD} \downarrow \quad \frac{\dfrac{4T_D}{4T_A}}{S} \times \frac{\dfrac{3T_B}{3T_F}}{S} = \frac{30}{1}$$

Using the ratios 5 : 1 and 6 : 1 as before, we have:

$$\frac{\dfrac{4 \times 5}{4 \times 1}}{24} \times \frac{\dfrac{3 \times 6}{3 \times 1}}{21} = \frac{30}{1} \quad \text{or} \quad \frac{\dfrac{20}{4}}{24} \times \frac{\dfrac{18}{3}}{21} = \frac{30}{1}$$

To correct the inequality of the sums 24 and 21, we now search for one number S which is divisible by both 24 and 21. The smallest such number is 168. Dividing 168 by 24 we get 7, so the first ratio becomes $\dfrac{7}{7} \times \dfrac{20}{4}$. Dividing 168 by 21 we get 8, making the second ratio $\dfrac{8}{8} \times \dfrac{18}{3}$.

The pattern then looks like this:

$$\frac{\dfrac{140}{28}}{168} \times \frac{\dfrac{144}{24}}{168} = \frac{30}{1}$$

To find the tooth numbers we factor out the coefficients 4 and 3 in the above ratios, obtaining the expression:

$$\frac{4 \times 35}{4 \times 7} \times \frac{3 \times 48}{3 \times 8} = \frac{30}{1}$$

which corresponds to the original array:

$$\frac{\dfrac{4T_D}{4T_A}}{S} \times \frac{\dfrac{3T_B}{3T_F}}{S} = \frac{30}{1}$$

This would indicate tooth numbers as follows:

$$T_D = 35, \quad T_A = 7, \quad T_B = 48, \text{ and } T_F = 8$$

These numbers satisfy the numerical conditions but there are no stock gears available with 35, 7, or 8 teeth, so we must use multiples of these numbers to bring them into the range of available stock sizes. Multiplying by 2, the expression becomes

$$\frac{\dfrac{4 \times 70}{4 \times 14}}{336} \times \frac{\dfrac{3 \times 96}{3 \times 16}}{336} = \frac{30}{1}$$

with all conditions satisfied when $T_D = 70$, $T_A = 14$, $T_B = 96$, and $T_F = 16$.

The sketch shown in Figure 7.20 completely describes the solution of this design problem.

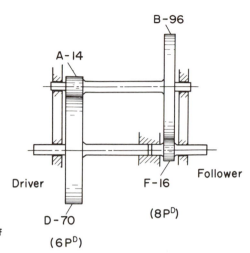

Figure 7.20 Reverted train with gears of different diametral pitch

7.10 Planetary, or Epicyclic, Gear Trains

All gears in the trains we have studied have been mounted in bearings which are secured in a stationary frame or case. The simplest form of this frame is a bar, such as A in Figure 7.21. If this bar is permitted to rotate about a fixed pin at O, gears D and F are still secured in mesh and F rolls about D just as a planet turns about the sun. A gear train in which one or more gears turn about a moving axis is called a *planetary*, or *epicyclic, gear train*.* These trains can produce large changes in speed using very few gears. They have many unique applications in differentials, computing devices, and other instruments.

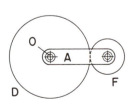

Figure 7.21 Gears on fixed axes

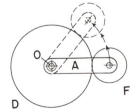

Figure 7.22 Epicyclic gear train

*The name *epicyclic* is derived from the fact that points on the planet gear describe epicycloids when in motion.

7.11 Speed Ratios of Epicyclic Trains

The simplest form of epicyclic train, shown in Figure 7.23, is composed of two gears, D and F, and arm A.* Gear D and arm A turn independently about fixed shaft O. Gear F turns about pin N, on A, and meshes with D. There are three rotating members whose speeds are of interest, gear D, gear F, and arm A. If the speeds of any two of these members are known, we can determine the speed of the third.

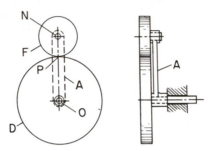

Figure 7.23 Two-gear epicyclic train

As it is very difficult to visualize the motion of planet gears like F, let us devise an equation which relates the speeds and the tooth ratio of the gears. This can be done by means of the familiar velocity analysis (Figure 7.24).

Assuming ω_A and ω_D to be known and both clockwise:

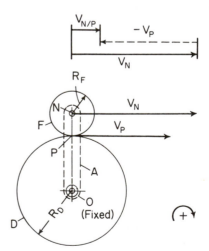

Figure 7.24 Velocity analysis of epicyclic train

*For convenience, only the pitch circles of D and F are shown in Figure 7.23. R_D and R_F are pitch radii of the gears.

$$V_N = \omega_A \times ON \quad \text{but} \quad ON = R_D + R_F$$

$$\therefore \quad V_N = \omega_A(R_D + R_F)$$

$$V_P = \omega_D \times R_D$$

(V_P on $D = V_P$ on F, since pitch point velocities are equal)

$$\omega_F = \frac{V_{N/P}}{PN} \quad \text{but} \quad V_{N/P} = V_N - V_P \quad \text{and} \quad PN = R_F$$

$$\therefore \quad \omega_F = \frac{V_N - V_P}{R_F} \quad \text{and} \quad \omega_F R_F = V_N - V_P$$

Substituting:

$$\omega_F R_F = \omega_A(R_D + R_F) - \omega_D R_D$$

Expanding:

$$\omega_F R_F = \omega_A R_D + \omega_A R_F - \omega_D R_D$$

So

$$\omega_F R_F - \omega_A R_F = \omega_A R_D - \omega_D R_D$$

Factoring out R_F and R_D:

$$R_F(\omega_F - \omega_A) = R_D(\omega_A - \omega_D)$$

So

$$\frac{\omega_F - \omega_A}{\omega_A - \omega_D} = \frac{R_D}{R_F}$$

The pitch radii R_D and R_F can be expressed in terms of tooth numbers through the diametral pitch formula $D = T/P^D$ or $2R = T/P^D$ or $R = T/2P^D$

$$R_D = \frac{T_D}{2P^D} \quad \text{and} \quad R_F = \frac{T_F}{2P^D}$$

The speed ratio equation then becomes

$$\frac{\omega_F - \omega_A}{\omega_A - \omega_D} = \frac{T_D}{T_F}$$

In the given gear train, the tooth ratio (t.r.) is $-T_D/T_F$. (It is negative, since D and F are mounted so as to turn in opposite directions relative to one another.) If we multiply both sides of the speed ratio equation by -1, it becomes:

$$\frac{-\omega_F + \omega_A}{\omega_A - \omega_D} = -\frac{T_D}{T_F}$$

$$\frac{-\omega_F + \omega_A}{\omega_A - \omega_D} = \text{t.r.}$$

in which $-T_D/T_F$ now equals the t.r. This equation becomes more logical if we multiply numerator and denominator of the left side by -1:

$$\frac{\omega_F - \omega_A}{\omega_D - \omega_A} = \text{t.r.}$$

All of the speeds on the left side are absolute speeds, and the tooth ratio is determined by the numbers of teeth on the gears and their relative sense of rotation. We can rationalize this equation if we note the similarity of the regular and epicyclic speed-ratio equations. The expression $\omega_F - \omega_A$ really is the speed of gear F relative to arm A, and $\omega_D - \omega_A$ is the speed of gear D relative to arm A. (As with linear velocities, *the difference of two absolute angular velocities is this relative velocity*.) So, whereas the regular train equation is

$$\frac{\omega_F}{\omega_D} = \text{t.r.}$$

the epicyclic equation is

$$\frac{\omega_{F/A}}{\omega_{D/A}} = \text{t.r.}$$

The difference lies in that, in the ω_F/ω_D term of the regular train equation, the velocities are absolute or relative to the earth (a stationary arm) and the equation could be written:

$$\frac{\omega_{F/\text{Earth}}}{\omega_{D/\text{Earth}}} = \text{t.r.}$$

whereas, in the epicyclic equation, the $\dfrac{\omega_{F/A}}{\omega_{D/A}}$ term states that the velocities are relative to the arm. By using the expanded form of the epicyclic equation:

$$\frac{\omega_F - \omega_A}{\omega_D - \omega_A} = \text{t.r.}$$

we introduce the absolute speed of the arm A into the expression. In the regular train equation, the speed of the earth is omitted since it is assumed to be zero. This comparison not only serves to show the logic behind these equations, but their similarity is an aid to remembering them.

Signs are very important in the angular velocity terms as well as in the t.r. *Like signs in speed terms denote the same directions of rotation.* In regular gear trains, the t.r. is determined by the numbers of teeth on the gears and a sign denoting the direction of rotation of the driver relative to the follower. In epicyclic trains the same holds true, except that we do not compare the absolute directions of rotation of driver and follower to establish the sign, because different directions of arm motion cause the follower gear to change its absolute sense of rotation. The sign of the t.r. in an epicyclic train is independent of any motion of the connecting frame or arm just as the t.r. of a regular train is independent of the stationary frame which holds the gears in mesh.

In each case, *the t.r. is an indication of the capability of meshing gears to turn relative to one another* and is not affected by the absolute speeds of the gears or motion of the connecting arm or frame. It is therefore convenient to assume the arm to be stationary when determining the t.r. of an epicyclic train. Arrows can then be sketched on the gear assembly to determine consistent directions of rotation, just as was suggested in the case of regular gear trains. *The t.r. does not depend on the absolute speeds or actual directions of rotation of the gears but only upon their ability to turn relative to one another.*

7.12 Analysis of Epicyclic Gear Trains

All epicyclic trains can be solved by the general equation

$$\frac{\omega_F - \omega_A}{\omega_D - \omega_A} = \text{t.r.}$$

There are four terms in the equation ($\omega_F, \omega_D, \omega_A$, and t.r.) so any three of these must be known in order to solve for the fourth. Let us try some sample problems.

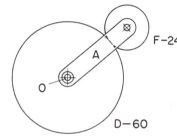

F–24

A

O

D–60   **Figure 7.25** Example 1

Example 1

Gear D in Figure 7.25 has 60 teeth and turns clockwise at 200 rpm. Gear F has 24 teeth. Arm A turns counterclockwise 100 rpm about fixed shaft O. The absolute speed of F is required. Substituting in the epicyclic equation (denoting clockwise rotation $+$):

$$\frac{\omega_F - (-100)}{+200 - (-100)} = -\frac{\overset{5}{\cancel{60}}}{\underset{2}{\cancel{24}}} \quad \text{so} \quad 2\omega_F + 200 = -1500$$

$\omega_F = -\frac{1700}{2} = -850$ rpm, counterclockwise (the negative sign denotes counterclockwise motion).

Example 2

A more practical form of epicyclic train is shown in Figure 7.26. The input is applied to gear N, which turns 3600 rpm about fixed axis L. The output

Figure 7.26 Example 2

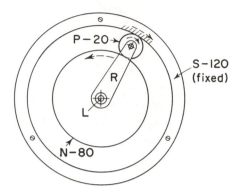

member is R, also turning about L, independently of N. Gear P meshes with N and S, which is a stationary annular. The angular speeed of R is required.

First, we identify the driver gear, the follower gear and the arm. The driver may be at either end of the train of three gears. If we select N as the driver, S will be the follower.* R is clearly the arm in this train—the output of a mechanism may be any rotating member (not necessarily a gear).

Now, we write our basic equation, substituting the proper members as identified above:

$$\frac{\omega_S - \omega_R}{\omega_N - \omega_R} = \text{t.r.}$$

Substituting:

$$\frac{0 - \omega_R}{3600 - \omega_R} = -\frac{\overset{2}{\cancel{80}}}{\underset{3}{\cancel{120}}}$$

(The sign of t.r. is established by the dotted arrows, and gear P is an idler.)

$$-3\omega_R = -7200 + 2\omega_R$$

$\omega_R = 1440$ rpm, in the same sense as gear N. Since ω_N was introduced in the equation as a positive quantity and the calculated value of ω_R is also positive, they must both turn in the same direction.

7.13 Epicyclic Speed Changer

With careful selection of tooth ratios, epicyclic trains can be designed to provide large speed changes with few gears. Figure 7.27 shows a speed reducer composed of only four standard gears of stock sizes. Gears J and K are keyed to the same shaft which is carried by the arm. The arm turns with the input

---

*The fact that S does not turn does not affect its function as a follower of the train. A driver or a follower gear may have any speed, including zero !

shaft B. J meshes with a stationary gear L, and K meshes with M, which turns
with the output shaft C. If B turns 2000 rpm, let us find the speed of C. Assuming L as the driver gear and M as the follower the general equation

$$\frac{\omega_F - \omega_A}{\omega_D - \omega_A} = \text{t.r.}$$

becomes:

$$\frac{\omega_C - \omega_B}{\omega_L - \omega_B} = \frac{T_L \times T_K}{T_J \times T_M}$$

(the sign of the t.r. is plus)

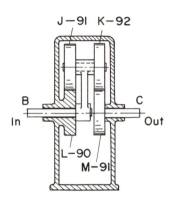

J−91 K−92

B

In   Out

C

L−90
M−91

Figure 7.27 Reverted epicyclic

$$\frac{\omega_C - 2000}{0 - 2000} = \frac{90}{91} \times \frac{92}{91} = \frac{8280}{8281}$$

$$8281\omega_C - (2000 \times 8281) = -(2000 \times 8280)$$

$$8281\omega_C = 2000\,(8281 - 8280)$$

$$\omega_C = \frac{2000}{8281} = 0.2415 \text{ rpm} \quad (\text{same sense as } B)$$

The ratio of the output speed to the input speed is:

$$\frac{\omega_C}{\omega_B} = \frac{\dfrac{2000}{8281}}{2000} = \frac{1}{8281}$$

a very large reduction of speed.

Just to illustrate the possibilities, let us join two epicyclic trains like the
one above in series, as shown in Figure 7.28. The output shaft C of the first

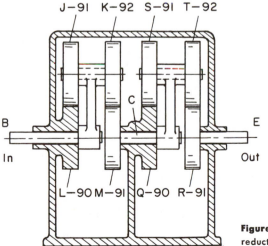

J−91 K−92 S−91 T−92

B   E

In   Out

C

L−90 M−91 Q−90 R−91

Figure 7.28 Series epicyclic for large
reduction

train becomes the input shaft of the second. If B turns 2000 rpm, let us find the speed of the final output shaft E.

Epicyclic trains in series like this must be treated separately—the two trains cannot be solved by one equation. Having calculated the speed of C in the first train, we can apply this figure directly in the second. With Q as the stationary driver gear, and C as the arm:

$$\frac{\omega_E - \omega_C}{\omega_Q - \omega_C} = \text{t.r.} \quad \text{so} \quad \frac{\omega_E - \dfrac{2000}{8281}}{0 - \dfrac{2000}{8281}} = \frac{8280}{8281}$$

$$8281\omega_E - \left(\frac{2000 \times 8281}{8281}\right) = -\left(\frac{2000 \times 8280}{8281}\right)$$

$$(8281)^2\omega_E = (2000 \times 8281) - (2000 \times 8280)$$

$$\therefore \quad \omega_E = \frac{2000}{(8281)^2} = 0.00002916 \text{ rpm, or 1 rev in 23.8 days!*}$$

7.14 Epicyclic Reversing Mechanism

A mechanism which permits reversal of sense of rotation while in motion has advantages over one which can only be reversed when at a standstill. The epicyclic reversing mechanism shown in Figure 7.29 has this feature and also provides different speeds in each direction.

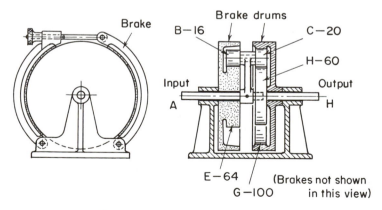

Figure 7.29 Reversing epicyclic train

*The rather spectacular speed change with this type of train is largely due to the tooth ratio of nearly 1. Longhand calculations are demanded in such solutions.

The input shaft A turns the arm which carries two planetary gears, B and C. These gears turn together on the same shaft which revolves freely in its bearings on the arm. Gear B meshes with a spur gear E which turns freely on shaft A. The large brake drum, shown stippled, is attached to E. Gear C meshes with an annular gear G and also with spur gear H, which turns the output shaft. The outer surface of G forms a second brake drum. A bearing for arm A is provided in H and the annular G is mounted on the output shaft H, but A, H, and G turn independently. Two pairs of brake shoes, shown in the left view of Figure 7.29, are provided for both E and G so that either gear may be prevented from turning.

Let us assume an input speed of 1000 rpm at A. If we apply the brake to G, preventing its rotation, and allow E to turn, the effective epicyclic train consists of gears G, C, and H only. Gears B and E do not contribute to the output speed of H. Considering G as the driving gear and H the follower, the epicyclic equation becomes:

$$\frac{\omega_H - \omega_A}{\omega_G - \omega_A} = \frac{T_G}{T_H}$$

Substituting:

$$\frac{\omega_H - 1000}{0 - 1000} = \frac{-100}{60}$$

Solving: $60\omega_H - 60{,}000 = 100{,}000$

$$\omega_H = \frac{160{,}000}{60} = 2667 \text{ rpm} \quad \text{(Same sense as } A\text{)}$$

Now, if we apply the brake to E and allow G to turn, the effective train consists of gears E, B, C, and H. Gear G does not contribute to the output speed of H. If we consider E the driving gear and H the follower, the equation is:

$$\frac{\omega_H - \omega_A}{\omega_E - \omega_A} = \frac{T_E}{T_B} \times \frac{T_C}{T_H}$$

Substituting:

$$\frac{\omega_H - 1000}{0 - 1000} = \frac{64}{16} \times \frac{20}{60}$$

Solving: $3\omega_H - 3000 = -4000$

$$\omega_H = -\frac{1000}{3} = -333\tfrac{1}{3} \text{ rpm} \quad \text{(Opposite sense to } A\text{)}$$

If both brakes are released there is no positive drive applied to H and it may remain stationary. Thus, in this device we have forward, neutral, and reverse speeds.

7.15 Epicyclic Selective Speed Transmission

The mechanism in Figure 7.30 consists of a series of three identical epicyclic trains. The driver gears, H, G, and E are all keyed to turn with the input shaft D. The annulars, C, B, and A are followers of the three trains. Annular B also serves as the arm of the first train, as it carries planet gear L. Annular A similarly serves as the arm of the second train, carrying planet gear K. The arm of the third train carries planet gear J and is attached to the output shaft F. The right end of shaft D is supported by bearings in F, but they turn independently. Similarly, the annulars all turn freely on shaft D which supports them. Brake bands (not shown in Figure 7.30) are fitted on each annular—so that each may be held stationary as required. Shaft D turns at constant speed of 1000 rpm.

First, let us hold annular A stationary and determine the speed of F. Applying the epicyclic equation

$$\frac{\omega_F - \omega_A}{\omega_D - \omega_A} = \text{t.r.}$$

to train 3, E is the driver, A is the follower, and F is the arm:

$$\frac{\omega_A - \omega_F}{\omega_E - \omega_F} = -\frac{T_E}{T_A} \quad \text{so} \quad \frac{0 - \omega_F}{1000 - \omega_F} = \frac{-36}{72}$$

And solving,

$$\omega_F = 333\tfrac{1}{3}\,\text{rpm} \quad (\text{same sense as } D)$$

(Examination will show that trains 1 and 2 also turn, but do not contribute to the speed of F.)

Now, if we apply the brake to annular B, while allowing A and C to turn, we can determine the speed of F. Train 1 does not affect the speed of F, so we

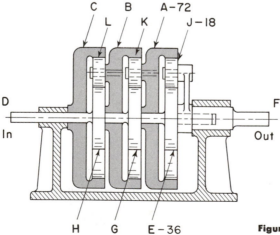

Figure 7.30 Selective speed series epicyclic

ignore it and start with train 2. Gear G is the driver, B the follower and A is the arm. The speed ratio formula becomes:

$$\frac{\omega_B - \omega_A}{\omega_G - \omega_A} = \text{t.r.} \quad \text{so} \quad \frac{0 - \omega_A}{1000 - \omega_A} = -\frac{1}{2}$$

And solving,

$$\omega_A = +\frac{1000}{3}\text{ rpm}$$

To finp ω_F we now turn to train 3, using the ω_A determined above:

$$\frac{\omega_A - \omega_F}{\omega_E - \omega_F} = \text{t.r.} \quad \text{so} \quad \frac{1000/3 - \omega_F}{1000 - \omega_F} = -\frac{1}{2}$$

And solving,

$$\omega_F = +555.5\text{ rpm} \quad \text{(in same sense as } D\text{)}$$

Lastly, we now hold annular C stationary and release A and B. To determine the speed of F we start with train 1:

$$\frac{\omega_C - \omega_B}{\omega_H - \omega_B} = \text{t.r.} \quad \text{so} \quad \frac{0 - \omega_B}{1000 - \omega_B} = -\frac{1}{2} \quad \text{and} \quad \omega_B = \frac{+1000}{3}\text{ rpm}$$

Now, in train 2, we apply the above value of ω_B:

$$\frac{\omega_B - \omega_A}{\omega_G - \omega_A} = \text{t.r.} \quad \text{so} \quad \frac{1000/3 - \omega_A}{1000 - \omega_A} = -\frac{1}{2} \quad \text{and} \quad \omega_A = +\frac{5000}{9}\text{ rpm}$$

Now, in train 3, we apply the above value of ω_A:

$$\frac{\omega_A - \omega_F}{\omega_E - \omega_F} = \text{t.r.} \quad \text{so} \quad \frac{5000/9 - \omega_F}{1000 - \omega_F} = -\frac{1}{2} \quad \text{and}$$

$$\omega_F = +\frac{19,000}{27} = +703.7\text{ rpm} \quad \text{(in same sense as } D\text{)}$$

Thus, by applying brakes to each annular in turn, we obtain three different output speeds without stopping the motor or shifting any gears in and out of mesh.*

7.16 Epicyclic Computing Mechanism

A *bevel gear differential* is shown in Figure 7.31. This type of train has many applications, but in this example it is used to add or subtract angular velocities. The given speeds are applied at the input shafts B and C in the same sense if they are to be added and in the opposite sense if subtracted. The speed

---

*A transmission of this type was used on the historic model T Ford and still has many current automotive applications.

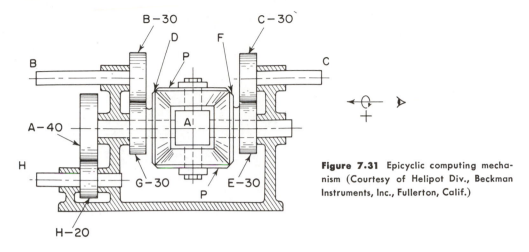

Figure 7.31 Epicyclic computing mechanism (Courtesy of Helipot Div., Beckman Instruments, Inc., Fullerton, Calif.)

of the output shaft *H* will be the sum or difference (the algebraic sum) of the given speeds.

The spur gear pairs *B-G*, *C-E*, and *A-H* are regular trains (mounted on fixed axes). Gear *G* turns with the miter gear *D*, and *E* turns with *F*. The planet gears *P* are identical in size with the meshing gears *D* and *F* and turn freely on the cross-shaped arm, which is secured to turn with gear *A*. *D* and *F* are supported by this arm but turn independently of the arm and of one another.

To observe the performance of this mechanism, let us apply a speed of +300 rpm to *B* and +100 to *C* and determine the speed of *H*. To define sense of rotation, we may assume clockwise rotation, seen looking from right toward the left, as plus. This convention must apply to all gears (except the planets) in the mechanism. *When a mechanism is composed of both regular and planetary gear trains, each type must be solved separately.*

As the tooth ratio of the regular train *B-G* is -1, $\omega_G = \omega_D = -300$. Similarly, the t.r. of *C-E* is -1 and $\omega_E = \omega_F = -100$.

In the planetary train, *D* may be the driving gear, *F* the follower, and *A* (turning gear *A*) the arm. The two planet gears *P* are kinematically equivalent. Only one is considered when establishing the t.r.—the other is added for balance.* The drive progresses from *D* through *P* to *F*, as shown in Figure 7.32. Since *P* is an idler and *D* and *F* are the same size, turning opposite directions relative to one another, the t.r. equals -1.

$$\frac{\omega_F - \omega_A}{\omega_D - \omega_A} = \text{t.r.}$$

So

$$\frac{-100 - \omega_A}{-300 - \omega_A} = -1 \quad \text{and} \quad \omega_A = -200 \text{ rpm}$$

In the regular train *A-H*, $\omega_H / \omega_A = \text{t.r.}$ so

$$\frac{\omega_H}{-200} = -\frac{\overset{2}{\cancel{40}}}{\underset{1}{\cancel{20}}} \quad \text{and} \quad \omega_H = +400 \text{ rpm}$$

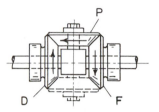

Figure 7.32 Bevel gear differential

which is the sum of $+300$ and $+100$ as predicted.

To check the subtraction operation, we may turn *B* at $+500$ rpm and *C* at -300 rpm. This will give *G* and *D* a speed of -500 and *E* and *F* a speed of $+300$. So in the epicyclic train:

$$\frac{-500 - \omega_A}{+300 - \omega_A} = -1 \quad \text{and} \quad \omega_A = -100 \text{ rpm}$$

In the *A-H* train:

$$\frac{\omega_H}{-100} = -\frac{2}{1} \quad \text{and} \quad \omega_H = +200 \text{ rpm}$$

which equals $+500 - 300$ as predicted.

7.17 Compound Epicyclic Train

The speed changer shown in Figure 7.33 is composed of two epicyclic trains which must be solved separately. Gear *D* is pinned to the input shaft *J* which turns 2000 rpm. *D* meshes with *E*, which is mounted on arm *A*. The

*Multiple planet gears are frequently used in epicyclic trains. The student must learn to recognize this repetition.

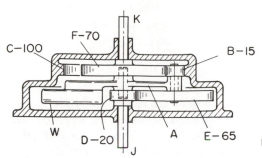

Figure 7.33 Compound epicyclic train

other planet gear B turns with E and meshes with the fixed annular C and gear F. Arm A carries a counterweight W and turns freely in bearings in gears D and F. F is pinned to the output shaft K, whose speed is to be determined.

The epicyclic train D-E-B-F is easily identified, in which gear D is the driver, F the follower, and A the arm.

$$\frac{\omega_F - \omega_A}{\omega_D - \omega_A} = \text{t.r.} \quad \text{so} \quad \frac{\omega_F - \omega_A}{2000 - \omega_A} = +\frac{20}{65} \times \frac{15}{70} = +\frac{6}{91}$$

$$91\omega_F - 91\omega_A = 12{,}000 - 6\omega_A$$

$$91\omega_F - 12{,}000 = 85\omega_A$$

$$\therefore \frac{91\omega_F - 12{,}000}{85} = \omega_A$$

The above equation does not give the speed of F but only relates it to ω_A. To find ω_A, we must examine a second epicyclic train, such as C-B-F, of which A is also the arm. Here C is the driving gear and F the follower.

$$\frac{\omega_F - \omega_A}{0 - \omega_A} = -\frac{100}{70} \quad \text{so} \quad 70\omega_F - 70\omega_A = 100\omega_A \quad \text{and} \quad \omega_A = \frac{70}{170}\omega_F$$

This does not give a solution for ω_F, but it does provide a second equation for ω_A, in terms of ω_F. The two expressions for ω_A may be equated:

$$\frac{91\omega_F - 12{,}000}{85} = \omega_A = \frac{70}{170}\omega_F$$

Solving for ω_F:

$$\omega_F = +214.3 \text{ rpm} \quad \text{(in the same sense as } J\text{)}$$

7.18 Kinematic Design of Epicyclic Gear Trains

The same considerations which governed the design of regular gear trains apply to epicyclics:

1. The overall speed ratio of the train must be obtained with the input and output shafts turning in the same or opposite directions as specified.
2. The least number of gears should be used.
3. The tooth numbers and diametral pitch must be consistent with the mounting arrangement and available space.
4. Stock gears should be used wherever possible.
5. The use of reasonably uniform gear ratios is desirable. (This frequently must be sacrificed to fulfill other more important conditions, however.)

Restrictions on the arrangement and relative size of the gears, space limitations, and mounting problems tend to complicate or alter the relative importance of the above rules when applied to epicyclic design. Dynamic problems are sometimes of primary importance, but they are too complex to be undertaken here and are reserved for the more advanced courses in applied mechanics. The purely kinematic problems, offered here, must first be understood before any dynamic refinements are attempted.

7.19 Design Procedure for Epicyclic Trains

As is the case in many design problems, the application of simple theory is so interspersed with interlocking choices and decisions that a well-planned program of attack is vital if we are to rise above blind hit-or-miss guesswork in epicyclic design. A certain measure of trial-and-error method cannot be avoided, but the trials must be intelligent choices and the errors must point the way to improvement. Trial and error should consist of "educated guesses" followed by "informative failures." A six-step procedure is recommended as follows:

1. *Identify output and input members of the train.*
   The input may be applied either at a moving gear or at the arm, so there are always two choices:
   a) Input at gear *D*
      Output at arm *A*
      Gear *F* stationary
   b) Input at arm *A*
      Output at gear *D*
      Gear *F* stationary
2. *Determine the tooth ratios of the required train.*
   This is done by substituting the given input and output speeds in the epicyclic equation (including signs for the sense of rotation specified) and solving for the value and sign of the t.r. Since there are two possible input members, as noted in 1 above, there will be two possible values of the t.r. from which to choose.

3. *Select the general type of train to be designed.*
   There are two general types of simple epicyclic trains, the *3-gear* and the *4-gear*, from which to choose. The type selected must be capable of producing the desired t.r. and specified sense of rotation. These types of trains are described in subsequent Articles 7.21 and 7.23.

4. *Select the tooth numbers of all gears.*
   These must provide the correct t.r. and fit the diametral pitch and space requirements.

5. *Sketch the train.*
   The solution is worthless without instructions as to the arrangement of the gears, arm, bearings, and tooth numbers. Input, output, and stationary members must be labeled.

6. *Check the design.*
   When the sketch is completed, it is only reasonable to see if the train satisfies the specifications.

7.20 Simple Epicyclic Trains

There appears to be an endless variety of epicyclic trains, but if we consider only simple epicyclics, excluding compound trains, series trains, and combinations of regular trains with epicyclics, there are but two basic types— the *three-gear train* and the *four-gear train*. All other simple trains may be recognized as variations of these two.

Three characteristics of simple epicyclic trains should be observed:

1. *The input and output shafts of all simple epicyclic trains must lie on a single straight line.*
   If these shafts are required to be parallel or at an angle with one another, regular trains must be added (thus forming a combination train).

2. *In simple epicyclic trains, either the input or the output shaft must turn with the arm.*
   The driver and follower gears of the train, used in establishing tooth ratio, must not be confused with the input or output members of the train as a whole. The arm will always be connected to either the input or the output shaft.

3. *In all simple epicyclic trains, either the driver or the follower gear must be stationary.*
   Two of the three speeds in the epicyclic equation must be known if the third is to be determined. Since simple epicyclics have only one input shaft, the other known speed becomes that of the fixed gear, namely, zero. If two input speeds are required (neither one zero) they must be supplied by a regular gear train added to the epicyclic. This constitutes a combination train—not a simple epicyclic.

7.21 The Basic Three-Gear Epicyclic Train

A basic three-gear train is shown in Figure 7.34. The arm is attached to the input shaft A, and gear B turns the output shaft. (These designations are interchangeable according to whether an increase or reduction in speed is required.) Gear C turns freely on the arm and meshes with both B and the stationary annular E. Figure 7.35 shows the equivalent train, using bevel gears.

A variation of the basic train in Figure 7.34 is shown in Figure 7.36. The same gears are used, but B is stationary and the annular E turns the output shaft. The input shaft turns arm A as before. Figure 7.37 shows the equivalent train with bevel gears. (In these diagrams all stationary members are cross-hatched, and moving gears that are shown in section are stippled.)

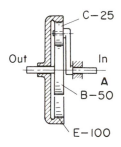

Figure 7.34 Basic three-gear epicyclic train

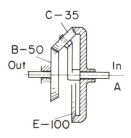

Figure 7.35 Bevel three-gear epicyclic train

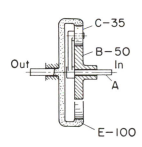

Figure 7.36

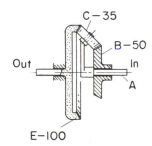

Figure 7.37

The tooth ratios of each of these trains are negative. There is no arrangement of a three-gear epicyclic in which the t.r. is positive except that shown in Figure 7.38. This is a special design of limited application in that gears D and F both mesh with the same planet gear and therefore can differ in size by only one or two teeth. The smaller of these must be mounted at an extended center distance (Article 6.27), which is not usual practice.

A common reason for using epicyclic trains is to obtain a relatively large

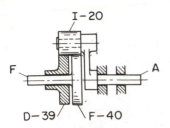

Figure 7.38 Special three-gear epicyclic with extended distance

speed change between input and output shafts with as few gears as possible. It is therefore useful to determine what factors influence the output-input ratio. In Figures 7.34 and 7.35 the larger of the two end gears in the train is fixed. In Figures 7.36 and 7.37 the smaller of the two end gears is fixed.

Let us apply an input speed of $+1$ rpm to A in each case and determine the comparative output speeds. In Figures 7.34 and 7.35, assuming B as follower gear,

$$\frac{\omega_B - 1}{0 - 1} = -\frac{100}{50} \quad \therefore \quad \omega_B = +3 \quad \text{(in the same sense as } A\text{)}$$

The output/input ratio is

$$\frac{\omega_{\text{out}}}{\omega_{\text{in}}} = +\frac{3}{1}$$

In Figures 7.36 and 7.37:

$$\frac{0 - 1}{\omega_E - 1} = -\frac{2}{1} \quad \therefore \quad \omega_E = +1.5 \quad \text{and} \quad \frac{\omega_{\text{out}}}{\omega_{\text{in}}} = +\frac{1.5}{1}$$

This shows that for a given tooth ratio, a larger speed change may be obtained by fixing the larger gear.

Now let us see what influence a change in tooth ratio has on the output speeds. Figure 7.39 shows a three-gear train similar to Figure 7.34, but with a smaller sun gear G, attached to the output shaft. Figure 7.40 shows the bevel gear equivalent.

Applying $+1$ rpm input at the arm as before, with G as the follower gear,

$$\frac{\omega_G - 1}{0 - 1} = -\frac{100}{20} \quad \therefore \quad \omega_G = +6 \text{ rpm} \quad \text{(same sense as } A\text{)}$$

The output/input ratio:

$$\frac{\omega_{\text{out}}}{\omega_{\text{in}}} = +\frac{6}{1}$$

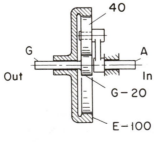

Figure 7.39

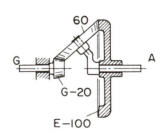

Figure 7.40

As this is much larger than the 3:1 obtained by the trains in Figures 7.34 and 7.35, we conclude that the larger the tooth ratio, the greater the speed change.* Note that, in all of the above examples, the t.r. is negative and the output/input ratio positive. All variations of the three-gear train have not been shown, but none will permit opposite rotation of input and output shafts.

We deduce from these experiments that the largest speed changes obtainable with three-gear epicyclic trains require the following:

1. Making the largest gear stationary.
2. Using the largest possible tooth ratio.

7.22 Design of a Three-Gear Epicyclic Train

Let us consider a design problem to illustrate procedure. A simple epicyclic train is required in which the output/input ratio is $\omega_{out}/\omega_{in} = 12/1$. Output and input shafts are to turn in the same direction, in fixed bearings. All gears are to be the same diametral pitch with the following stock sizes available (in spur, annular, or bevel gears): all tooth numbers from 12 to 24; even numbers from 24 to 128; every fourth number from 128 to 160. We must use the following procedure:

1. *We identify the output and input members of the train.*
   There are two choices:
   a) Input to gear D
      Output to arm A
      Gear F stationary
   b) Input at arm A
      Output at gear D
      Gear F stationary
2. *We calculate the two tooth ratios,* substituting the above identifications in the epicyclic formula: $\dfrac{\omega_F - \omega_A}{\omega_D - \omega_A} = \text{t.r.}$

In case a) with the input at D this yields:

$$\frac{0 - 12}{1 - 12} = +\frac{12}{11} = \text{t.r.} \quad (\text{No. 1})$$

---

*The meaning of "large," or "small," as applied to any ratio, does not necessarily apply to the quotient of the numerator divided by the denominator. The magnitude of a ratio is always obtained by dividing the larger number (numerator or denominator, as the case may be) by the smaller number of the ratio. For example, a ratio of 1:5 is larger than 4:1. The positive or negative sign does not enter into consideration. (A ratio of $-1:3$ is equal in size to $+1:3$.)

In case b) with the input at A:

$$\frac{0-1}{12-1} = -\frac{1}{11} = \text{t.r.} \quad (\text{No. 2})$$

3. *We select the type of train.*

Tooth ratio No. 1 is positive and cannot be achieved by a 3-gear train. Tooth ratio No. 2 is negative, and since it is not too high a ratio to be

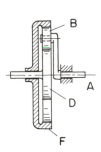

Figure 7.41 Basic three-gear epicyclic

attained with a single pair of gears, let us decide upon a 3-gear train of the basic type shown in Figure 7.41. This requires that the input be at arm A, the output at gear D, and that gear F be a stationary annular. Gear B is the planet gear which functions as an idler, not contributing to the value of the t.r., but is an important factor in establishing its sign. Using the available gears, the largest possible tooth ratio would be

$$\frac{T_D}{T_F} = -\frac{12}{160} = -\frac{1}{13\frac{1}{3}}$$

Since this is larger than the required t.r. (No. 2) of $-1:11$, the selection of the 3-gear train is justified.

4. *Tooth numbers must be selected.*

Since B is an idler gear, D and F establish the (negative) t.r. of $1:11$. Gear D could have 12 teeth and F have 132 teeth. The size of planet gear B is now determined by the space requirements. The pitch diameters of D, B and F are related as follows:

$$\frac{D_F}{2} = D_B + \frac{D_D}{2}$$

And since $D^P = T/P^D$,

$$\frac{T_F}{2P^D} = \frac{T_B}{P^D} + \frac{T_D}{2P^D} \quad \text{or} \quad T_B = \frac{T_F - T_D}{2}^*$$

And substituting:

$$T_B = \frac{132 - 12}{2} = 60$$

6. *A sketch of the train is shown,* in Figure 7.42, giving tooth numbers and bearing locations, and designating input and output shafts.

7. *The design is checked.* Referring to Figure 7.42, we may insert the

---

*All three gears must have the same diametral pitch, so P^D cancels.

input speed of 1 rpm at arm A and zero speed for gear F in the epicyclic formula and solve for ω_D, the output:

$$\frac{0-1}{\omega_D-1} = -\frac{\frac{1}{\cancel{12}}}{\frac{\cancel{132}}{11}}$$

$$-\omega_D + 1 = -11$$

$$\omega_D = +12$$

which is the required output speed and in the same sense as the input at A.

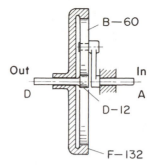

Figure 7.42 Sketch of design of epicyclic train

7.23 The Basic Four-Gear Epicyclic Train

An example of the four-gear epicyclic is shown in Figure 7.43. The input shaft A is connected to the arm and the output shaft to gear C. Gears G and E are carried on the arm and turn together, G meshing with the fixed gear B and E meshing with C. (This forms a reverted train in which B and G must be $5P^D$, while E and C are $3P^D$ to give equal center distances.) The t.r. is positive, $+14:1$, in this train.

Figure 7.44 shows gears of the same tooth numbers and arrangement as in Figure 7.43, except that B is a fixed annular. (All gears can have the same P^D in this case.)

This train has a negative t.r. of $-14:1$.

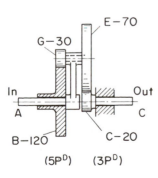

$(5P^D)$ $(3P^D)$

Figure 7.43 Basic four-gear epicyclic train

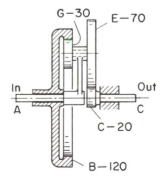

Figure 7.44 Four-gear epicyclic with fixed annular

Figure 7.45 is equivalent to Figure 7.43 except that the small gear C is fixed, and B is the output gear, providing a t.r. of $+14:1$.

Figure 7.46 is equivalent to Figure 7.44 with the small gear C fixed and the annular B as the output gear. The t.r. here is $-14:1$. We note that it is possible to obtain either a positive or negative t.r. if a four-gear basic train is used.

Now to determine what factors influence the output/input ratios of four-gear trains. Let us first consider the relative size of the fixed gear, keeping the t.r. constant. If we apply an input of $+1$ rpm to the arm in each case, we can compare the output speeds.

In the train in Figure 7.43 (assuming C as follower gear for purposes of determining t.r.):

$$\frac{\omega_C - 1}{0 - 1} = \frac{120 \times 70}{30 \times 20} \quad \omega_C = -13 \text{ rpm} \quad \text{(sense opposite to } A)$$

The output/input ratio is:

$$\frac{\omega_\text{out}}{\omega_\text{in}} = -\frac{13}{1}$$

In the train in Figure 7.44 (with gear C as follower):

$$\frac{\omega_C - 1}{0 - 1} = -\frac{14}{1} \quad \omega_C = +15 \text{ rpm} \quad \text{(same sense as } A)$$

$$\frac{\omega_\text{out}}{\omega_\text{in}} = +\frac{15}{1}$$

In Figure 7.45 (with C as follower):

$$\frac{0 - 1}{\omega_B - 1} = +\frac{14}{1} \quad \omega_B = +\frac{13}{14} \text{ rpm} \quad \text{(same sense as } A)$$

$$\frac{\omega_\text{out}}{\omega_\text{in}} = +\frac{0.93}{1}$$

In Figure 7.46 (with C as follower):

$$\frac{0 - 1}{\omega_B - 1} = -\frac{14}{1} \quad \omega_B = +\frac{15}{14} \text{ rpm} \quad \text{(same sense as } A)$$

$$\frac{\omega_\text{out}}{\omega_\text{in}} = +\frac{1.07}{1}$$

From these results we see that (with the t.r. held constant) the output/input ratio was much larger in the trains where the large gears were fixed (Figures 7.43 and 7.44) than where the small gears were stationary (Figures 7.45 and 7.46). Therefore, if a large output/input ratio is wanted, the large gears should be fixed.

Now let us determine the influence of tooth ratio upon output/input ratio. To do this we will increase the t.r. in the trains shown in Figures 7.43 and

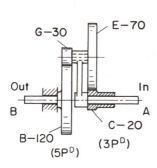

Figure 7.45 Four-gear train with gears of different diametral pitch.

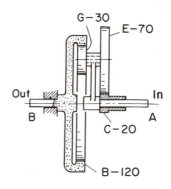

Figure 7.46 Four-gear train with moving annular

7.44, using the same input speed of 1 rpm, and observe the effect on the output speed.

Figure 7.47 shows a train like that in Figure 7.44 except that a 75-tooth gear J has been substituted for E and a 15-tooth gear K used in place of C. If K is the follower gear, the epicyclic equation will be:

$$\frac{\omega_K - 1}{0 - 1} = \frac{120}{30} \times \frac{75}{15} \quad \left(\text{in which the t.r.} = -\frac{20}{1}\right)$$

$$\omega_K = +21 \text{ rpm, (sense same as } A)$$

The output/input ratio:

$$\frac{\omega_{\text{out}}}{\omega_{\text{in}}} = +\frac{21}{1}$$

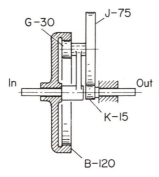

Figure 7.47

We note that in Figure 7.44 the t.r. $= -14:1$ and the output/input ratio was $+15:1$. Thus we see that an increase in t.r. caused as increase in output/input ratio.

Figure 7.48 shows a train like that in Figure 7.43 except that the 75- and 15-tooth gears J and K have replaced gears E and C. Assuming K as follower gear, the equation is:

$$\frac{\omega_K - 1}{0 - 1} = +\frac{20}{1}$$

$$\omega_K = -19 \text{ rpm} \left(\text{sense opposite to } A\right)$$

$$\frac{\omega_{\text{out}}}{\omega_{\text{in}}} = -\frac{19}{1}$$

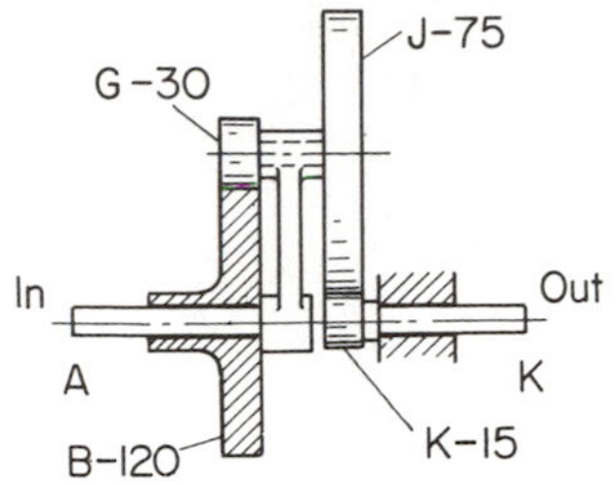

Again, we have increased the t.r. from $+14 : 1$ in Figure 7.43 to $+20 : 1$ and note a corresponding increase in output/input ratio from $-13 : 1$ to $-19 : 1$. The larger t.r. has increased the output/input ratio.

From these experiments we deduce that the largest speed changes obtainable with four-gear epicyclic trains require the following:

Figure 7.48

1. Making the largest gear stationary.
2. Using the largest possible tooth ratio.

7.24 Design of a Four-Gear Epicyclic Train

Suppose it is required to design a simple epicyclic train with an output/input ratio of $\omega_{\text{out}}/\omega_{\text{in}} = 1/55$. The output and input shafts must turn in opposite directions in fixed bearings. All gears are to have the same diametral pitch with the following stock sizes available (in spur, annular, or bevel gears): all tooth numbers from 12 to 24; even numbers from 24 to 128; every fifth number from 25 to 125; every fourth number from 128 to 160. Mating gears are to be mounted on correct center distances (separation is not permitted). The procedure is as follows:

1. *First, we identify the output-input members of the train.*
   There are two choices, each of which will give different tooth ratios:
   a) Input at F
      Output at A
      Gear D fixed
   b) Input at A
      Output at F
      Gear D fixed
2. *Second, the two tooth ratios resulting from these identifications can be calculated* by substituting the specified velocities in the epicyclic formula,

$$\frac{\omega_F - \omega_A}{\omega_D - \omega_A} = \text{t.r.}$$

Since opposite senses of rotation of input and output members is specified, we will insert a negative input value (-55) and a positive output value ($+1$).

In case a) with the input at F:

$$\frac{-55 - 1}{0 - 1} = +\frac{56}{1} = \text{t.r.} \quad \text{(No. 1)}$$

In case b) with the input at A:

$$\frac{+1 + 55}{0 + 55} = +\frac{56}{55} = \text{t.r.} \quad \text{(No. 2)}$$

3. *Next, the type of train must be selected.*
   A 3-gear train cannot be used, since both of the t.r. values above are positive, and the standard form of the 3-gear train always has a negative t.r. (separation is not permitted here, so the special form of the 3-gear train is ruled out). Accordingly, we must use that form of the 4-gear train which has a positive t.r., as shown in Figure 7.49.

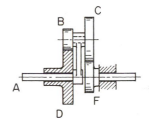

Figure 7.49 Basic four-gear epicyclic

4. *Now, tooth numbers must be determined.*
   Since the chosen tooth numbers must yield the correct t.r., the first step is to decide which of the two t.r. values will provide the best gear-train design. Ordinarily this decision is based upon experimenting with both t.r. values to see which one is most compatible with the tooth numbers of available gears. A solution which utilizes smaller gears (low tooth numbers) is preferred to that involving large gears if interference is avoided. Also a solution in which the planet gears (carried by the arm) are relatively small is desirable.

Let us first investigate t.r. No. 1, which is $+56:1$. In a 4-gear train, there are two pairs of gears contributing to the tooth ratio. Referring to Figure 7.49:

$$\text{t.r.} = \frac{T_D}{T_B} \times \frac{T_C}{T_F} = +\frac{56}{1}$$

First we must factor 56 into two numbers which are preferably nearly equal. Here the obvious selection is 8×7, and accordingly T_D/T_B may be $8:1$ and T_C/T_D may be $7:1$ as far as numerical ratios are concerned. Regardless of available tooth numbers, it is best to work with small numbers at this point. Second, since we are using a reverted train, the center distances for each pair of gears must be equal. Here all gears are of the same diametral pitch so, if $T_D + T_B = T_C + T_F$, this requirement will be satisfied. Using the array suggested for the design of reverted trains,

MULTIPLY

$$\text{ADD} \downarrow \quad \dfrac{\dfrac{T_D}{T_B} \times \dfrac{T_C}{T_F}}{N \; = \; N} = \text{t.r.} \quad \text{we have} \quad \dfrac{\dfrac{8}{1} \times \dfrac{7}{1}}{9 \; \neq \; 8} = \dfrac{56}{1}$$

If we let $N = 72$ (the product of 9 and 8) this number is divisible into 9 parts of 8 each and 8 parts of 9 each, so in the pattern:

$$\dfrac{8 \times 8}{1 \times 8} \times \dfrac{7 \times 9}{1 \times 9} = \dfrac{\dfrac{64}{8}}{72} \times \dfrac{\dfrac{63}{9}}{72} \qquad \text{there are no 9-tooth gears so we multiply all through by 2}$$

$$\dfrac{\dfrac{128}{16}}{144} \times \dfrac{\dfrac{126}{18}}{144} = \dfrac{56}{1} \qquad \text{we find that all of these tooth numbers are available}$$

So $T_D = 128$, $T_B = 16$, $T_C = 126$, and $T_F = 18$.

5. *Next, we sketch the train,* giving all neccessary data, as shown in Figure 7.50.

6. *To check this solution,* with repuired input speed of F equal to -55, we solve for the output at A in the epicyclic formula:

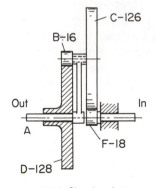

B–16

C–126

Out   In

A

F–18

D–128

$$\dfrac{-55 - \omega_A}{0 - \omega_A} = \dfrac{\overset{8}{\cancel{128}}}{\underset{1}{\cancel{16}}} \times \dfrac{\overset{7}{\cancel{126}}}{\underset{1}{\cancel{18}}} = \dfrac{56}{1}$$

$$-55 - \omega_A = -56\omega_A \quad \text{and} \quad \omega_A = +1$$
$$\text{(as specified)}$$

Figure 7.50 Sketch of design of four-gear epicyclic train

Now, let us investigate t.r. No. $2 = +56 : 55$, in which the input is at A and output at F. Since this t.r. is also positive, we refer to the 4-gear form in Figure 7.49 again. The ratio $56 : 55$ can be factored into several pairs of fractions, such as:

$$\dfrac{7}{5} \times \dfrac{8}{11} \quad \text{or} \quad \dfrac{56}{1} \times \dfrac{1}{55} \quad \text{or} \quad \dfrac{28}{11} \times \dfrac{2}{5}$$

To satisfy the reverted train requirements:

MULTIPLY

$$\text{ADD} \downarrow \quad \dfrac{\dfrac{T_D}{T_B} \times \dfrac{T_C}{T_F}}{N \; = \; N} = \text{t.r.} \quad \text{we may have} \quad \dfrac{\dfrac{7}{5} \times \dfrac{8}{11}}{12 \; \neq \; 19}$$

or

$$\frac{\dfrac{56}{1} \times \dfrac{1}{55}}{57 \neq 56} \quad \text{or} \quad \frac{\dfrac{28}{11} \times \dfrac{2}{5}}{39 \neq 7}$$

In the first case, the least common multiple of 12 and 19 is their product 228, so the transformed array would be:

$$\frac{\dfrac{133}{95} \times \dfrac{96}{132}}{228 = 228}$$

Using the above figures as tooth numbers is not possible because a 133-tooth gear is not available and the terms of the two ratios are not divisible by any one number. Also all gears are very large. We discard this combination.

Trying $\frac{56}{1} \times \frac{1}{55}$, the L. C. M. of 57 and 56 is 3192. The transformed array:

$$\frac{\dfrac{3136}{56} \times \dfrac{57}{3135}}{3192 = 3192}$$

again contains ratios that cannot be reduced to available gear sizes (57 is a prime number).

The third possibility, $\frac{28}{11} \times \frac{2}{5}$, involves the L. C. M. of 39 and 7, which is 273 and the corresponding array

$$\frac{\dfrac{196}{77} \times \dfrac{78}{195}}{273 = 273}$$

again cannot be reduced to the range of available tooth numbers.

Since this investigation has failed to yield any solution, we should abandon it. (In fact experience indicates that tooth ratios, in which the numerator and denominator are 2-digit numbers and are also consecutive numbers, rarely yield a solution if separation is forbidden.)

7.25 Design with Extended Mounting (Separation)

If it were permissible to use separation in the problem of the previous article, a simple solution is possible when the input is applied to the arm A and the output to gear F. This yields the $+56 : 55$ t.r. which we investigated unsuccessfully in the example above (where separation was not acceptable.) Since separation is permitted, a three-gear train of the type shown in Figure 7.51 may be used, as this has a positive t.r. $= T_D/T_E \times T_E/T_F$ (T_E cancels out, since E functions as an idler). The tooth ratio,

$$\frac{T_D}{T_F} = \frac{56}{55}$$

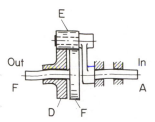

Figure 7.51 Three-gear epicyclic with separation

and T_E may be any convenient number as far as t.r. is concerned. Since gears of 55 and 56 teeth are available, these may be used for gears D and F. Gears D (56 teeth) and E will be mounted at the normal center distance. Gears F (55 teeth) and E will be mounted at this same distance, which is greater than normal for them. For the most compact design, E should be the smallest gear which will run with F at this mounting distance without causing interference or intermittent contact. The number of teeth on E would be established by studies like those conducted in Articles 6.18 and 6.24. In this case it will be found that E can have 20 teeth without any risk of interference or intermittent contact. The solution of this problem is shown in Figure 7.52. Aside from the presence of separation, this design is far more desirable than the four-gear train, since only three gears are used with a shorter arm.

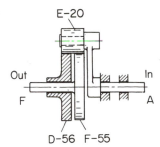

Figure 7.52 Sketch of alternate design

PROBLEMS

7.1. A four-gear speed changer unit is shown Figure P7.1 composed of helical gears. Assume that this unit has the tooth numbers noted on the diagram. Shaft D is connected to the motor and carries the 15-tooth pinion. The 16-tooth pinion and the 60-tooth gear both turn with the intermediate shaft and, the 48-tooth gear turns the output shaft F. If the motor runs at 1800 rpm, determine the speed of F. State direction of rotation relative to D.

7.2. A commercial speed changer is shown in Figure P7.2 composed of 6 herringbone gears with tooth numbers given on the sketch. All gears turn with the shafts upon which they are mounted. If input shaft D turns 3600 rpm, determine the speed of the output shaft F. Report sense relative to D.

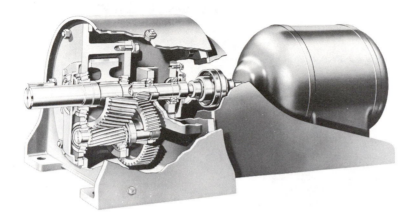

Figure P7.1 (Courtesy of The Falk Corp., Milwaukee, Wis.)

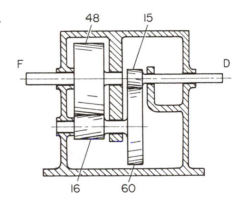

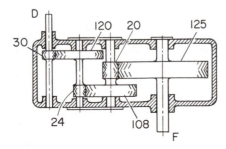

Figure P7.2

7.3. A four-stage spur gear speed changer of stock design is shown in Figure P7.3. All gears turn with the shafts upon which they are mounted. All shafts turn in fixed bearings. Tooth numbers of the 8 gears are noted. If an input speed of 1800 rpm is applied to shaft D, what is the speed of shaft F? Do they turn in the same or opposite sense?

Figure P7.3 (Courtesy of Abart Gear & Machine Co., Chicago, Ill.)

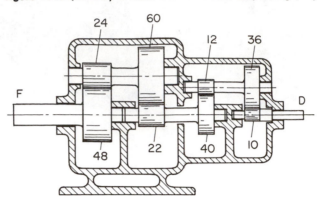

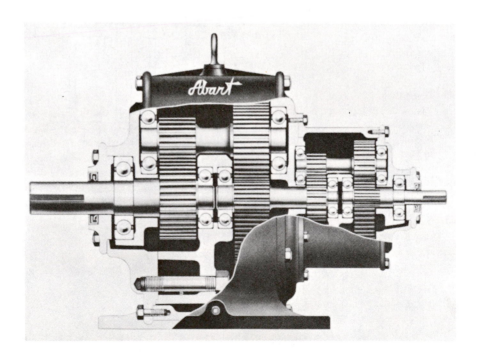

7.4. The gear train shown in Figure P7.4 has 2 output shafts F and G. All bearings are fixed. Gears turn freely on shafts shown dotted, but in all other cases, gears are attached to the shafts upon which they are mounted. If the input shaft D turns 1000 rpm, determine the speeds of F and G. Do F and G turn in the same or opposite directions?

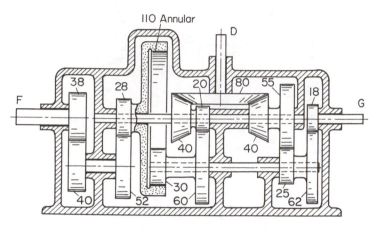

Figure P7.4

7.5. A shaft-mounted speed reducer is shown in Figure P7.5. The housing is mounted on bearings which carry the hollow output shaft F and is prevented from turning by the tie rod. The hollow shaft F is keyed to the shaft of the driven machine upon which it is mounted. Input shaft D turns pinion A which in turn drives gear B. Gear B is keyed to the intermediate shaft, and pinion C is cut solid on the intermediate shaft. C drives E, which is keyed to the hollow shaft F. If D is driven at 1800 rpm and the gears have the tooth numbers shown, find the speed of shaft F.

7.6. A six-gear reverted train is shown in Figure P7.6. Gears B and C turn together, and E and G turn together. Gear A has 50 teeth and a diametral pitch of 6. B has 16 teeth and a $P^D = 8$. C has 75 teeth and a $P^D = 10$. E has 10 teeth. If shaft D turns 100 rpm, find the speed of F and state sense relative to D.

7.7. Design a regular gear train in which the ratio of the output to the input speed is exactly 405:1, with both shafts turning in the same direction. Use as few gears as possible selected from the following stock sizes: every tooth number from 12 to 24; every even number from 24 to 128; every fifth number from 25 to 125. All gears are to be mounted on fixed axes and speed ratios of each pair are to be as uniform as possible. Sketch the train, giving all tooth numbers and labeling input and output shafts.

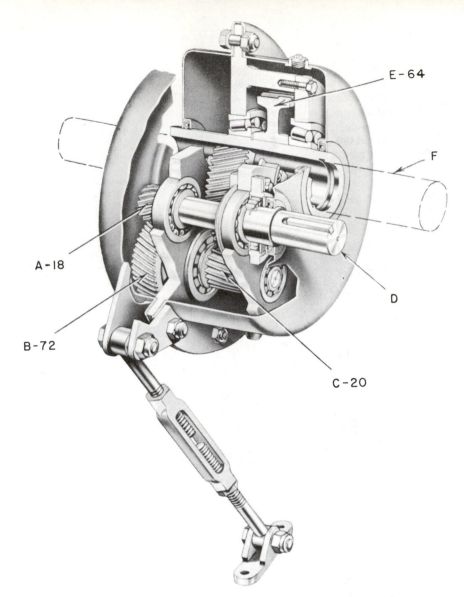

Figure P7.5 (Courtesy of The Falk Corp., Milwaukee, Wis.)

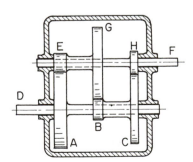

Figure P7.6

7.8. Design a speed reducer in which the ratio of output to input speeds is exactly 1 : 176. The driver and follower shafts are to be at right angles to one another, as shown in Figure P7.8. They need not lie in the same plane, nor is direction of rotation restricted. Use as few gears as possible, selecting from the following stock sizes (in spur, bevel, and annular gears): every tooth number from 12 to 20; every even number from 20 to 100; every fifth number from 20 to 120. All gears are to be of the same diametral pitch and mounted on fixed axes. Worm gears are not to be used. The unit is to be as compact as possible, with reasonably uniform speed ratios in each gear pair. Sketch the train in a case, labeling all tooth numbers and indicating input and output shafts.

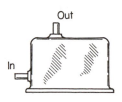

Figure P7.8

7.9. Design a reverted gear train in which the output to input speed ratio is 38 : 1. Select from the following stock gears: every tooth number from 12 to 24; every even number from 24 to 160; every third number from 100 to 160. Sketch the train, showing bearings and tooth numbers and labeling output and input shafts.

7.10. Design a reverted gear train for a reduction in speed of 20 : 1. The input gear and its mate are to be 10 diametral pitch, and the other pair are to be 8 diametral pitch. Use only stock gears as follows: every tooth number from 12 to 24; every even number from 24 to 160. Make a sketch of the train, giving tooth numbers, diametral pitches, and distance between shafts. Label input and output shafts.

7.11. Design a speed reducer having six gears mounted on two parallel shafts, as shown in Figure P7.11. The output to input speed ratio is to be 1 : 120. Gears *A* and *B* are 10 diametral pitch. Gears *C* and *E* are 8 diametral pitch. Gears *G* and *H* are 6 diametral pitch. Use only stock gears of the following tooth numbers: every number from 16 to 24; every even number from 24 to 100; every fifth number from 25 to 100; every fourth number from 100 to 172. Sketch the train, labeling tooth numbers and input and output shafts.

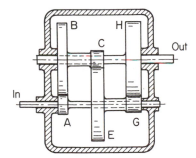

Figure P7.11

7.12. Design a selective-speed drive in which the input shaft turns at a constant speed of 3600 rpm. The output shaft is to have available speeds as follows: 600,

2000, and 2700 rpm, all in the same direction. These shafts are to turn in fixed bearings and are to be parallel to one another. Use only the following stock gears: every tooth number from 12 to 30; every even number from 30 to 100; every fifth number from 100 to 160. Changing speeds with the gears in motion is not required. Sketch the mechanism in position for the 600-rpm output speed, giving tooth numbers and labeling input and output shafts.

7.13. Design a two-stage worm gear speed reducer of the type shown in Figure P7.13. Input and output shafts must be parallel and must turn in opposite directions with a reduction ratio of 1:1500. Stock worms are available with single, double, triple, and quadruple threads, right-hand or left-hand. Worm gears for the worms above may be purchased in every tenth tooth number from 20 to 100. Sketch the reducer, showing arrangement, tooth numbers, and specifications and labeling the input and output shafts.

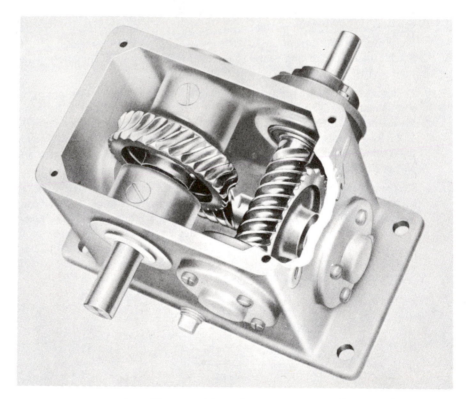

Figure P7.13 (Courtesy of Abart Gear & Machine Co., Chicago, Ill.)

7.14. A simple three-gear epicyclic train is shown in Figure P7.14. Annular G is fixed. Gear J turns freely on K and meshes with both G and H. The input shaft is attached to K and the output to H. If the input speed is 100 rpm, determine the output speed. Do input and output shafts turn alike or opposite?

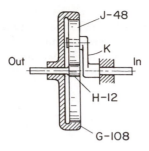

Figure P7.14

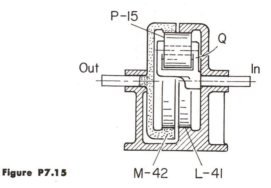

Figure P7.15

7.15. In Figure P7.15, annular L is cut on the frame of the machine, and annular M turns the output shaft. Gear P meshes with both M and $L*$ and turns freely on a shaft supported by Q, which is attached to the input shaft. If the input speed is 1800 rpm, determine the output speed and report direction of rotation relative to the input.

7.16. The bevel gear epicyclic shown in Figure P7.16 has a fixed annular C. The input shaft turns gear E. Two identical planet gears B, turning freely on A, are used to balance the mechanism. The planet gears mesh with C and E. The output is connected to A. If the input speed is 1400 rpm, determine the output speed and sense of rotation relative to input.

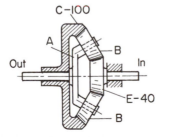

Figure P7.16

*P and L are mounted at a slightly extended center distance.

7.17. The speed changer shown in Figure P7.17 combines a regular with an epicyclic train. Gears S and P turn with the input shaft. Gear T turns on a fixed stud and meshes with S and annular M. R meshes with P and M, which turns in fixed bearings, independently of W. The output shaft is attached to W. If the input speed is 3600 rpm, find the output speed and sense of rotation relative to the input.

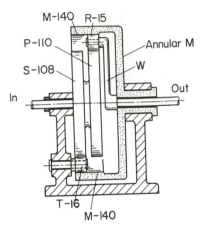

Figure P7.17

7.18. In the four-gear epicyclic train in Figure P7.18 annular gear U is stationary. T is attached to the output shaft and gear S to the input. Gears Q and R are keyed to the same shaft, which turns freely in T. What is the output to input speed ratio? Indicate directions of rotation.

Figure P7.18

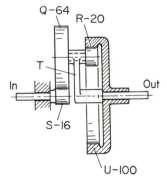

7.19. The speed changer shown in Figure P7.19 combines a regular with an epicyclic train. The input is at gear A and the output at gear H. R and gear B turn together in fixed bearings. Gear E is attached to the case. Gears C and G turn together in the bearing in R. If the input speed is 3600 rpm, what is the output speed? State relative directions of rotation.

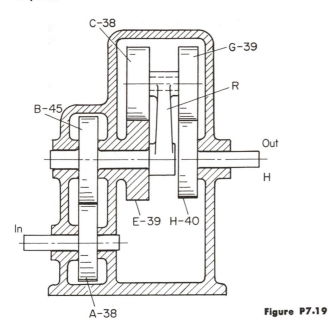

Figure P7.19

7.20. The bevel train shown in Figure P7.20 has the input at gear L and the output at N. Gears P and M turn together. What is the output to input speed ratio? Do the shafts turn in the same or opposite directions?

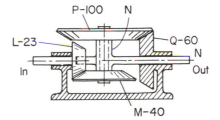

Figure P7.20

7.21. A selective four-speed drive is shown in Figure P7.21. Input or output may be applied at V-belt pulleys attached to B, A, and E. Brakes (not shown) may be applied to the outer surfaces of B, A, or E, so as to hold any of these stationary as desired. An annular gear B is cut on the inner surface of pulley B. Gear C turns freely on a stud carried by A and meshes with gears B and E. Gear E and pulley E are attached to the same shaft, which is supported by fixed bearings (stippled) at the right and by bearings in the center of B at the left. This shaft also supports A, but both B and A turn independently upon it. B turns in fixed bearings at the left (stippled).

If the input speed is constant at 1000 rpm and is always in the same direction, determine the output speeds under the following conditions (name the output member in each case and state its direction of rotation relative to the input):

(a) *B* stationary, input at *A*.

(b) *A* stationary, input at *B*.

(c) *A* stationary, input at *E*.

(d) *E* stationary, input at *B*.

Figure P7.21

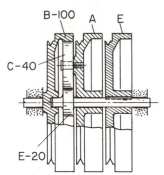

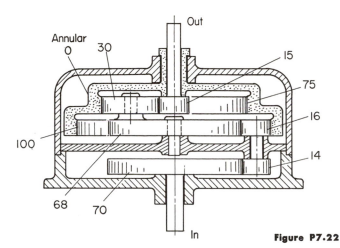

Figure P7.22

7.22. In the speed changer shown in Figure P7.22, the 14- and 16-tooth gears are attached to the same shaft. The 30-tooth gear turns freely on a stud which is mounted in the rim of the 68-tooth gear. Two annular gears are cut in the same piece *O* (shown in section), which turns freely in its bearing in the case. The output shaft turns freely in *O*. All portions of the case are stationary. If the input speed (applied to the 70-tooth gear) is 1800 rpm, determine the output speed (at 15-tooth gear) and report relative directions of rotation.

7.23. The speed reducer shown in Figure P7.23 contains nine gears. The 42- and 40-tooth gears are cut on the same piece and are freely attached to member *M*. The 12- and 38-tooth gears are keyed to the same shaft. The 36-tooth gear and 120-tooth annular are cut on the same casting. The 54-tooth gear meshes with both the 120- and 12-tooth gears. If the input speed is 3000 rpm, find the output speed and relative direction of rotation.

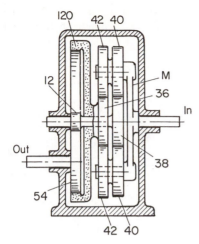

Figure P7.23

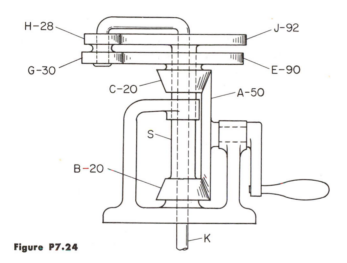

Figure P7.24

7.24. A hand-driven *centrifuge* is shown in Figure P7.24. Gear *A* is attached to the crank. Gears *B* and *J* are fastened to the hollow shaft *S*. Gears *C* and *E* are on the same piece and rotate freely on shaft *S*. Gears *G* and *H* are also cut on one piece. The arm *K* turns freely at all points. If the crank is turned at 40 rpm, determine the speed of the centrifuge bowl (not shown) which is attached to *K*.

7.25. A *compound epicyclic train* is shown in Figure P7.25, which is composed of bevel gears. Gear *G* is attached to the input shaft and gear *L* to the output. Gears *J* and *K* are fastened to the same shaft, which is supported by the X-shaped member *M*. *M* is mounted to turn freely in bearings in *G* and *L*.

The large bevel gear *H* is stationary. If the input speed is 1000 rpm, what is the output speed and in which direction does the output shaft turn relative to the input?

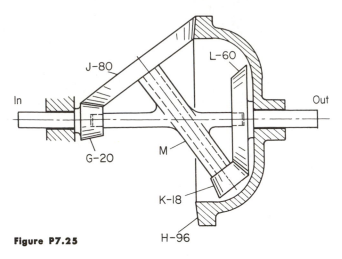

Figure P7.25

7.26. Design an epicyclic gear train for an output to input speed ratio of 9:1, in which output and input shafts turn in the same direction. Use as few gears as possible, selecting from the following stock sizes of spur, bevel, or annular gears: every tooth number from 14 to 24; even tooth numbers from 24 to 144. Separation is not permitted. Sketch the required train, giving all tooth numbers and labeling output and input shafts.

7.27. Design an epicyclic gear train having an output to input speed ratio of 11:1, with output shaft turning opposite to input. Use as few gears as possible (without separation), selecting only from the following sizes (spur, bevel, or annular): every tooth number from 14 to 24; even tooth numbers from 24 to 144. Sketch the train, giving tooth numbers and labeling output and input shafts.

7.28. Design an epicyclic train having an output to input speed ratio of 1:27 with output and input shafts turning in the same direction. Use as few gears as possible (without separation) selecting from the following stock sizes (spur, bevel, and annular): every tooth number from 14 to 30; even numbers from 30 to 144; every fifth number from 30 to 160. Sketch the train, giving tooth numbers and indicating output and input shafts.

7.29. Design an epicyclic train with an output to input ratio of 1:27, with output and input shafts turning in opposite directions. Separation is permitted in effort to use as few gears as possible. Stock gears are available (spurs, bevels, and annulars) in the following sizes: every tooth number from 14 to 70; even numbers from 70 to 160. Sketch the train, giving tooth numbers and indicating output and input shafts.

7.30. Design the best gear train that you can in which the output to input speed ratio is exactly 145 : 1. Output and input shafts must turn in the same direction. Assume that stock gears (spur, bevel, or annular) are available as follows: all tooth numbers from 10 to 30 and even numbers from 30 to 180. Assume that all of these gears will run together properly. Sketch the train, noting tooth numbers and labeling input and output shafts.

7.31. Design a *combination* regular and epicyclic gear train with an output to input speed ratio of 1 : 250, in which the input shaft turns in the same direction as the output shaft. Use as few gears as possible (without separation), chosen from the following sizes (spur, bevel, or annular): every tooth number from 12 to 24; even numbers from 24 to 144; every fifth number from 25 to 120. Sketch the train, showing arrangement and tooth numbers and indicating input and output shafts.

7.32. Design a *two-stage series-type* of epicyclic train, in which the output to input speed ratio is 1 : 1600, with input and output shafts turning in the same direction. Use as few gears as possible (without separation) chosen from the following sizes (spur, bevel, or annular): every tooth number from 12 to 70; even numbers from 70 to 160; every fifth number from 70 to 160. Sketch the train, showing arrangement, giving all tooth numbers and labeling input and output shafts.

7.33. Assume that you have only 4 spur gears consisting of one 78-tooth gear, two 79-tooth gears, and one 80-tooth gear, all of the same diametral pitch. Design a gear train which will give the greatest reduction in speed between the input and output shafts. It is not required that all of the gears be used. It is not important whether the input and output shafts turn in the same or opposite sense. It is intended that the reducer will be run continuously in one direction without reversing. Sketch your design, giving tooth numbers on the sketch and labeling input and output shafts. Report the input speed when the output shaft turns at 1 rpm.

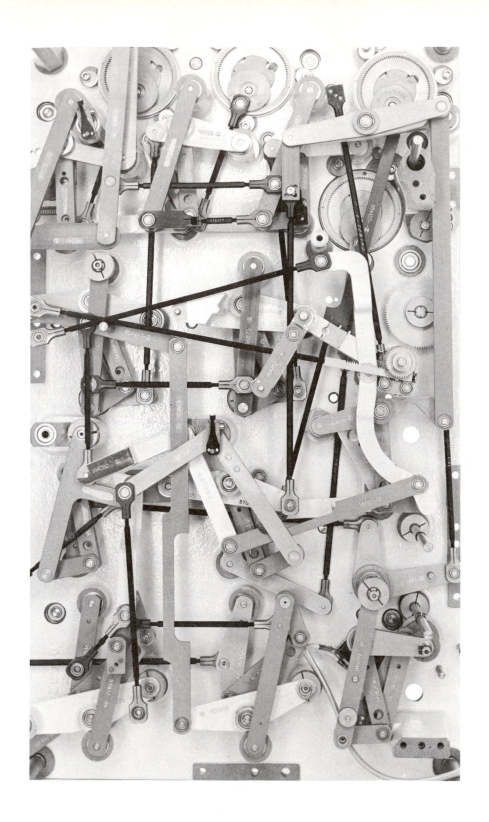

304

8

Linkages

The linkage might be called the universal mechanism, since almost any conceivable motion can be produced by this device. We have already observed many applications of linkages to produce various displacement paths, velocities, and accelerations. In the majority of cases, the linkage is the simplest mechanism which can be devised to provide a given motion. (Perhaps the most significant exception is the cam.) It is most unfortunate that, in general, the design of a linkage to produce a given motion is often very difficult.

In our early work in velocity and acceleration, we confined our study to a single position of a linkage, one which enabled us to obtain instantaneous values. Later we recognized that most machines are designed for continuous operation and we learned to use graphical calculus as a tool to study complete cycles of motion. While it is theoretically possible to differentiate a displacement-time curve graphically to obtain velocity and then differentiate that velocity-time curve to determine the acceleration graph, it has been found that the results of this second differentiation are usually very inaccurate, owing to the magnification of errors by two successive approximations. In this chapter, we propose to develop a fast and accurate method of obtaining velocities throughout an entire cycle of motion. The resulting velocity-time curve can then be differentiated graphically to obtain reliable acceleration values in one step, or it can be integrated to obtain displacements. To this end we will concentrate upon the study of velocities in linkages during complete cycles of motion to round out our abilities in motion analysis.

8.1 The Basic Four-Bar Linkage

The four-bar linkage derives its renown from the fact that the members of a three-bar linkage are incapable of relative motion and a linkage composed of more than four consecutive bars has indeterminate motion with a single input. Though it may assume many forms, often with little resemblance to the usual representation, a four-bar linkage consists of two members in pure rotation about fixed axes, called the *driving crank* and *follower crank;* a *coupler** in combined motion, which joins the moving ends of the cranks; and a fixed *frame*, which establishes the relative position of the stationary crank centers.

Several forms of the basic four-bar linkage are shown in Figures 8.1 to

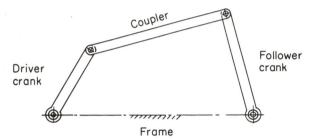

Figure 8.1 Basic four-bar linkage

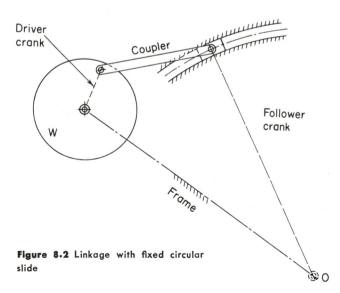

Figure 8.2 Linkage with fixed circular slide

---

*The term *connecting rod* is very commonly used for this member, but *coupler* is preferred here for the specific designation of the linkage element.

8.6.* In Figure 8.2, a block sliding in a fixed circular slot is used to avoid a long follower crank. Point O is the center of the slot, and the dotted line indicates the imaginary crank. Wheel W becomes the driving crank. A circular cam with swinging roller follower in Figure 8.3 is really a four-bar linkage. In Figure 8.4, the eccentric E is a coupler, and strap S is the driving crank. In Figure 8.5, the center of the curved slot is at point P on body M. The member M serves as the driving crank and also establishes the length of the coupler. Figure 8.6 shows a linkage with but two physical members. Centers of curvature of the contacting surfaces are at O and T.

The members of the linkage can best be identified by their motion, regardless of their physical form. The frame must be stationary, the cranks in pure rotation about fixed axes, and the coupler in combined translation and rotation.

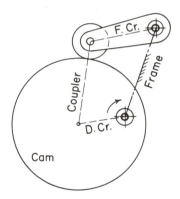

Figure 8.3 Circular cam and follower as a linkage

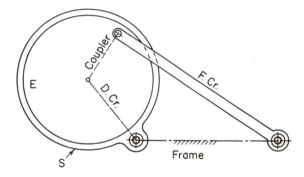

Figure 8.4 Linkage with eccentric

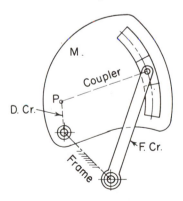

Figure 8.5 Linkage with moving slide

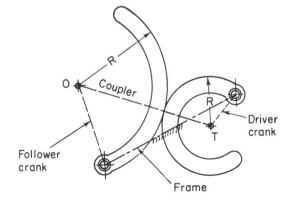

Figure 8.6 Sliding contact linkage

*In all simplified linkage representation the fixed axes will be shown by two concentric circles ⊕ . Crosshatching denotes stationary surfaces.

8.2 Angular Speed Ratio of Cranks

In the velocity analysis of linkages, the angular speed ratio of the follower crank to the driver crank is a most significant value. We can readily determine this, using the familiar velocity vector method, but, for studies of complete motion cycles, this involves drawing too many vectors, so let us attempt to simplify.

Figure 8.7 shows a typical four-bar linkage with driver crank *AB*, follower crank *CD*, and coupler *BC*. *A* and *D* are fixed axes. The linear velocity of *B* is perpendicular to *AB*, and the velocity of *C* is perpendicular to *CD*. The in-

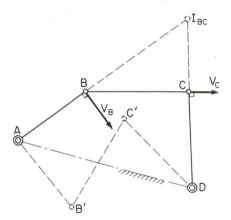

Figure 8.7 Velocity analysis using instant center of coupler

stant center of rotation of *BC* lies at *I*, the intersection of perpendiculars to V_B and V_C, drawn from *B* and *C*. As these velocities are directly proportional to the distances of *B* and *C* from *I*:

$$\frac{V_C}{V_B} = \frac{IC}{IB}$$

But

$$V_C = \omega_{CD} \times CD \quad \text{and} \quad V_B = \omega_{AB} \times AB$$

Substituting:

$$\frac{\omega_{CD} \times CD}{\omega_{AB} \times AB} = \frac{IC}{IB} \quad \text{and} \quad \frac{\omega_{CD}}{\omega_{AB}} = \frac{AB}{CD} \times \frac{IC}{IB}$$

This equation for the speed ratio only requires drawing two lines to locate I_{BC}; no vectors need be drawn. *AB* and *CD* are constants. There is only a practical objection to the use of this equation: the instant center *I* will not be accessible in all positions of the mechanism. In the position *AB′C′D* (shown dotted), *I* falls entirely beyond reach, and neither *IB′* nor *IC′* could be measured.

This difficulty can be overcome by substituting two other lines for *IB* and

IC, which are in the same ratio to one another. If we draw a line from *A*, parallel to the coupler *BC*, meeting *CD* at *K*, then

$$\frac{CK}{AB} = \frac{IC}{IB} \quad \text{(see Figure 8.8)}$$

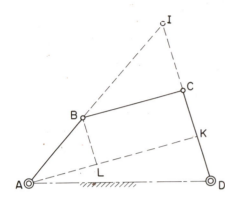

Figure 8.8 Velocity analysis using similar triangles

To prove this, we draw a line from *B*, parallel to *CK*, meeting *AK* at *L*. Then triangle *IBC* is similar to *BAL*, since their corresponding sides are parallel. Now

$$\frac{IC}{IB} = \frac{BL}{AB} \quad \text{(In similar triangles corresponding sides are proportional.)}$$

But, in the parallelogram *BLKC*, *BL* = *CK*, so

$$\frac{IC}{IB} = \frac{CK}{AB}$$

Now, substituting *CK/AB* for *IC/IB* in the original equation:

$$\frac{\omega_{CD}}{\omega_{AB}} = \frac{AB}{CD} \times \frac{CK}{AB}$$

and canceling *AB*:

$$\frac{\omega_{CD}}{\omega_{AB}} = \frac{CK}{CD}$$

This is simpler than the original equation, since only one line (*AK*) need be drawn and only one measurement (*CK*) is required (*CD* is a constant length in all positions).

If point *K* becomes remote, as it is in the position shown in Figure 8.9, we note that *I* is accessible, so the original equation

$$\frac{\omega_{CD}}{\omega_{AB}} = \frac{AB}{CD} \times \frac{IC}{IB}$$

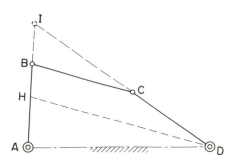

Figure 8.9 Crank speeds are (inversely) proportional to crank and intercept

can be used. Another scheme would be to draw DH parallel to BC, instead of AK, and by reasoning similar to that above, $IC/IB = CD/BH$. Then

$$\frac{\omega_{CD}}{\omega_{AB}} = \frac{AB}{CD} \times \frac{CD}{BH}$$

and canceling CD:

$$\frac{\omega_{CD}}{\omega_{AB}} = \frac{AB}{BH}$$

in which AB is constant and only BH need be measured.

Thus, *to obtain the speed ratio of the cranks of a four-bar linkage, we draw a parallel to the coupler through either fixed center and measure the intercept on the opposite crank between the coupler and the parallel. The ratio of the crank speeds is (inversely) equal to the ratio of this intercept to the full length of the same crank.* As with all ratios, it is most important to get this ratio "right side up." Here we have a sort of inverse ratio. If crank length CD is used, it appears in the denominator of the ratio of lengths, since ω_{CD} appears in the numerator of the speed ratio. If crank length AB is used, it appears in the numerator, since ω_{AB} is in the denominator. A crank speed and the length of that crank always occupy opposite terms of the ratios:

$$\frac{\omega_{CD}}{\omega_{AB}} = \frac{CK}{CD} \qquad \frac{\omega_{CD}}{\omega_{AB}} = \frac{AB}{BH}$$

Given the speed of the driver crank, we can now write an equation for the speed of the follower crank of any recognizable four-bar linkage, which will apply in all positions. For example, if the cam in Figure 8.10 turns at a constant speed of 100 rpm, the speed of the follower crank will be:

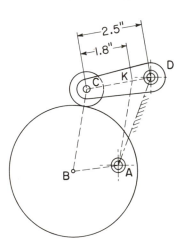

Figure 8.10 Angular speed of cam follower

$$\omega_{CD} = \omega_{AB} \times \frac{CK}{CD}$$

Since ω_{AB} and CD are both constants, ω_{AB}/CD may be called a constant factor F. In this problem, $\omega_{AB} = 100$ rpm, and crank $CD = 2.5$ in. So $F = 40$. If CK measures 1.8 in. in the position shown,

$$\omega_{CD} = 40 \times 1.8 = 72 \text{ rpm}$$

8.3 The Crank and Slider Linkage

The linkage shown in Figure 8.11 is a form of four-bar linkage which has so many applications that it deserves special attention. If the follower crank in Figure 8.2 were lengthened until it was infinitely long, the circular fixed slot would become straight, thus forming a *crank and slider linkage* like Figure 8.11. The follower crank would extend from C perpendicular to the slot to infinity, and the frame, in order to meet the follower crank, would have to extend from A to infinity in the same direction (perpendicular to the slot).

The crank and slider mechanism is used in piston engines, pumps, saws, and other types of reciprocating machinery.

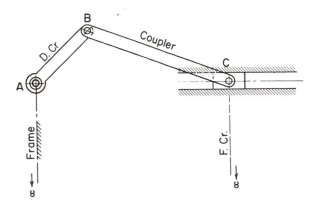

Figure 8.11 Crank and slider linkage

8.4 Velocity Analysis of Crank and Slider

In this mechanism, the follower crank does not rotate, so it is the linear velocity of the slider which is of interest. Let us work out an equation for this which will avoid drawing velocity vectors.

In Figure 8.12 we know that the velocity of B is perpendicular to AB and that the velocity of C lies along its path, the center line of the fixed slot. The instant center of the coupler BC will lie at I, the intersection of perpendiculars to V_B and V_C. Then

$$\frac{V_C}{V_B} = \frac{IC}{IB} \quad \text{and} \quad V_C = V_B \times \frac{IC}{IB}$$

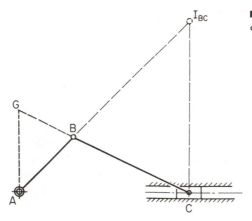

Figure 8·12 Velocity analysis of crank and slider linkage

But

$$V_B = \omega_{AB} \times AB \quad \text{so} \quad V_C = \omega_{AB} \times AB \times \frac{IC}{IB}$$

in which ω_{AB} and AB are constants. The same objection is raised to this equation as in the former case: the instant center is inaccessible in many positions of the mechanism. We effect a remedy by finding an alternate ratio equivalent to IC/IB. If we draw a line from A parallel to IC to meet BC (extended) at G, triangle AGB will be similar to triangle ICB, since their corresponding sides are parallel. Then $IC/IB = AG/AB$ (corresponding sides are proportional), and substituting in the former equation:

$$V_C = \omega_{AB} \times AB \times \frac{AG}{AB} \quad \text{or} \quad V_C = \omega_{AB} \times AG$$

Again, we have eliminated one measurement, so V_C equals a constant (ω_{AB}) multiplied by the length AG. Since line AG is parallel to IC and IC is perpendicular to the fixed slot, we note that AG is always drawn perpendicular to the direction of the slot. Also it must be noted that in the equation, $V_C = \omega_{AB} \times AG$, a linear velocity is equated to the product of an angular velocity and a length. As in the equation $V = \omega \times r$, ω_{AB} must be in radians per second or radians per minute.

An isosceles linkage is a special form of crank and slider mechanism in which AB equals BC, as in Figure 8.13(a). With this linkage, point G will be quite as inaccessible as I in positions such as that shown in Figure 8.13(b). In a case like this, we can draw a parallel to BC intersecting AB (extended) at H and the perpendicular to the guides at J and substitute JC/HB for IC/IB. These ratios can be shown to be equal by drawing HK parallel to JC [Figure 8.13(a)] which forms triangle HKB, similar to ICB. Then $HK/HB = IC/IB$, but $HK = JC$, so $JC/HB = IC/IB$. The equation for V_C will then be:

$$V_C = \omega_{AB} \times AB \times \frac{JC}{HB}$$

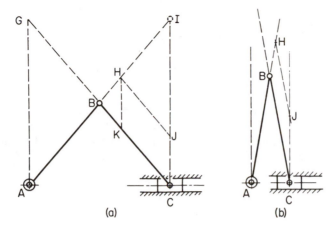

Figure 8.13 Velocity of slider of isosceles linkage

A parallel line *HJ* may be drawn in any convenient position so *JC* and *HB* will always be within reach, as shown in Figure 8.13(b).

8.5 Velocities of Points on the Coupler

If a velocity analysis of any point on a coupler is required, an equation can be devised for determining this graphically. Figure 8.14 shows a four-bar linkage in which the velocity of point *E* on the coupler is to be studied while driver crank *AB* makes a complete revolution.

Using the instant center of coupler *BC*, $V_E/V_B = IE/IB$, so

$$V_E = V_B \times \frac{IE}{IB}$$

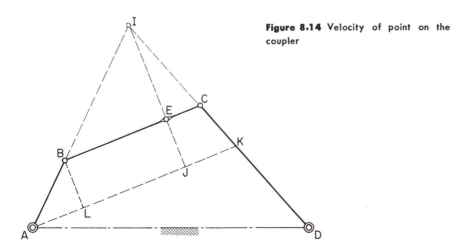

Figure 8.14 Velocity of point on the coupler

But

$$V_B = \omega_{AB} \times AB$$

$$\therefore V_E = \omega_{AB} \times AB \times \frac{IE}{IB}$$

in which ω_{AB} and AB are constants and ω_{AB} is in radians per unit of time.

Since I is inaccessible in some positions, we must find two alternate measurements in the same ratio as IE/IB. A line may be drawn through A, parallel to the coupler BC, meeting CD at K. If IE is extended to meet AK at J and BL drawn parallel to IE, triangle IBE will be similar to BAL. In these triangles $IE/IB = BL/AB$, but $BL = EJ$, so:

$$\frac{IE}{IB} = \frac{EJ}{AB}$$

Substituting in the original equation,

$$V_E = \omega_{AB} \times \cancel{AB} \times \frac{EJ}{\cancel{AB}} \quad \text{or} \quad V_E = \omega_{AB} \times EJ$$

(ω_{AB} in radians per unit of time). If I is inaccessible, this equation requires that we locate point J by other means than extending IE.

It can be shown that $JK/AK = EC/BC$, as follows: In the similar triangles IEC and IJK, $EC/JK = IC/IK$, and in the similar triangles IBC and IAK, $BC/AK = IC/IK$. Then $EC/JK = BC/AK$, since they are both equal to IC/IK.

$$\therefore \frac{EC}{BC} = \frac{JK}{AK} \quad \text{(multiplying both sides by } JK/BC)$$

Thus point J may be located by dividing line AK so that the ratio of JK to AK is equal to the ratio of EC to BC.

This division of line AK may be done very rapidly by the use of *proportional dividers*, shown in Figure 8.15. This instrument may be so adjusted that when the large opening equals BC, the short opening, on the opposite end, equals EC. With the pivot in this position, the large end may be set equal to

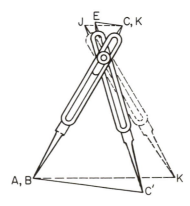

Figure 8.15 Proportional dividers
(Courtesy of Keuffel & Esser Co.)

any length of *AK* with the small end automatically giving the corresponding length of *JK*. Since the proportion of *BC* to *EC* remains the same, whatever position the linkage assumes, one setting of the pivot is correct for the entire problem.

If this instrument is not available, a proportional scale can easily be made. A piece of tracing paper with three parallel lines drawn upon it is shown in Figure 8.16. The distances between the parallel lines are equal (or in proportion) to *BE* and *EC*. The lines should be labeled *A*, *J*, and *K*. To find *J*, we place the paper over the drawing so that line *A* crosses point *A* and line *K* crosses point *K*; then line *J* will cross the line *AK* of the drawing at the proper point *J*—which may be pricked through the tracing paper with a needle point.

Now that we have a method for locating point *J*, *EJ* can be measured and the equation for the linear velocity of *E* ($V_E = \omega_{AB} \times EJ$) can be used in all positions of the linkage. Here again ω_{AB} must be in radians per unit of time.

If point *E* on the coupler does not lie between *B* and *C*, the procedure is the same. To devise an equation for V_E (in Figure 8.17) we first draw *BE* and

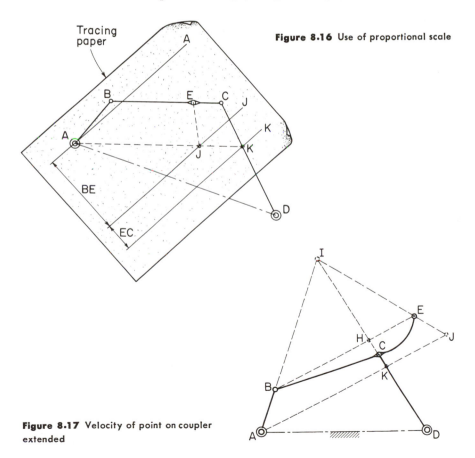

Figure 8.16 Use of proportional scale

Figure 8.17 Velocity of point on coupler extended

then extend *CD* to meet *BE* at *H*. Next *AK* is drawn parallel to *BE*, meeting *CD* at *K*. *J* will fall on *AK* (extended) in such a location that $BH/BE = AK/AJ$ (use proportional dividers or a proportional scale on tracing paper, as in the previous problem). *EJ* may now be drawn and measured.

$$V_E = \omega_{AB} \times EJ$$

If the point *E*, whose velocity is required, is on the coupler of a *crank and slider linkage*, the same method applies. In the example in Figure 8.18, we first draw *BE*, extending it to meet the follower crank extended at *H*. (The follower crank is an imaginary line through *C* perpendicular to the slot, as explained in Article 8.3). Next *AK* is drawn parallel to *BE*, meeting the follower crank at *K*. Point *J* is located on *AK* so that $BH/BE = AK/AJ$ by proportional division, and *EJ* is drawn and measured. Then $V_E = \omega_{AB} \times EJ$ as before.

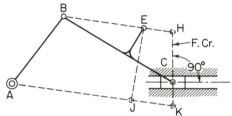

Figure 8.18 Velocity of point on coupler extended

8.6 Linkage Chains

Just as gears are combined in trains, linkages may be connected in series to form *linkage chains*. All linkages, however complex, can be shown to be composed of one or more four-bar units, either of the conventional form or the crank and slider type, depending upon whether the output member is to be in translation or rotation. Figure 8.19 shows a simple linkage chain, in

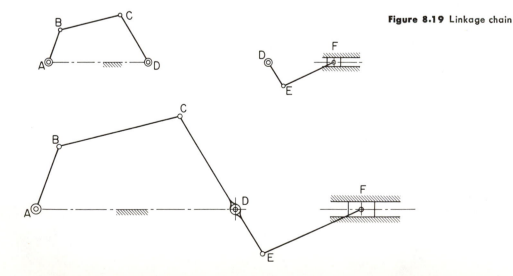

Figure 8.19 Linkage chain

which *ABCD* may easily be identified as the conventional type and *DEF* as a crank and slider.

Equations can be devised for graphical determination of the velocity of the output member in terms of the given velocity at the input and certain measured distances.

8.7 Velocity Analysis of Linkage Chains

Example 1 (Figure 8.20)

In the linkage *ABCD* we draw *AK* parallel to *BC* and measure *CK*. Then (as in Article 8.2)

$$\frac{\omega_{CD}}{\omega_{AB}} = \frac{CK}{CD}$$

We note that, since they are on the same body,

$$\omega_{DE} = \omega_{CD} \quad \text{so} \quad \omega_{DE} = \omega_{AB} \times \frac{CK}{CD}$$

In the crank and slider linkage *DEF*, if we draw *DG* perpendicular to the fixed slot meeting the coupler *EF* (extended) at *G*, then (as in Article 8.3)

$$V_F = \omega_{DE} \times DG$$

If we substitute the value of ω_{DE} above in this equation, $V_F = \omega_{AB} \times CK/CD \times DG$, in which ω_{AB}/CD is a constant,* so V_F is equal to this constant factor multiplied by the product of *CK* and *DG*.

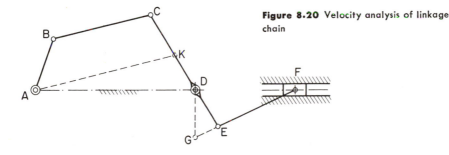

Figure 8.20 Velocity analysis of linkage chain

Example 2 (Figure 8.21)

In the linkage train shown, ω_{AB} is given, and it is required to devise an equation for V_F. The conventional four-bar linkage *ABCD* is easy to recognize, while the second linkage is not as obvious. It is not always necessary to identify the four-bar linkages in order to derive an equation. In some cases this identification is extremely difficult.

*ω_{AB} must be in radians per unit of time.

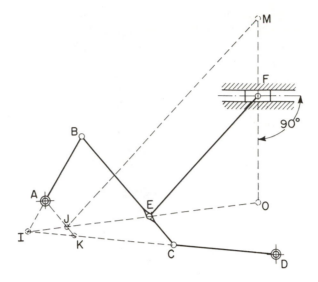

Figure 8.21 Velocity analysis of linkage chain

Therefore let us attack this problem without reference to previous velocity equations. If we locate the instant center of BC, by extending AB and CD to meet at I, then, since E is a point on BC, $V_E/V_B = IE/IB$. As I may become inaccessible, we can draw AK parallel to BC intersecting IE at J, and it has been proved (see Article 8.5) that $IE/IB = JE/AB$.* Substituting in the above equation:

$$\frac{V_E}{V_B} = \frac{JE}{AB}$$

Then

$$V_E = V_B \times \frac{JE}{AB} \quad \text{but} \quad V_B = \omega_{AB} \times AB$$

So

$$V_E = \omega_{AB} \times \cancel{AB} \times \frac{JE}{\cancel{AB}} = \omega_{AB} \times JE$$

Next, we locate O, the instant center of EF. The V_E is perpendicular to IE and V_F in the direction of the fixed slot, so O is at the intersection of IE (extended) and a perpendicular to the slot from F. Then $V_F/V_E = OF/OE$ and $V_F = V_E \times OF/OE$. If we substitute the value of V_E, obtained above, in this equation, then $V_F = \omega_{AB} \times JE \times OF/OE$, which includes three measurements (JE, OF and OE) and the constant ω_{AB}.

This equation can be simplified and the difficulty of an inaccessible O removed. Let us draw a parallel to EF from J and extend it to meet OF ex-

*When I is inaccessible, JE must be found by dividing AK so that $AK/AJ = BC/BE$ using one of the methods of proportional division described in Article 8.5.

tended at M. Now OF/OE can be proved equal to FM/JE (Article 8.5), so we substitute it in the above equation for V_F and:

$$V_F = \omega_{AB} \times \cancel{JE} \times \frac{FM}{\cancel{JE}} \quad \text{or} \quad V_F = \omega_{AB} \times FM$$

This is a simpler equation to use, since only FM need be measured. The construction is not greatly abbreviated, however. To get FM we would proceed as follows (assuming I and O to be inaccessible):

1. Extend CD to meet a parallel to BC through A at point K.
2. Locate J on AK, by dividing AK so that $AK/AJ = BC/BE$.
3. From J draw a parallel to EF to meet a perpendicular to the slot drawn from F at M.
4. Draw FM.

While there are many other linkages for which we could devise velocity equations, the methods explained in the examples above may be applied to all others. It is important that the student learn to apply his basic knowledge of velocity relationships and ingenuity rather than to attempt to memorize formulas or follow worked out examples.

8.8 Procedure for Derivation of Velocity Equations

A step-by-step plan may guide the student to an orderly process of devising these velocity equations. The use of instant centers, substitution in similar triangles, and elimination of terms through cancellation are the principal techniques.

1. *Sketch velocity vectors* on mechanism drawing, showing instant centers of all members in plane motion.
2. *Relate linear velocities* of significant points through ratios of distances to instant centers.
3. *Write a preliminary equation* for the desired value in terms of given data.
4. *Find, or construct, triangles which will be similar* to those which involve distances to instant centers. Look for more accessible triangles, or triangles that have a side of constant length, or triangles that have a common side.
5. *Substitute ratios of sides of these new similar triangles* for the ratios of distances to the instant centers in the original triangles.
6. *Make cancellations and separate the constant terms* of the equation. (The velocity equations of most mechanisms can be reduced to contain but one, two, or three variables.)
7. *Indicate clearly on the drawing* just what construction lines must be drawn and what distances are to be measured in order to solve the equation.

8.9 Linkages with Moving Slides

All of the linkages in the previous articles have had pin joints or sliders moving in fixed guides. In some linkages members are restrained by sliders on *moving guides*. Equations for graphical analysis of velocities can be written for this type of linkage also, so that complete cycles of motion can be studied.

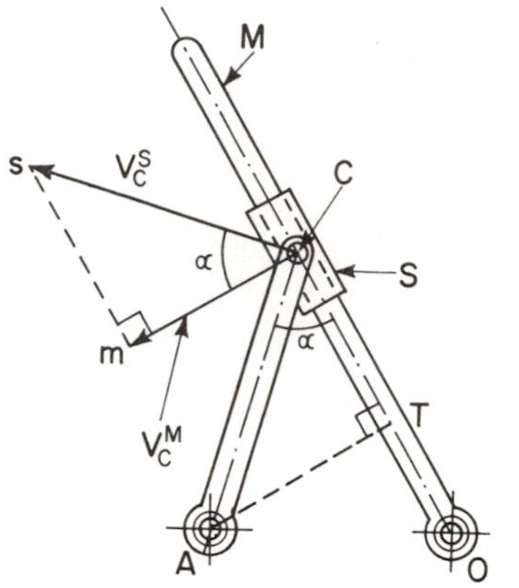

A basic form of the moving slide linkage is shown in Figure 8.22. The straight bar M swings about O. The driving crank AC is pinned at C to a sleeve S, which slides freely on M. If the angular speed of AC is known, let us write an equation for ω_M. As with velocity vector analysis, we can assume that there are two points C, one on sleeve S and one inside the sleeve on bar M. Since there is sliding contact, V_C^M and V_C^S have the same effective component perpendicular to sliding (i.e., perpendicular to bar M). The effective component of V_C^S perpendicular to M is equal to V_C^M, since velocities of all points on M are perpendicular to OM. These vectors are sketched on Figure 8.22, showing that they form a right triangle Cms in which the angle between the vectors is labeled α. As we do not wish to use vectors, let us construct another right triangle similar to Cms. AC is perpendicular to Cs and bar M is perpendicular to Cm, so the angle between AC and M equals α. If we draw a line from A perpendicular to M, intersecting M at T, triangle ACT will be similar to triangle Cms (corresponding angles are equal). Therefore

$$\frac{Cm}{Cs} = \frac{CT}{AC} \quad \text{or} \quad \frac{V_C^M}{V_C^S} = \frac{CT}{AC}$$

But

$$V_C^M = \omega_M \times OC \quad \text{and} \quad V_C^S = \omega_{AC} \times AC$$

So

$$\frac{\omega_M \times OC}{\omega_{AC} \times AC} = \frac{CT}{AC} \quad \text{and} \quad \omega_M = \frac{\omega_{AC} \times AC \times CT}{OC \times AC}$$

$$\text{or} \quad \omega_M = \omega_{AC} \times \frac{CT}{OC}$$

The *Scotch yoke* shown in Figure 8.23 is another example of a linkage with a moving slide. The block B is pinned to AC at C and slides freely in the slot in Y. If ω_{AC} is given, let us evaluate the velocity of the yoke Y. Considering the two points C on the block and on yoke Y, V_C^B and V_C^Y have equal

Figure 8.22 Linkage with moving slide

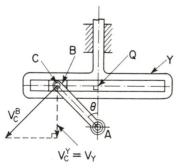

Figure 8.23 Scotch yoke

amounts of velocity perpendicular to the slot. Since V_C^Y is directed perpendicular to the slot, it is equal to the effective component of V_C^B in that direction. If we draw a perpendicular to the slot through A, the triangle ACQ will be similar to the triangle formed by the two vectors (corresponding sides are perpendicular), so $V_C^Y/V_C^B = CQ/AC$. As the yoke is in translation, all points upon it have equal velocities, so $V_C^Y = V_Y$. Also $V_C^B = \omega_{AC} \times AC$, so

$$\frac{V_Y}{\omega_{AC} \times \cancel{AC}} = \frac{CQ}{\cancel{AC}} \quad \text{and} \quad V_Y = \omega_{AC} \times CQ$$

(where ω_{AC} is in radians per unit of time).

The motion of the yoke in this mechanism is called *simple harmonic* or *sinusoidal motion*. We note in the equation above that

$$V_Y = \omega_{AC} \times AC \times \frac{CQ}{AC}$$

In the triangle ACQ, we note that CQ/AC is the sine of angle θ, which is the angular displacement of the crank AC from the vertical position.* Therefore $V_Y = \omega_{AC} \times AC \times \sin\theta$, and since ω_{AC} and AC are constants, it may be said that V_Y varies as the sine of θ (hence *sinusoidal*).

Figure 8.24 shows a linkage in which the slot on the body N is circular with center at point P (similar to Figure 8.5). Body N swings about fixed axis O. Block B is pinned to AC at C and slides freely in the curved slot.

Figure 8.24 Velocity of member with circular slide

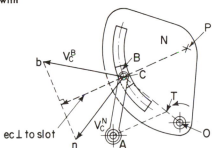

Let us evaluate the angular velocity of N in terms of ω_{AC} and certain measured distances. If we sketch in the vectors for the velocities of points C on the block B and on N, we see that both have the same effective component perpendicular to the slot (in the radial direction PC); V_C^B and V_C^N form two sides of the triangle Cbn. To construct a similar triangle, we note that AC is perpendicular to V_C^B and CO is perpendicular to V_C^N. If we draw a line from A parallel to PC, it will be perpendicular to the third side of the triangle bn

---

*In a right triangle the sine of an angle is equal to the side opposite the angle divided by the hypotenuse.

and will meet CO at T. Thus triangle ACT will be similar to triangle Cbn (corresponding sides are perpendicular). Therefore

$$\frac{Cn}{Cb} = \frac{V_C^N}{V_C^B} = \frac{CT}{AC} \quad \text{but}$$

$$V_C^N = \omega_N \times OC \quad \text{and} \quad V_C^B = \omega_{AC} \times AC$$

So

$$\frac{\omega_N \times OC}{\omega_{AC} \times AC} = \frac{CT}{AC} \quad \text{and} \quad \omega_N = \omega_{AC} \times \frac{CT}{OC}$$

8.10 Quick-Return Linkages

In machine tools and production machinery, there is frequent demand for a long, slow, cutting or feed motion of fairly uniform velocity in one direction, followed by a fast return stroke in which no work is done. The linkage shown in Figure 8.22 is adapted to such requirements and, along with other devices of this sort, is called a *quick-return mechanism*.

A typical example, the *oscillating-beam quick-return linkage*, is shown in Figure 8.25. The slotted beam M swings about fixed axis O driven by the constant velocity crank AC, to which the sliding block B is pinned. A crank and slider linkage OPR changes the oscillating motion into translation of the block R in fixed guides.

Figure 8.25 Oscillating beam quick return

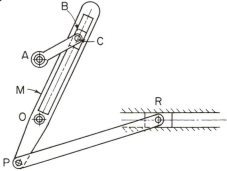

R is in its extreme positions when AC is perpendicular to the slot, as shown in the simplified sketch in Figure 8.26. If the crank rotates clockwise, starting from position AC, to position AC', it turns through $270°$, while R moves from right to left, from R to R'. Continuing clockwise, the crank now swings from AC' to AC, an angle of $90°$, while R moves left to right from R' to R again. If crank AC turns at constant speed, it will take three times as

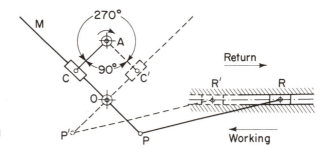

Figure 8.26 Time ratio of working and return strokes

long to traverse 270° as it does to turn 90°. Therefore R travels from right to left at a slow pace and returns from left to right in one-third the time. The working stroke of R will then be from right to left and the fast return from left to right.

Let us derive an equation for the velocity of R, so that a velocity-time plot can be made in preparation for an acceleration study. The velocity of C on M is equal to the effective component of the velocity of C on B in the direction perpendicular to sliding, as shown in Figure 8.27. If AT is drawn perpendicular to M, triangle ACT will be similar to the vector triangle and $V_C^M/V_C^B = CT/AC$. Then, since $V_C^M = \omega_M \times OC$ and $V_C^B = \omega_{AC} \times AC$, we find

$$\frac{\omega_M}{\omega_{AC}} = \frac{CT}{OC} \quad \text{and} \quad \omega_M = \omega_{AC} \times \frac{CT}{OC}$$

In the crank and slider linkage OPR, $\omega_{OP} = \omega_M$, and if we draw OS perpendicular to the fixed slides to meet PR, $V_R = \omega_M \times OS$. Substituting the value of ω_M above:

$$V_R = \omega_{AC} \times \frac{CT}{OC} \times OS$$

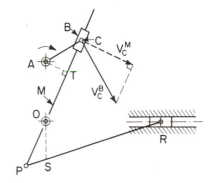

Figure 8.27 Velocity analysis of quick return

In this equation ω_{AC} must be in radians per unit of time. Point T is located by a perpendicular to the beam M through the driving crank center A, and OS is perpendicular to the fixed slide meeting the coupler PR at S. Two lines must be drawn and three measurements made.

If this equation is applied in a series of positions of the linkage and the resulting values of the velocity of R plotted against time, the curve will look like that in Figure 8.28. The velocity during the working stroke is relatively low and uniform, while that of the return stroke (plotted as negative in Figure 8.28) attains a high peak. This curve can be differentiated graphically obtaining an acceleration-time graph.

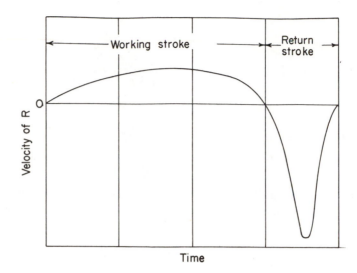

Figure 8.28 Velocity-time curve of quick return slider

Other types of quick-return linkages are shown in Figures 8.29–8.31. In the *Whitworth* (Figure 8.29), both cranks make complete revolutions, providing a compact mechanism for a long stroke of *R*. The oscillating-beam type in Figure 8.30 is used on the Atlas shaper. Here the beam also serves as the coupler to the crank and slider linkage. Figure 8.31 shows a pin-jointed quick return composed of the conventional four-bar linkage and a crank and slider in series. Both cranks make complete revolutions. This is a simple mechanism, but is not as easily adjusted to different strokes and speeds as the others.

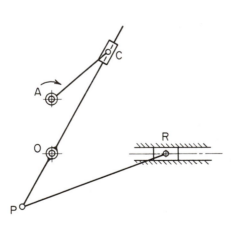

Figure 8.29 Whitworth quick return linkage

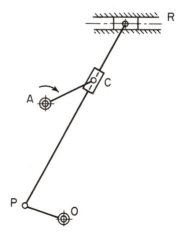

Figure 8.30 Quick return linkage on Atlas shaper

Figure 8.31 Drag-link quick return linkage

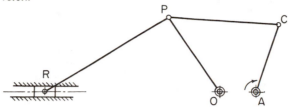

8.11 Design of a Quick-Return Linkage

Suppose it is required to design a quick-return mechanism having a working stroke of 6 in. and a time ratio of the working to the return stroke of 2:1. The driving crank is 1.6 in. long.

Let us first try an oscillating-beam type of mechanism, as shown in Figure 8.32. Here, crank *AC* turns at constant speed about *A*, causing the sleeve to slide along *M*. A pin on the output block at *F* is carried in a yoke on *M*. The extreme positions of *M* are reached when *AC* is perpendicular to *M*, as shown in Figure 8.33. To achieve the 2:1 time ratio the 360° of crank circle must be divided into two angles whose ratio is 2:1. The two positions of

Figure 8.32 Oscillating bean with yoke

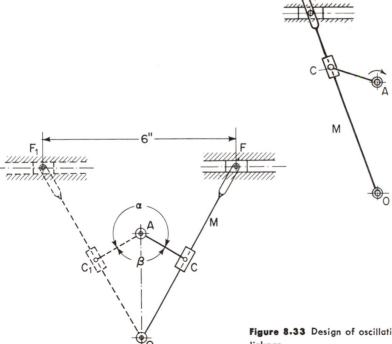

Figure 8.33 Design of oscillating beam linkage

crank AC in Figure 8.33 form sides of these angles, hence angle $\alpha = 240°$ and angle $\beta = 120°$. Since the figure is symmetrical about the center line OA angles OAC and OAC_1 are each equal to $\frac{120}{2}$ or $60°$ and angles AOC and AOC_1 equal $30°$ (triangle ACO is a right triangle).

Starting at A we can lay out the crank circle of 1.6 in. radius and draw the two positions of M tangent to this circle at $30°$ with the center line. These will intersect the center line at O. Lines OC and OC_1 are extended until FF_1 equals the 6-in. stroke. Since the short side and the hypotenuse of the $30°$–$60°$ right triangle ACO are in the ratio of 1:2, the center distance AO will equal twice AC, or 3.2 in. The triangle OFF_1 is equilateral, so OF will equal 6 in.

The same problem may be solved using a Whitworth quick-return linkage, as shown in Figure 8.34. The extreme positions of F are attained when the mechanism assumes the positions shown in Figure 8.35. The crank positions AC and AC_1 are $120°$ apart, since angle $\alpha = 120°$ as before. Angles OAC and OAC_1 are both $60°$, so triangle ACO is a $30°$–$60°$ right triangle in which AO equals $1.6/2 = 0.8$ in. The stroke of F equals twice the length of crank OP, so OP must be 3 in. long. Coupler PF may be any convenient length greater than OP.

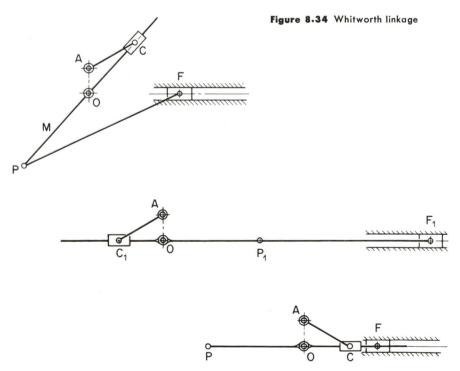

Figure 8.34 Whitworth linkage

Figure 8.35 Design of Whitworth linkage

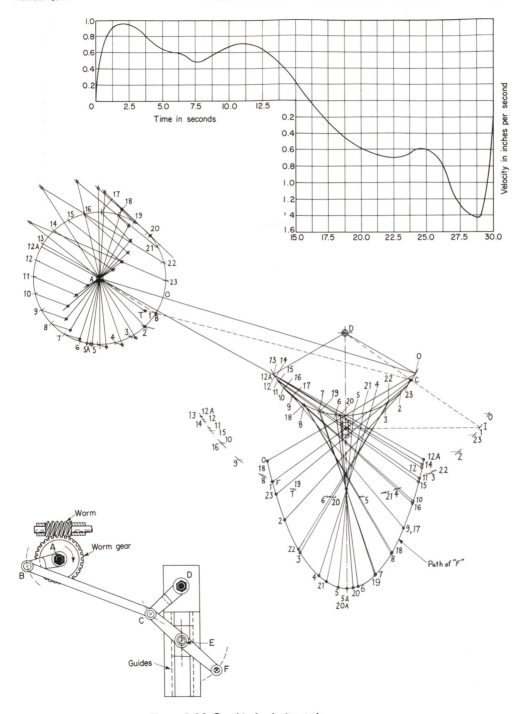

Figure 8.36 Graphical velocity study

8.12 Velocity Equations for All Mechanisms

The technique of simplified graphical velocity analysis is not limited to linkages. Velocity equations can be devised for any mechanism or connected system of mechanisms: friction and rolling contact drives, cams, and gear trains. To employ this method, one must know velocity-vector relationships but no vectors need be drawn.

Since the mechanism being studied must be drawn in a series of consecutive positions, it is important to "skeletonize" the drawing into the simplest equivalent graphical representation so as to save layout time. "Stick-figure" representations are adequate, and oftentimes the path of a significant point is sufficient without drawing the member upon which it is located in every position.* (Fig. 8.36.)

8.13 Velocity Equation for a Cam and Follower

In Figure 8.37 a circular cam C whose center is at O turns counterclockwise about a fixed axis at A with a constant angular velocity, ω_c. Roller R is kept in pure rolling contact with the cam at all times by a spring (not shown). R turns freely about pin E on follower F, which is constrained to translate up and down in fixed guides.

Let us devise an equation for the velocity of F in terms of the angular velocity of C and certain accessible measured distances. This equation is to be used to determine the velocity of F graphically for a series of positions so as to provide a complete velocity study throughout the motion cycle.

Point P on line EO is the point of contact of cam C and roller R in the typical position shown. The velocity of P on the cam equals $\omega_c \times AP$ and is directed perpendicular to AP. With pure rolling contact, the velocity of P on C will be equal to the velocity of P on R and therefore will also be perpendicular to AP.

The instant center of roller R is at I, which is the intersection of AP

---

*Those who have access to specially equipped computers, and who can properly program the machine, will be able to produce multiposition configurations of a mechanism, record the necessary measurements, and perform the computations without doing any "hand-drafting" at all. This would save hours of drafting time and tremendously increase precision in the solution yet would enable one to employ the relatively simple graphical methods rather than some intricate mathematical process. This facility does not render what we have studied obsolete, however, for the computer is only a tool. The user must know the basic theory and must devise the method equation before he can program the computer to do his drafting for him and enjoy the advantages of this labor-saving technique. Everything we have learned here is necessary and essential no matter which technique of execution is eventually employed in the solution of a problem.

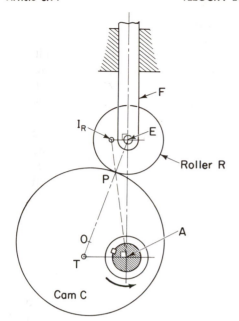

Figure 8.37 Velocity of rolling cam follower

extended (perpendicular to V_P) and a line through E perpendicular to the vertical V_E.

Now $V_P = AP \times \omega_C$, and on roller R

$$\frac{V_E}{V_P} = \frac{IE}{IP} \quad \text{so} \quad V_E = \frac{IE}{IP} \times AP \times \omega_C$$

If we draw EP, extended to intersect a line through A, perpendicular to the path of E, at point T, triangle ATP will be similar to triangle IEP. (Corresponding sides are parallel.) In these triangles:

$$\frac{IE}{IP} = \frac{AT}{AP}$$

and, substituting in the above equation:

$$V_E = \frac{AT}{AP} \times AP \times \omega_C \quad \text{or} \quad V_E = AT \times \omega_C$$

To use this equation for values of V_E, we need only draw EO and AT, perpendicular to the path of E, intersecting EO at T. Then AT is measured and multiplied by the constant ω_C. No vectors are needed.

8.14 Velocity Equation for an Epicyclic Gear Train

Gear trains are usually analyzed by computations using the speed and tooth ratio formulas developed in Chapter 7. Sometimes the graphical method is more enlightening and instructive than the numerical method. In any event, it is always desirable to have an alternative method to check one's work or just to find out what is going on.

Figure 8.38 shows a *compound epicyclic gear train.* The gears are represented by pitch circles for simplicity and can be assumed to be drawn exactly to scale. The input (ω_D) is applied at gear *D*, which turns about fixed axis *O* and is held in mesh with planet gear *B* by arm *A* (shown dotted) at pitch point *P*.

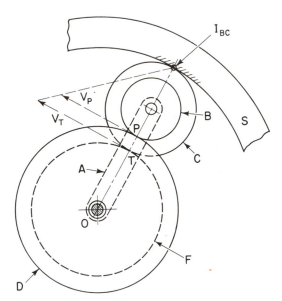

Figure 8.38 Velocity of compound epicyclic train

Planet gears *B* and *C* are keyed to the same shaft so that they turn together. Gear *C* meshes with a stationary annular gear *S* and also with the output gear *F*, which turns about axis *O*, independently of gear *D*. Arm *A* also turns independently of gears *D* and *F*.

Having given the angular velocity of *D*, let it be required that we derive an equation for the angular velocity of *F* in terms of ω_D and certain measured distances. Pitch circles are kinematically equivalent to cylinders rolling without slip, so the velocity of *P* will be identical on gears *D* and *B*:

$$V_P = \omega_D \times OP$$

The point *I* on gear *C* will for the same reason have the same velocity on gear *C* and annular *S*. Since *S* is stationary, point *I* will have zero velocity and therefore will be the instant center of the rigid body composed of gears *B*, *C*, and their connecting shaft.

At the pitch point *T*, gears *C* and *F* will have identical velocities. On the rigid body of gears *B* and *C*, with the instant center at *I*:

$$\frac{V_T}{V_P} = \frac{IT}{IP} \quad \text{and} \quad V_T = \frac{IT}{IP} \times OP \times \omega_D$$

$\omega_F = V_T/OT$ and substituting the above values:

$$\omega_F = \frac{IT}{IP} \times \frac{OP}{OT} \times \omega_D$$

This equation is entirely composed of constant terms, so it holds good for all positions of the mechanism. Also it may be noted that $IT = D_C$, $IP = D_C + D_B/2$, $OP = D_D/2$, $OT = D_F/2$. So in terms of pitch circle diameters, the equation could be:

$$\omega_F = \frac{2D_C}{D_C + D_B} \times \frac{D_D}{D_F} \times \omega_D$$

and since $D = T/P^D$, if all gears were of the same diametral pitch:

$$\omega_F = \frac{2T_C}{T_C + T_B} \times \frac{T_D}{T_F} \times \omega_D \quad \text{(in terms of tooth numbers)}$$

So this graphics-based equation could be applied without drawing anything. The vectors shown are only to explain the derivation.

PROBLEMS

Although several design problems are included, the principal object of the linkage analysis problems is to develop the art of devising brief and usable equations which may be employed for complete studies of each linkage, throughout its motion cycle.

The derivation of each equation is required, accompanied by a sketch showing necessary measurements.

In these problems velocity vectors are not to be used in the solutions. They may be sketched as a guide to the derivation of velocity equations, or a vector study may be made to check an equation, but a vector solution alone is not sufficient.

8.1. In the four-bar linkage of Figure P8.1 crank TS turns at constant speed of 50 rpm, counterclockwise. Determine the following:

(a) Angular speed ratio of the follower crank QR to driver crank TS in the position shown.

(b) Angular speeds of QR when ST is at 90° with SR.

(c) Angle of oscillation of QR (angle between extreme positions).

(d) The over-all time ratio of the clockwise to the counterclockwise stroke of QR.

(e) The maximum angular velocity of QR during its counterclockwise stroke (draw the linkage in the position of maximum speed of QR).

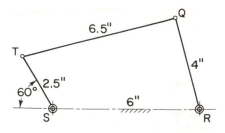

Figure P8.1

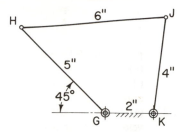

Figure P8.2

8.2. A four-bar linkage called a *drag link* is shown in Figure P8.2.

(a) When crank *HG* turns 180° (starting in the position shown), through what angle does *JK* turn?

(b) When *HG* turns 360°, through what angle does *JK* turn?

(c) What is the speed ratio of the cranks when *H* is on line *GK* extended to the left of *G*? Give the speed ratio when *H* is on line *GK* extended to the right.

(d) If *HG* turns at a constant speed of 60 rpm, what is the maximum angular velocity of *JK* (draw the linkage in the position in which ω_{JK} is maximum).

8.3. In the crank and slider linkage shown in Figure P8.3, *OL* turns at a constant speed of 35 rpm.

(a) Devise an equation for the linear velocity of *M* in terms of the angular velocity of *OL* and a measured distance which is accessible in all positions of the mechanism.

(b) Draw the mechanism at every 15° position of *OL* for 180°, starting with *M* at the left end of the stroke.

(c) Using the above equation, obtain values of the velocity of *M* in each position.

(d) Plot a curve of the velocity of *M* (ordinates) versus time (abscissae) at suitable scales (use velocity values obtained above and time intervals for 15° displacement of *OL*).

(e) By graphical differentiation of the velocity curve determine an acceleration curve for *M* and plot it on the same graph as the velocity above.

Figure P8.3

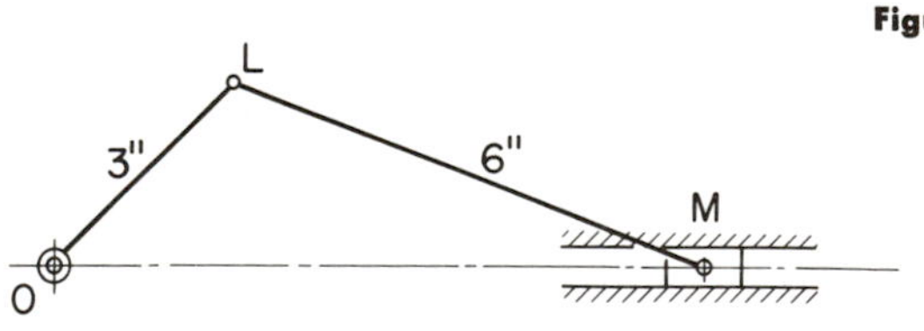

8.4. The four-bar linkage in Figure P8.4 has a driving crank *OP*, which turns clockwise at a constant speed of 60 radians/min.

(a) Devise an equation for the angular velocity of RS in terms of ω_{OP} and certain accessible measurements on the layout.

(b) Using this equation, determine the angular velocity of RS at each $15°$ position of OP, starting at OP_0 and continuing to OP_1.

(c) Plot a curve of ω_{RS} (ordinates) versus time (abscissae) at suitable scales.

(d) Differentiate this curve graphically to obtain values of the angular acceleration of RS.

(e) Plot the angular acceleration of RS on the same graph as the angular velocity above.

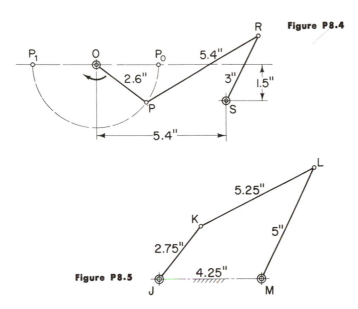

Figure P8.4

Figure P8.5

8.5. In the four-bar linkage shown in Figure P8.5, the driving crank JK turns clockwise at a constant speed of 60 radians/min. It is desired to study the angular acceleration of follower crank LM during its counterclockwise motion.

(a) Lay out the linkage and determine the extreme positions of LM.

(b) Write an equation for the angular velocity of LM in terms of ω_{JK} and certain accessible measured distances.

(c) Divide the angle turned by JK (during the counterclockwise motion of LM) into 12 equal parts and, using the above equation, find the angular velocity of LM at each of the 12 stations.

(d) Plot a curve of ω_{LM} (ordinates) versus time (abscissae) at suitable scales.

(e) Differentiate this angular velocity curve graphically to obtain values of angular acceleration of LM.

(f) Plot a curve of the angular acceleration of LM on the same graph as the angular velocity curve.

8.6. In the linkage in Figure P8.6, bar PS is one continuous member 7 in. long. OR is the driving crank which turns clockwise at 10 radians/sec.

(a) Derive an equation for the velocity of P in terms of ω_{OR} and any required measurements.

(b) Using the above equation, determine V_P in the position shown. Indicate direction by a sketched vector.

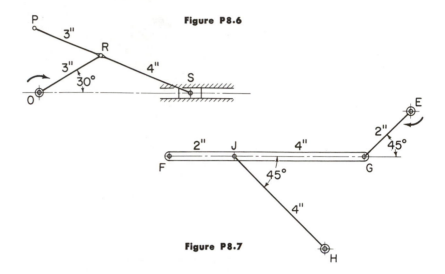

Figure P8.6

Figure P8.7

8.7. In the linkage in Figure P8.7, bar FG is one continuous member 6 in. long. The driving crank EG turns clockwise 60 radians/min.

(a) Derive an equation for the velocity of F in terms of ω_{EG} and any required measurable distances.

(b) Using this equation determine V_F in the position shown. Indicate direction.

8.8. In Figure P8.8, CE is the driving crank which turns clockwise 3.5 rpm. DAE is a rigid triangular member.

(a) Derive an equation for V_A in terms of ω_{CE} and accessible measured distances.

(b) Using this equation, determine the value of V_A when the mechanism is in the position shown. Indicate direction.

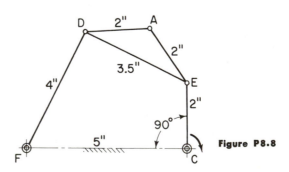

Figure P8.8

8.9. In the linkage shown in Figure P8.9, *JK* is the driving crank, which turns at 7 rpm. *OL* is a continuous rigid member, 4 in. long.

(a) Devise an equation for the angular velocity of *PS* in terms of ω_{JK} and any accessible measured distances as required. Make it as brief as possible.

(b) Using this equation, determine the angular velocity of *PS* in the position shown.

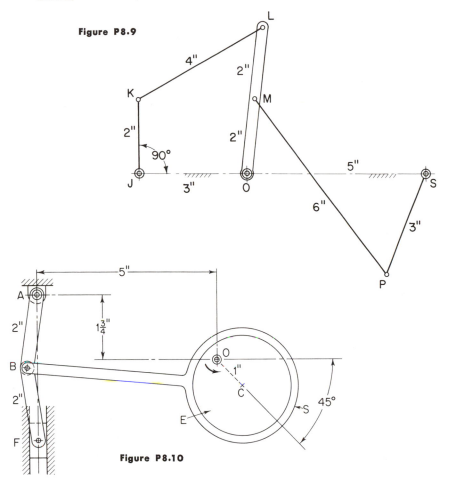

Figure P8.9

Figure P8.10

8.10. The *toggle linkage* shown in Figure P8.10 is driven by the eccentric *E* with center at *C*, which turns counterclockwise about fixed shaft *O* at 3.5 rpm. *E* turns freely inside of the eccentric strap *S*, which carries a rigidly attached extension bar pinned to the toggle linkage at *B*. Measurement *BC* equals 6 in.; other measurements are given on the drawing.

(a) Write an equation for the velocity of *F* in terms of the angular velocity of the eccentric *E* and any accessible measured distances required.

(b) Using this equation, determine V_F when in the position shown.

8.11. The linkage shown in Figure P8.11 is driven by the crank *EF*, which turns at 70 rpm. Bar *FB* is one rigid member, 4 in. long, carrying pin *C* at its midpoint. *E, O* and *A* are fixed axes.

(a) Derive an equation for the angular velocity of *OP* in terms of ω_{EF} and certain measurable distances. Make it as brief as possible.

(b) Using the above equation, determine ω_{OP} when the linkage is in the position shown.

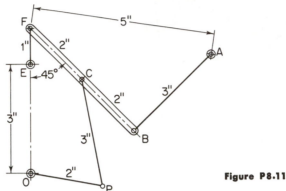

Figure P8.11

8.12. The linkage in Figure P8.12 has fixed axes at *A* and *D*. The driving crank is *AB*. Bar *EC* is one rigid 4-in. member carrying pin *F* at its midpoint.

(a) Devise an equation for the angular speed ratio of the follower crank *DE* to the driver crank *AB*. Make it as brief as possible and try to employ measurements which will be accessible in all positions of the linkage.

(b) Using the above equation, determine the value of the angular speed ratio of ω_{DE} to ω_{AB} when the linkage is in the position shown.

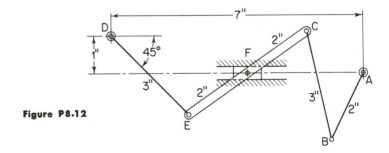

Figure P8.12

8.13. A *drag-link mechanism* is shown in Figure P8.13. *P* and *Q* are fixed axes. The driving crank is *QT*, which revolves at constant speed of 17.5 rpm, clockwise. An additional link *SR* drives the output slider *R* in fixed guides, located as shown. Dimensions in inches are given on the drawing. In all drag-link mechanisms, both cranks make complete revolutions. This linkage is designed so that the time required for the stroke of R in one direction is much less than for its return in the opposite direction. Investigate this linkage graphically, and determine the following information:

(a) What is the length of the stroke (distance between extreme positions) of the slider R?

(b) Is the fast stroke of R from left to right or from right to left?

(c) What is the overall time ratio of the fast to the slow stroke of R?

(d) Derive an equation for the velocity of R in terms of ω_{QT} and certain accessible distances.

(e) Using this equation, determine the velocity of R in inches per second when it is in the position shown, first during the fast stroke and then during the slow stroke.

Figure P8.13

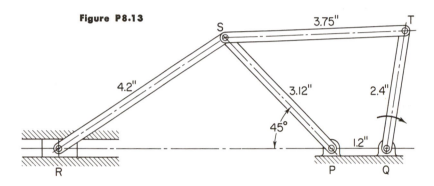

8.14. In the design of a new machine, it is required that a lever arm M oscillate back and forth about a fixed axis O with a particular motion described below.

1. The angle of oscillation (between extreme positions) is to be exactly $90°$.
2. The length of the lever arm M from the axis of rotation O to a point P on its centerline is to be 6 in.
3. The total time ratio of the forward to the return stroke is to be 5:3.
4. The mechanism which propels the lever is to be driven by a shaft which revolves at a constant speed of 25 rpm about a fixed axis S. The driving crank is to be labeled ST for purposes of identification.

(a) Design a pin-connected linkage which will provide the required motion to the lever. Space limitations demand that the complete mechanism operate in a $9'' \times 12''$ (or smaller) rectangle. Make the linkage as simple as you can. It must be capable of continuous operation without jamming. The solution should consist of a full-size assembly drawing of the linkage with dimensions of all members, location of fixed axes, and orientation of the angle of oscillation of arm M relative to the line of centers of these axes.

(b) Make a cardboard model of this linkage to demonstrate that it satisfies the conditions and operates without jamming.

Next, it is desired to investigate the angular velocity and angular acceleration of arm M. To do this, proceed as follows:

(c) Devise an equation for the angular velocity of OP (arm M) in terms of the constant angular velocity of the input crank ST and certain measurable distances. This is to facilitate obtaining values of ω_M without drawing vectors.

(d) Using this equation, determine ω_M in radians/sec for a series of positions throughout the fast or return stroke. (Obtain at least 9 values of ω_M at equal time intervals throughout this stroke.)

(e) Make a plot of ω_M against time at scales which will give a well-proportioned curve (see Article 5.7).

(f) Using graphical differentiation, determine the maximum angular acceleration of arm M in radians/sec/sec during this stroke.

8.15. In Figure P8.15 bar M is a straight rigid member which turns about fixed axis O. The sleeves are pinned to cranks at R and P and slide freely on M. Crank LR turns at 10 rpm about L.

(a) Write an equation for the angular speed ratio of TP to LR.

(b) Using this equation, determine ω_{TP} in the position shown.

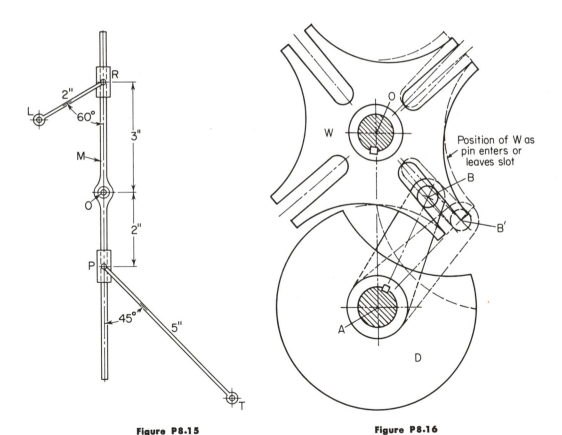

Figure P8.15   **Figure P8.16**

8.16. A well-known intermittent motion device called a *Geneva mechanism* is shown in Figure P8.16. The lower shaft turns in fixed bearings at *A*. This is the input shaft, and it revolves at constant speed. The partly circular disk *D* and arm *AB* are both keyed to turn with this input shaft. The arm carries a pin *B* as shown. The upper shaft turns about a fixed axis at *O* and is the output shaft which turns with intermittent motion. The slotted *Geneva wheel W* turns this shaft. The number of slots may vary, but they are always equally spaced. When the drive is set in motion, the pin *B* enters one of the slots in the wheel *W*. At the moment of engagement the radial centerline *AB* must be at 90° with the center line of the slot so as to avoid impact as the wheel starts to turn. As the arm *AB* turns, the pin causes the wheel to turn as long as it is engaged in the slot. The disk *D* is cut away so as to clear the slotted lobe of *W* during this motion. At the instant the pin leaves the slot, the circular part of disk *D* engages the matching circular surface between the slotted lobes of the Geneva wheel, thereby locking the wheel in a stationary position while the pin *B* completes its circuit about the drive shaft. The slots are so spaced that this pin will enter the next slot when it comes again to the starting position. As the pin enters the slot the circular surfaces of *D* and *W* must separate so as to again allow *W* to turn. Disk *D* and arm *AB* turn 60 rpm, counterclockwise. The center distance *OA* = 4.24 in.* *AB* = 3 in. In the position where *W* is just starting to turn and the pin is at *B'* (shown dotted) angle *OB'A* = 90° and angle *OAB'* = 45°.

(a) Write an equation for the angular velocity of Geneva wheel *W* in terms of ω_{AB} and certain measurable distances.

(b) Using this equation and an equivalent simplified representation of the mechanism, obtain values of ω_W during 45° of the rotation of *AB* starting from the position *AB'* shown dotted. Use 9 positions of the mechanism at 5° intervals of the displacement of *AB*.

(c) Plot ω_W versus time at suitable scales.

(d) By graphical differentiation of this curve, determine the maximum angular acceleration of wheel *W* during this typical interval of its motion.

8.17. On a production stamping machine, specified lengths of flat strip stock must be fed into the cutting dies at regular intervals. The stock is supplied to the manufacturer wound upon reels. The feed motion is always in the same direction, and there is a specified ratio between the time the stock is in motion to the time it is held stationary during the stamping operation. To accomplish this automatic feed process, a *Geneva mechanism* is to be designed in which the input shaft turns at constant speed while the output shaft alternately turns through a suitable angle, then dwells (remains stationary) for a given interval before repeating an identical rotation. The output shaft always turns in the same direction. Figure P8.17 shows several typical Geneva mechanisms. Problem 8.16 describes the operation of another Geneva shown in Figure P8.16.

*Actually this center distance equals 4.2426 in., but it can be laid out as 4.24 without introducing objectionable inaccuracy in the graphical solution.

Figure P8.17 (Courtesy of Tangen Drives, Inc., Clearwater, Florida.)

The specified motion characteristics are as follows: The time ratio of the motion of the output shaft (Geneva wheel) to its dwell period is 3:7. The time for each dwell period is to be exactly 2.1 sec. The distance between centers of the input and output shafts is to be 4 in. Both shafts are to be $\frac{7}{8}$ in. in diameter. Design the mechanism, and make a full-size assembly drawing. Two views are required, and dimensions are to be omitted. Also determine and report:

(a) Angle of rotation of the output shaft for each motion.

(b) Angular speed of input shaft in rpm.

(c) Maximum angular velocity of the output shaft in radians per sec.

Derive an equation for the angular velocity of the output shaft in terms of the constant angular velocity of the input shaft and certain accessible measured distances shown and labeled on the drawing.

Using this equation, plot and draw a curve of angular velocity of the output shaft versus time. Select suitable scales and plot only the typical interval from the time the pin enters the slot until it reaches the line of centers of the input and output shafts.

By graphical differentiation, determine the angular acceleration of the output shaft at sufficient intervals to plot a curve, using the same time scale as was used for the angular velocity curve.

8.18. A *quick-return mechanism* is shown in Figure P8.18. Driving crank *OP* turns about fixed axis *O* at 1 radian/min, counterclockwise. The slotted bar *S* swings about fixed shaft *T* and has a gear quadrant (2.5-in. pitch radius) cut on its lower end. This quadrant drives rack *R*.*

(a) How long is the stroke of *R* (distance between extreme positions)?

(b) What is the time ratio of the slow to the fast stroke?

(c) Derive an equation for V_R in terms of ω_{OP} and some measurable distances.

(d) Applying this equation, determine V_R when *S* is in the position shown during the slow stroke.

(e) Similarly determine V_R when *S* is in the same position during the fast stroke.

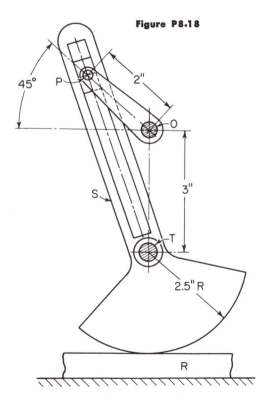

Figure P8.18

*The teeth on the quadrant and rack are omitted here for simplicity.

8.19. A quick-return mechanism is shown in Figure P8.19. The driving crank MP is 2 in. long and turns 60 rpm about fixed axis M. The slotted beam is $8\frac{1}{2}$ in. long from T to R. Link OT is 2 in. long and swings about fixed axis O. The block at R travels in horizontal fixed guides.

(a) Determine the length of the stroke of block R.

(b) Determine the time ratio of the slow to the fast stroke.

(c) Write an equation for V_R in terms of ω_{MP} and some accessible measured distances.

(d) Using the above equation find the value of V_R in the position shown during the slow stroke.

(e) Find V_R in position shown during the fast stroke.

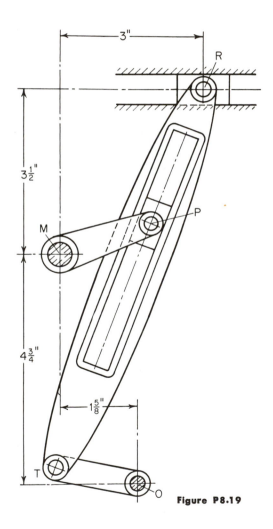

Figure P8.19

8.20. Design a quick-return mechanism having a stroke of 7 in. The ratio of the slow to the fast stroke is to be 2.6:1. Any form of linkage may be used. Make a drawing of the mechanism, giving all measurements and mounting dimensions.

8.21. Design a Whitworth quick-return linkage having a stroke of 5 in. The ratio of the slow to the fast stroke is to be 7:2. Make a drawing of the mechanism giving all measurements and mounting dimensions.

8.22. A *cyclic gear intermittent motion drive* is shown in Figure P8.22. Points O and R are fixed axes. The 3 gears are represented only by their pitch circles for simplicity. The input is at the shaft at O and the output at R. Gear A turns about O with the input shaft at a constant speed of 35 rpm causing gear C to alternately rotate through a given angle and dwell. C always turns in the same direction. Gears A and B are held in mesh at pitch point S by the link NP connecting their centers. Gear B meshes with gear C at their pitch point T, and they are held in mesh by link PR. Gear C turns independently of link PR and is keyed to the output shaft. $OR = 6\frac{1}{4}$ in., and $ON = 0.8$ in. The pitch diameters of the gears are $A = 2\frac{1}{4}$ in., $B = 3\frac{3}{4}$ in., and $C = 4\frac{3}{4}$ in. Derive an equation for the angular velocity of gear C in terms of ω_A and certain accessible measured distances which may be used for any position of the mechanism. Using this equation, determine ω_C when the mechanism is in the position shown.

Figure P8.22

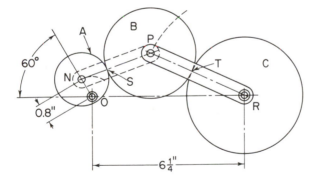

8.23. A *4-gear epicyclic train* is shown in Figure P8.23. The gears are represented by their pitch circles. The input is at shaft *S*, and the output is at shaft *F*, which turns with arm *A*. Shaft *S* turns gear *D* at 1000 rpm. *D* meshes with gear *B*. The planet gears *B* and *C* turn together and are carried by the arm *A*. Gear *C* meshes with the stationary annular *N*. The pitch diameters of the gears are as follows: *D* = 5 in., *B* = 3 in., *C* = 4 in., and *N* = 12 in. Derive an equation for the angular speed of shaft *F* in terms of the angular speed of the input shaft *S* and the pitch diameters of the gears. This equation will hold for any position of the mechanism. Using this equation, determine the angular speed of shaft *F*. Check your equation by solving for the speed of *F*, using the epicyclic formula.

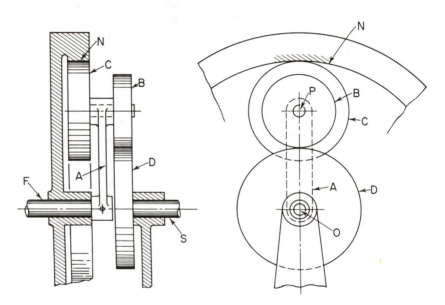

Figure P8.23

8.24. A *plate cam* with a roller follower on a swinging arm is shown in Figure P8.24. Cam J turns clockwise at a constant speed of 70 rpm about fixed axis O in contact with roller R. (Contact is assured by a spring not shown.) R turns about pin E in the follower arm F, which oscillates about a fixed axis at G. Cam J is symmetrical about center line OH, and its surface is composed entirely of circular arcs: a radius of $2\frac{1}{4}$ in. about O, two radii of $4\frac{1}{2}$ in. about points A and D, which are diametrically opposite to one another on the first arc of $2\frac{1}{4}$-in. radius, and an arc of $1\frac{1}{2}$-in. radius about H, tangent to the two $4\frac{1}{2}$-in. radius arcs at B and C. $OG = 4\frac{3}{4}$ in., and $GE = 3$ in. Roller R is 2 in. in diameter. In the given position $OE = 3\frac{13}{16}$.

(a) Derive an equation for the angular velocity of arm F in terms of the angular velocity of cam J and certain accessible measured distances when the roller contacts the cam surface between points C and D as shown in Figure P8.24.

(b) Draw the mechanism in the position where the roller R contacts the cam surface at K between points B and C. Derive another equation for ω_F similar to the one above for this range of positions.

(c) Using these equations, solve for ω_F in the given position and when contact is at K.

Figure P8.24

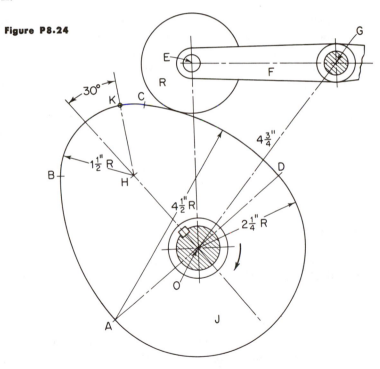

9

Cams

Of all mechanisms, the cam is the most versatile. A very ordinary input motion can be transformed into a wide variety of intricate output motions by some form of this simple two-member device. Perhaps the outstanding virtue of the cam is that, unlike the linkage, its design is straightforward and easy.

Cams are made in various forms with different types of followers, but generally, a cam is a specially shaped rotating body which transmits motion to a follower through sliding or rolling contact. The shape of the contacting surface of the cam can assume an infinite variety of contours which can be rationally designed to produce a desired follower motion. Although it is often necessary to study the acceleration of a cam follower resulting from a given cam, our principal concern here is with design rather than analysis.

9.1 Geometric Design of Cams

Although some high-speed cams are generated from analytical designs, the more common applications are less critical and the contour of the cam can be defined by a large-scale, precise, geometric layout.

Before the actual layout can be attempted, a program of the required displacements of the follower, at successive time intervals, must be available. Whatever the source or nature of the motion of the follower, it must be stated in terms of displacement and time to be usable in the cam-design process. Since cams usually turn at uniform speed, the angular increments of cam displacement correspond to specific time intervals. The follower displacements can be tabulated for successive time intervals, or these displacements can be presented graphically in the form of a graph of follower displacements versus cam (angular) displacements. This graph is called a *displacement diagram* by cam designers. In the illustrations of geometric design, we will assume that follower displacements are available for each angular position of the cam.

It is fairly obvious that the total time required for a complete cycle of follower motion must equal the time allotted to one revolution of the cam. This relationship establishes the angular speed of the cam and guarantees identical motion of the follower in each cycle.

9.2 Design of Radial Cam with Pointed Follower

If the follower moves in translation along a straight line which (if extended) passes through the axis of rotation of the cam, the cam is described as *radial*. A radial cam with pointed follower is the simplest to design. This type of follower is best suited to intricate motions, since the point is sensitive to abrupt changes in cam contour. The point is subject to wear and produces large stress connections, however, so its use is limited to light output loads. Figure 9.1 shows this assembly with a compression spring to maintain contact between cam and follower.

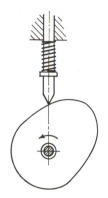

Figure 9.1 Radial cam with pointed follower

A *displacement schedule* is shown in Figure 9.2 with positions numbered for reference. All displacements are measured from the zero position, and follower displacements are upward.

No.	θ-cam in degrees	S-follower in inches
0	0	0
1	15	.1
2	30	.25
3	45	.4
4	60	.5 and 0
5	75	0
6–24	90–360	0

Figure 9.2 Cam-follower displacement schedule

Let us design a radial cam to turn counterclockwise about point A in Figure 9.3, with the initial position of the follower at 0. First we draw the *reference line* $A0$ and measure the follower displacements from 0 along $A0$ extended, numbering each position. Next we lay out the cam displacements

(15° increments) from the reference line, numbering them in clockwise order.*

If we rotate the cam counterclockwise through a 15° angle, line 1 on the cam will lie upon the reference line. In this same interval of time the point of the follower is to rise from 0 to point 1, so there must be a point on the cam surface at point 1 to support it. If we mark this point 1 on line 1 and then rotate the line back (clockwise) to its original position, we will have one point on the cam surface plotted. The simplest way to plot this point is to swing an arc about *A* through point 1 on the reference line around to line 1 on the cam.

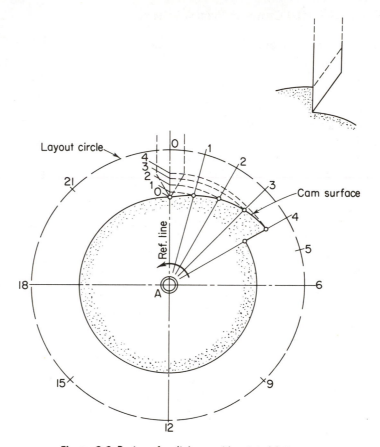

Figure 9.3 Design of radial cam with pointed follower

---

*Accuracy is vital in this layout, since the solution is a drawing. To obtain precise angles, a process of division of a large circle is recommended, first into 180° by the reference line, then into 90° by a "constructed" perpendicular, then into 45° by bisected angles, and finally into 15° using the dividers. Ordinary small protractors are unsatisfactory, but the protractor on a good drafting machine could be used instead of the above process.

Other points on the cam are transferred from the reference line to the corresponding cam lines in the same manner. Then, to define the cam surface, we draw a smooth curve through these points on the cam lines 0-1-2-3-4, as shown in Figure 9.3, thus ensuring a smooth lift motion of the follower.

In the displacement schedule in Figure 9.2, we note that two different displacements are given for the 60° cam angle. While we cannot place the follower in two positions at one time, we can provide for a sudden drop by making the cam radial between these two points. To clear the cam, the follower must be designed as shown in Figure 9.3.

The schedule indicates that the follower is to "dwell" at the zero displacement level while the cam completes its revolution. This will, of course, make the cam contour circular between line 4 and line 24 (the reference line).

9.3 Design of Cam for Roller Follower

A roller follower, as shown in Figure 9.4, will decrease friction, wear, and stress concentrations developed by the pointed follower and will permit heavier output loads. This is achieved with the sacrifice of sensitivity and simplicity, as will be seen.

To design a cam with a roller follower, using the same displacement schedule as in Figure 9.2, we will consider that the given displacements are to apply to the center of the roller and that point 0 is the initial position of the roller center. First, we design a theoretical cam curve called a *pitch line* (similar to the term *pitch circle* used in gearing), which represents the cam surface for a pointed follower with the point at the center of the roller. The actual cam contour of the former example will coincide with the pitch line of this roller follower cam, which is constructed in the same manner, as shown in Figure 9.5. To describe the real cam contour, we simply draw in the roller arcs with centers along the pitch line and then draw a smooth curve tangent to these arcs.

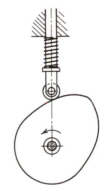

Figure 9.4 Radial cam with roller follower

This curve defines the cam surface from point 0 to point 4. Then we note that, in order to provide for the sharp drop at the 60° line, we will have to remove some of the cam surface which supported the roller when its center was at *P*. If this is done, the roller will swing around the corner *S* (Figure 9.5) and its center will never reach point *P*. If its center does not reach point *P*, the follower will never attain the maximum displacement specified in the schedule. This is an example of the lack of sensitivity of the roller as compared with the pointed follower.

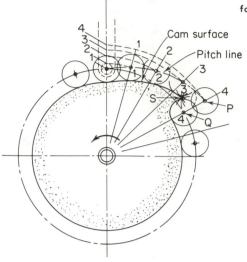

Figure 9.5 Design of cam for roller follower

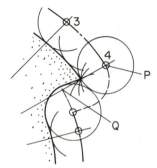

Figure 9.6 Actual motion of roller at a cusp

If the maximum displacement must be attained, the sharp drop will have to be delayed until the center reaches point *P*, as shown in Figure 9.6. It is a question as to whether it is more important to attain point *P* or point *Q*. With a roller follower one cannot attain both.

9.4 Design of Cam with Sliding Circular Follower

A circular follower, which is nonrolling, is simpler and lighter than the roller follower. The example in Figure 9.7 shows that the arc of the contacting surface may be large (to reduce stress concentrations) without the bulk of a large roller. The larger the arc, the less sensitive is the follower to abrupt changes in cam contour, of course.

The design of a cam for this type of follower is identical with that of the roller follower, since we lay out the displacements from the center of the follower arc, as shown in Figure 9.7.

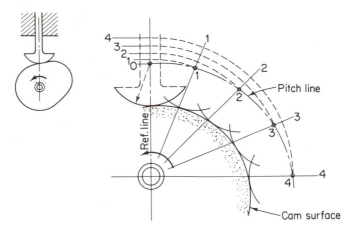

Figure 9.7 Cam with circular translating follower

9.5 Design of Cam with Flat Translating Follower

If the slope of a cam surface is too steep with respect to the path of the follower, both the pointed and roller followers will jam. This cannot occur with a flat follower, as the *pressure angle* (between the path and the common normal at the contact point) is constant and can be zero or very small. The flat follower is simple in design, but it is the least sensitive of all to cam curvature. It can only be used for very simple motions. Figure 9.8 shows an example with zero pressure angle.

When designing a cam for a flat follower, the displacements are laid out

Figure 9.8 Design of cam with flat translating follower

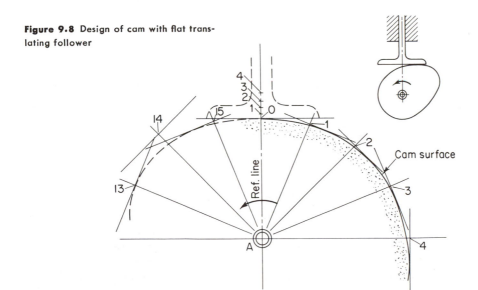

on the reference line as before and numbered corresponding to the cam angles, as shown in Figure 9.8. As it is impossible to predict at what point on the follower contact will take place, we draw lines representing the follower surface at 90° with each cam line, making the radial distance $A1$ on the cam line equal to $A1$ on the reference line, etc. This forms an envelope of follower surface lines to define the cam surface.

The cam surface is a smooth curve drawn tangent to each follower line successively. If it is impossible that this curve be drawn tangent to any follower line, as in the case of line 14 in Figure 9.8, then the flat follower may have to be abandoned.

9.6 Design of Cam with Swinging Roller Follower

In all of the previous examples, the follower has been in translation. Rotating arms are also used with pointed or roller followers, as shown in Figure 9.9. Since the paths of these types of followers are arcs, they cannot follow the radial reference line.

Figure 9.9 Cams with swinging followers

A typical example is shown in Figure 9.10. The cam is to turn counterclockwise about axis A, with the center of the roller follower starting at 0 and following the arc about B, the center of rotation of the arm. The reference line is line $A0$. It is assumed that angular displacements of the follower are available and have been laid out from line $B0$, locating the stations of the roller center along its path. The corresponding cam angles have been laid out *starting from the reference line $A0$* and numbered in clockwise order to correspond with the roller stations.

To visualize the design process, let us turn the cam counterclockwise through an angle of 20°. Line 1 will then lie upon the reference line. During this same time interval, the center of the roller will have moved from 0 to point 1 on its path. Since this point is not on the reference line, the corresponding point on the cam will not fall on cam line 1, which coincides with the reference line in its rotated position. If we draw an arc about A through point 1 and across the reference line, the distance from point 1 to the reference line ($1a$) can be measured as a chord of this arc. If we now rotate the cam

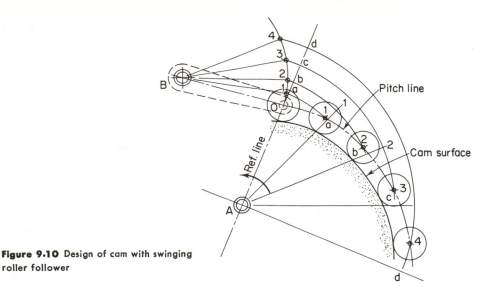

Figure 9.10 Design of cam with swinging roller follower

line 1 back to its original position and draw the same arc across it, we can use the same chord (1*a*) to locate point 1 on the cam. Point 1 on the cam must have the same location relative to line 1 as point 1 on the path of the follower has relative to the reference line. It is important that these chordal offsets be measured in the same direction in each case. From the reference line to point 1 on the follower we measure counterclockwise, so from cam line 1 to point 1 on the cam we must also measure counterclockwise.

All other points on the *pitch line* of the cam are laid out in this same manner, as shown in Figure 9.10. Equal offsets are labeled alike: 1*a*, 2*b*, 3*c*, etc. The pitch line curve is drawn through these points, roller arcs are added, and the cam surface is drawn as a smooth curve tangent to the roller arcs.

A cam with translating follower whose path does not follow the reference line is designed by this same method of offsets, as in Figure 9.11.

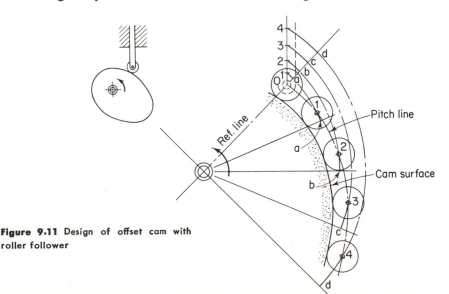

Figure 9.11 Design of offset cam with roller follower

353

9.7 Design of Cam with Swinging Flat Follower

In Figure 9.12 a cam with an oscillating flat follower is shown. In this example the cam turns counterclockwise about axis A and the follower turns about the fixed pin B. The contacting surface of the follower is a horizontal radial line through B. Several consecutive positions of the follower are shown, bearing the same numbers as the lines of the corresponding cam angles. The reference line may be drawn in any position, but in this case it is convenient to draw it perpendicular to the starting (horizontal) position of the follower surface. Note that the cam angles are laid out starting from the reference line.

As cam line 1 turns 30° to lie upon the reference line, the follower surface swings upward until it crosses the reference line at point 1. We must construct an envelope of these lines to define the cam surface, just as we did in the former flat follower layout. In other words line $B1$ must be drawn in the same relative position to line $A1$ as it now bears, relative to the reference line. To do this we reconstruct the triangle $A1B$ (formed by AB, $B1$, and the reference line) on the cam line $A1$, making sure that the triangle is laid out on the same side of line $A1$ as it is with respect to the reference line. To do this we swing an arc about A through point 1 on the reference line to intersect the cam line $A1$. We next draw a circle about A, with radius AB (labeled the "b" circle in Figure 9.12). Now, with a radius equal to $B1$ and center at point 1 (on cam line $A1$), we describe an arc intersecting the "b" circle at point b. This creates the triangle $A1b$ (stippled), identical with triangle $A1B$ on the reference line. The side $b1$ (extended) represents the follower surface relative to cam line $A1$.

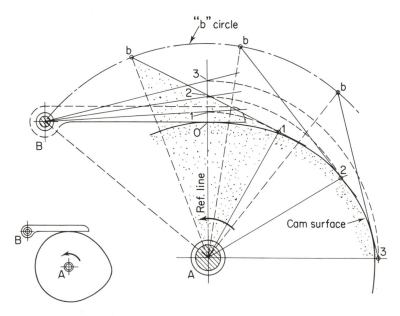

Figure 9.12 Design of cam with swinging flat follower

In a similar manner we construct triangle *A2b* identical to *A2B* and triangle *A3b* identical to *A3B*, and so on. (Once the construction is understood, the triangles need not be drawn.) The side *Ab*, common to each triangle, is defined by the "*b*" circle. Other sides are obtained from the reference line position (*A2* on the reference line is equal to *A2* on cam line 2, and side *2B* equals *2b*, etc.).

We now have an envelope composed of lines *B0*, *b1*, *b2*, *b3*, etc. The actual cam surface is a smooth curve drawn tangent to each of these in succession. As in the former case of flat follower design, if one of the *b*-lines lies outside of the envelope, the flat follower will have to be redesigned or abandoned.

9.8 Positive Motion Grooved-Plate Cams

In all of the previous cam assemblies it is assumed that contact between cam and follower will be maintained by springs. In high-speed cams, a large spring force is required to guarantee continual contact. This spring force, added to the output load, increases stresses at the contact points. In positive motion cams the follower is restrained by contact with the cam in two directions, so no spring is required.

A grooved-plate cam with a roller follower is shown in Figure 9.13. A groove slightly larger than the roller is milled into a circular plate with its center on the pitch line. The roller is constrained in this groove as the cam turns. In order to roll, the roller may contact only one side of the groove at any given time, so a small clearance must be provided. In certain positions

Figure 9.13 Grooved-plate positive motion cam (Courtesy of Eonic, Inc., Detroit, Michigan).

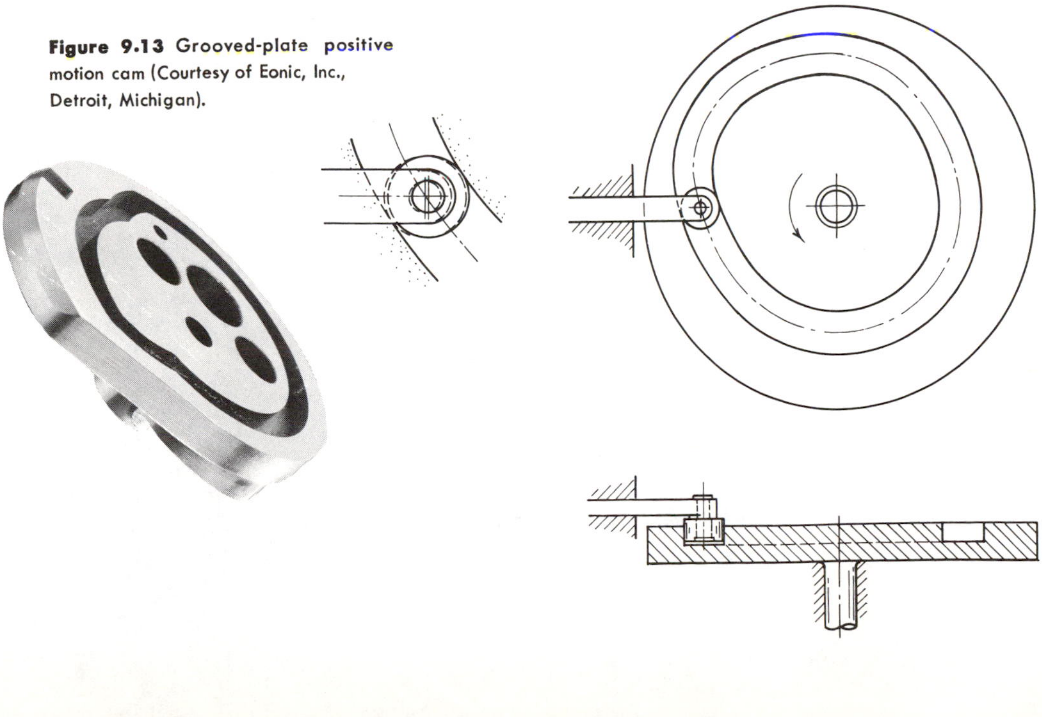

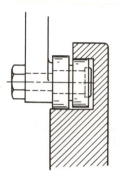

Figure 9.14 Dual roller follower to eliminate backlash

this clearance becomes magnified, as shown in Figure 9.13, so that considerable backlash is developed, permitting the follower motion to be quite inaccurate.

This inaccuracy can be overcome by using two rollers of slightly different diameters bearing on an eccentric groove, as shown in Figure 9.14. This really adds up to the manufacture of two cams, so the cost is greatly increased.

The design of all grooved-plate cams follows the same method as any roller follower design except that cam surfaces are drawn tangent to the roller on both sides.

9.9 Positive Motion Cylindrical Cams

Another type of grooved cam, shown in Figure 9.15, is the cylindrical cam. This one has a translating roller follower and matching groove. A conical roller must be used to accommodate different surface velocities at the top and bottom of the groove.

The design of this type of cam becomes the layout of the pitch (center) line of the groove on the developed outer surface of the cylinder, as shown in Figure 9.16. The circumference C equals the cam diameter multiplied by π.

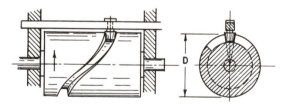

Figure 9.15 Cylindrical positive motion cam

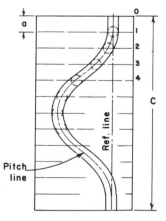

Figure 9.16 Design of cylindrical cam

The cam lines are parallel to one another (perpendicular to C) and divide the circumference into the same number of parts as there are angular divisions of the cam or follower displacements. The displacements of the follower are laid out from a convenient reference line which is parallel to C (toward the left in Figure 9.16). The pitch line is a smooth curve drawn through these plotted points, so that actually this layout is a displacement diagram. The groove may be drawn with sides tangent to roller circles centered on the pitch line, as shown.

9.10 Positive Motion Plate Cams

Two *plate cams of matched design* fastened together can be used as a positive motion mechanism for either translating or rotating followers.

Figure 9.17 shows such a device with a translating follower, carrying two rollers at a fixed distance apart. Cams C and E turn together about fixed shaft A, C always in contact with roller F and E maintaining contact with G. Cam C is designed first to provide the required motion of F, just as any radial cam with roller follower. Then cam E is designed to maintain contact with roller G, which travels at the fixed distance from F.

A similar idea is shown for an oscillating follower in Figure 9.18. The cams C and E are attached together and turn about fixed shaft A. The followers F and G are mounted on a bell-crank lever which turns on fixed axis B. F maintains contact with cam C and G with cam E at all times. Cam C is designed to give the required motion of F, and then E is designed for the corresponding motion of follower G. This device holds backlash at a minimum.

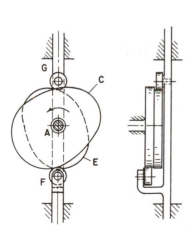

Figure 9.17 Matched cams for translating follower

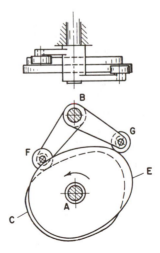

Figure 9.18 Matched cams for swinging follower

9.11 Combined Cams for Irregular Follower Paths

The design layouts described in the previous articles enable us to propel followers along any given path, but we depend upon some auxiliary mechanism to define the path. We have used fixed guides for translation and oscillating arms for rotation. An isosceles linkage might be used to guide a follower on an elliptical path, while a cam propels it along that path at a specified velocity, as in Figure 9.19.

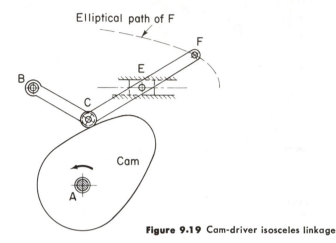

Figure 9.19 Cam-driver isosceles linkage

We have noted the difficulty of designing a linkage to guide a point along any random path. By the use of two synchronized cams we can solve this problem directly without catalogs, calculus, or conjecture. Figure 9.20 shows a simple mechanism driven by two cams in which pin P describes a figure 9 during each motion cycle. Cam C turns about fixed axis A driving a roller follower which causes bar FJ to slide back and forth horizontally in fixed guides. The link JP swings about J and is supported by cam E, which contacts the roller at P. E turns about fixed axis B at the same speed as C.

To design this mechanism, the figure 9 was first divided into 24 segments and each cam divided into twenty-four 15° (360/24) angles. Cam C provides only the horizontal displacements, which are projected from the figure 9 onto a horizontal line and numbered. Cam C is now designed for these horizontal displacements like any radial cam. Since the path of P leaves the reference line PB, cam E is designed as an offset cam to support the follower P at each station along the figure.

In a similar fashion these two cams could guide point P along any prescribed irregular path at any desired rate of speed.

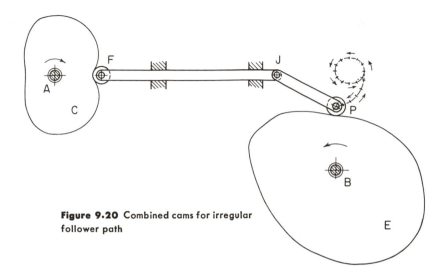

Figure 9.20 Combined cams for irregular follower path

9.12 Limitation in Cam Design

It has been pointed out that, if the *pressure angle* (between the path of the follower and the normal to the cam surface at the contact point) is too large, a roller or pointed follower will fail to climb the cam contour and will jam. For example, in Figure 9.21 with the follower in the position shown, the maximum pressure angle ϕ is 45° and the system is about to jam.* One remedy

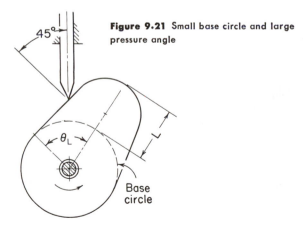

Figure 9.21 Small base circle and large pressure angle

*The Greek letter ϕ (phi) is used to denote pressure angles of cams in the text.

for this difficulty is to increase the size of the cam or, more specifically, to increase the size of the *base circle*. (The circle of radius equal to the minimum cam radius is called the *base circle*.) This makes the *lift* of the cam a smaller fraction of the maximum radius and decreases the pressure angles. The *lift* is the total distance that a follower rises from its lowest position, or L in Figure 9.21. The cam displacement during lift is the *lift angle* θ_L.

Figure 9.22 shows a cam having the same lift L as in Figure 9.21, which is effective in the same lift angle θ_L, but with the base circle enlarged. We note that the pressure angle is now only 30° and the system will not jam.

As it is desirable to keep the base circle as small as possible for a compact design, it is necessary that we devise some means of predicting the minimum permissible base circle.

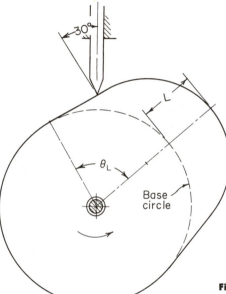

Figure 9.22 Enlarging base circle decreases pressure angle

9.13 Maximum Pressure Angle and Minimum Base Circle

Experience shows that in a well-designed cam, the maximum pressure angle should not exceed 30°. With this as a criterion we can devise a way to determine the minimum base-circle radius.

A *displacement diagram* (Figure 9.23) is a graph of the follower displacements versus cam displacements. If a wedge-shaped translation cam were made, with this curve as its upper contour, and passed beneath a vertical follower, the follower would have the same motion as if a rotating cam were used. Figure

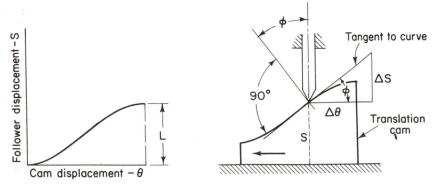

Figure 9.23 Displacement diagram and translation cam

9.23 shows that the pressure angle ϕ on such a cam would be bounded by the slope of the curve and a perpendicular to the path of the follower. In trigonometric terms, the tangent* of angle ϕ would be $\Delta S/\Delta\theta$ in the right triangle shown.

If these same displacements were plotted on a conventional rotation cam, the same triangle can be drawn, as shown in Figure 9.24. The pressure angle ϕ lies between the tangent to the cam contour and a perpendicular to the follower path as before. The altitude of the triangle is ΔS as before, but the base is now equal to $r\Delta\theta$, since it is the straight length of an arc of radius r and the subtending angle $\Delta\theta$.† So the tangent of ϕ now equals $\Delta S/r\Delta\theta$. But, as shown in Figure 9.24, radius r equals the radius of the base circle R^B plus the linear displacement (S) of the follower from the base circle to its present level. If $r = R^B + S$, then

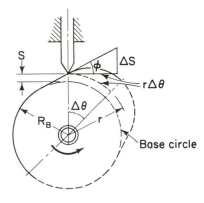

Figure 9.24 Minimum base circle for a given pressure angle

$$\tan\phi = \frac{\Delta S}{(R^B + S)\Delta\theta} \quad \text{or} \quad \frac{\Delta S/\Delta\theta}{R^B + S}$$

This equation can be used to calculate the minimum radius of the base circle for a given cam, if we use the proper values. The maximum permissible pressure angle is 30°, so tan ϕ becomes tan 30°, or 0.577. The value of $\Delta S/\Delta\theta$ may be obtained graphically from the steepest slope of the displacement diagram of the lift motion only,‡ as a follower cannot jam when descending.

> *The tangent of an angle of a right triangle equals the side opposite the angle divided by the adjacent side.
> †An arc length is equal to its radius multiplied by the subtending angle in radians.
> ‡This maximum value of $\Delta S/\Delta\theta$ is obtained as in graphical differentiation in Chapter 5.

The value of S will be measured from the displacement diagram at the point where $\Delta S/\Delta\theta$ is maximum (where the slope is steepest) (see Figure 9.23). In the equation above, solving for R^B:

$$R^B = \frac{\Delta S^{\max}/\Delta\theta}{0.577} - S$$

This will give the minimum radius of the base circle without danger of jamming the follower. In other words it establishes the proper distance between the axis of the cam and the lowest point on the path of the follower (AP in Figure 9.25). This is an important layout measurement in the design of a cam.

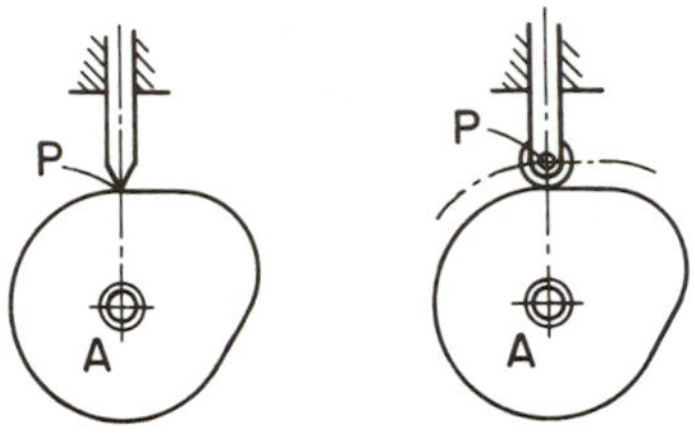

Note that, in the case of a roller follower mechanism, the base circle radius is measured from the axis of the cam shaft A to the center of the roller follower P, when the follower is in its lowest position. It is not the minimum radius of real cam surface.

Figure 9.25

9.14 Lift Motions for Cam Followers

When a designer is establishing the displacement schedule for a cam design, the kind of motion selected for the lift portion of the travel of the follower is important. The total displacement is usually specified by the application, but the lift motion is often a matter of choice. Three motions are in general use.

9.15 Uniformly Accelerated Motion*

We have studied this motion in Chapter 4, Article 4.2. For cam design purposes, we are mainly interested in the displacement equation, which is $S = V^0\Delta T + a(\Delta T)^2/2$. If the follower starts from rest, $V^0 = 0$ and $S = a(\Delta T)^2/2$. It is convenient to know that if progressive equal time intervals are substituted in this equation, the successive corresponding values of S will be in the ratio of $1 : 3 : 5 : 7$ and so on.

If $\Delta T = 1$, $S_1 = a/2$.

If $\Delta T = 2$, $S_2 = 2a$ and $S_2 - S_1 = 3a/2$.

If $\Delta T = 3$, $S_3 = 9a/2$ and $S_3 - S_2 = 5a/2$.

*This is sometimes called parabolic motion, since its displacement-time curve is composed of two parabolas. It is also called gravity motion, as a falling body has uniform acceleration.

Therefore, as a point travels along a line from A to B to C to D, the ratio of distances AB to BC to CD will be equal to $a/2$ to $3a/2$ to $5a/2$, or $1:3:5$, etc.

When designing a lift motion for a follower, we wish to accelerate for half the lift, then decelerate for the remainder of the lift. For example, if the lift of a follower totaled 3.6 in. and we wished to determine the displacements for six equal time intervals with uniform acceleration and deceleration, we would divide 3.6 into six parts in the ratio of $1:3:5:5:3:1$. These units total 18, so the unit of displacement will be $3.6/18 = 0.2$ in. and the successive displacements are 0.2, 0.6, 1, 1, 0.6, and 0.2. If desired, this division may be done graphically, as shown in Figure 9.26. If AC is the total lift, we draw an oblique line from A at any convenient angle. Using any convenient unit, we lay out from point A on this line 1 unit, 3 units, 5 units, etc,, in the required order. At the end of the group (point B), we draw a straight line to C and lines parallel to BC from the other divisions. These parallels will intercept distances on AC in the proper ratios.* The displacement diagram for this motion is shown in Figure 9.26.

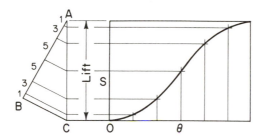

Figure 9.26 Displacement diagram for uniformly accelerated and decelerated motion

If extreme accuracy is demanded, total displacements from the starting point may be obtained from the *displacement table* in Figure 9.27. This is a nondimensional table listing 20 equal divisions of the lift angle (total cam displacement during the lift). The first column lists cam angles divided by the lift angle. If these numbers are each multiplied by the lift angle to be used in the particular application, they will specify the successive cam displacements. Similarly, the second column lists follower displacements divided by the total lift L. To obtain the successive follower displacements, we multiply the tabulated numbers by the total lift.

For example, if the lift angle in a cam being designed is 90° and the lift is 3 in., to find the first displacement we multiply 90 by 0.05 (from the table),

*The proof of this, by use of similar triangles, has been given at several other places in this text.

obtaining a cam angle of 4.5°. The corresponding follower displacement is
3 × 0.005 (from the table), or 0.015 in. Thus, we find this type of table
adaptable to any lift or lift angle. If 20 stations are not required, 10 may be
used by omitting intermediate tabulations.

DISPLACEMENT TABLE FOR CAMS

Cam Displacements	Uniform Acceleration	Simple Harmonic	Cycloidal
$\dfrac{\theta}{\text{Lift Angle}}$	$\dfrac{S}{\text{Lift}}$	$\dfrac{S}{\text{Lift}}$	$\dfrac{S}{\text{Lift}}$
0	0	0	0
.05	.005	.006	.0008
.10	.02	.024	.006
.15	.045	.055	.021
.20	.08	.096	.049
.25	.125	.146	.091
.30	.18	.206	.148
.35	.245	.273	.221
.40	.32	.346	.306
.45	.405	.422	.401
.50	.500	.500	.500
.55	.595	.578	.599
.60	.68	.655	.694
.65	.755	.727	.779
.70	.82	.794	.852
.75	.875	.854	.909
.80	.92	.905	.951
.85	.955	.946	.979
.90	.98	.976	.994
.95	.995	.994	.9992
1.00	1.000	1.000	1.000

θ = Cam displacement S = Follower displacement
Lift angle = Total cam displacement during lift   Lift = Total follower displacement

Figure 9.27 Displacement table for uniformly accelerated, simple harmonic,
and cycloidal motion.

9.16 Simple Harmonic Motion*

This is the accelerated and decelerated motion as produced by the Scotch
yoke (Article 8.9). Since it has a smoother acceleration curve than uniformly
accelerated motion, it is often used as the lift motion for cam followers. The
displacements for a given lift are easily determined graphically, as shown in

*Also called sinusoidal motion.

Figure 9.28. A semicircle is drawn with the given lift *AC* as a diameter. The arc is divided into as many equal divisions as there are stations in the layout. Perpendiculars to the diameter from these divisions of the semicircle intercept the proper displacements on the lift line *AC*. The displacement diagram for this class of motion is shown in Figure 9.28.

If greater accuracy is required, the table in Figure 9.27 also lists follower displacements for corresponding cam angles in simple harmonic motion. The use of this table is described in Article 9.15. It is much easier than solving the displacement equation for each station analytically.

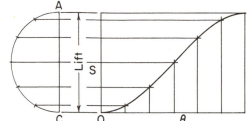

Figure 9.28 Displacement diagram for simple harmonic motion

9.17 Cycloidal Motion

Cycloidal motion is so named because the displacements of the follower are the same as those of a point on a circle as it rolls along a straight line, describing a path called a *cycloid*. It is the best motion for high-speed cam followers, since the acceleration curve is continuous and smooth, with zero acceleration at the start and end of the lift motion. This eliminates shock and vibration and gives a dynamic performance far superior to harmonic or uniformly accelerated motion. A high degree of precision is demanded in the manufacture of a cycloidal cam, with the result that it is the most costly design of the three.

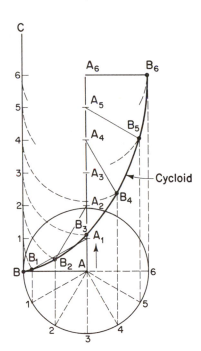

Figure 9.29 Layout of a cycloid

The layout of half a cycloid is shown in Figure 9.29. The generating circle rolls upward along line *BC*. The path of point *B* on the circle is a cycloid. Successive positions of the radial line *AB* at each 30° position of the circle are shown. Segment 1-2 on line *BC* equals arc 1-2 on the circle, etc.

A displacement diagram for cycloidal lift motion is shown in Figure 9.30. The complete cycloid, generated by one revolution of the circle, is drawn at the left. During this revolution, the center of the circle will travel a distance equal to the circumference ($2\pi r$). This distance is made equal to the lift *L*.

Since $2\pi r = L$,

$$r = \frac{L}{2\pi}$$

Here the positions of the generating point *B* are shown for each 30° of rotation of the generating circle. The vertical displacements of point *B* are projected horizontally to the follower displacement diagram at the right, thereby making the follower displacements equal to those of point *B*.

Since the generating circle will always be small as compared to the displacement curve, it is practically impossible to obtain sufficiently accurate displacements for a cycloidal cam design from a layout. The precision demanded for cycloidal motion requires that displacements be calculated from an equation. The above layout is offered only as a graphic description of the displacement curve. The table of displacements in Figure 9.27 offers accurate calculated values for cycloidal motion. This table must be used to obtain displacements when designing a cycloidal cam, as small inaccuracies in the design will destroy all the advantages of this motion. The use of this table is explained in Article 9.15.

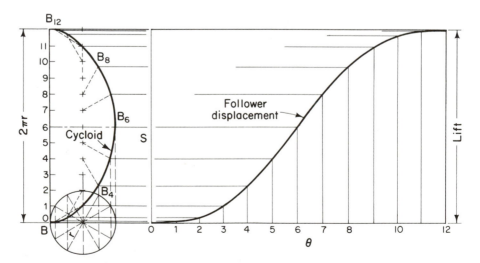

Figure 9.30 Displacement diagram for cycloidal motion

9.18 Acceleration Analysis of Cams

If the lift motion of a cam cannot be one of those described in the previous article, a plot of the acceleration of the follower will be needed in order to design a return spring and make a dynamic study. If an equation of the follower displacement in terms of time can be written, this can be differentiated analytically to obtain velocity and the velocity equation can be differentiated in turn to obtain acceleration. With an irregular motion this equation is very difficult to devise, so we usually use a graphical method. Two successive graphical differentiations rarely produce very accurate results. If we can plot a velocity curve by more accurate methods, this can be differentiated graphically to obtain a reliable acceleration curve.

9.19 Graphical Velocity Analysis of Cams

If the cam contour is drawn, we can use the same method employed in linkage analysis to obtain an equation for graphical velocity study. A cam C with roller follower F is shown in Figure 9.31. The motion of F will be the same if we use the equivalent system of a pointed follower (with the point at the roller center P), driven by the pitch line contour of the cam, since this is the method of design.

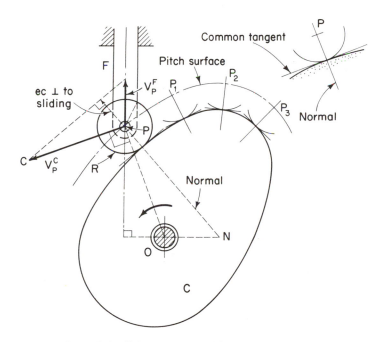

Figure 9.31 Velocity analysis of follower of a given cam

The velocity of point P on the pitch cam will be perpendicular to OP and equal to $\omega_c \times OP$. Since there is sliding contact between the assumed pointed follower and the pitch surface of the cam, the velocity of P on C will have the same effective component perpendicular to sliding (i.e., along the normal to the pitch surface) as the velocity of point P on follower F, shown by the sketched vectors. These vectors form a triangle Pcf in which side cf is parallel to the tangent (or perpendicular to the normal to the pitch surface at P).

If a line is drawn through O, perpendicular to the path of F, to meet the normal extended at N, triangle OPN will be similar to triangle Pcf (corresponding sides are perpendicular). Therefore:

$$\frac{Pf}{Pc} \quad \text{or} \quad \frac{V_P^F}{V_P^C} = \frac{ON}{OP} \quad \text{and} \quad V_P^F = \frac{ON}{OP} \times V_P^C$$

Substituting $\omega_c \times OP = V_P^C$, we have:

$$V_P^F = \frac{ON}{OP} \times \omega_c \times OP \quad \text{or} \quad V^F = ON \times \omega_c$$

ω_c must be in radians per unit of time, and ON is drawn perpendicular to the path of F to meet the normal to the pitch line through P.

To establish the direction of the normal, a common tangent to the real cam surface and the roller arc is drawn by eye.* The normal is drawn perpendicular to this tangent through P. Roller arcs may be drawn at the several station points P_1, P_2, P_3, etc., around the cam, as shown in Figure 9.31, to avoid redrawing the cam in the several positions.

The values of the velocity of F obtained by this method are plotted versus time at suitable scales, and the velocity curve is drawn. This velocity curve is then differentiated graphically (as described in Chapter 5) to obtain the required acceleration graph.

PROBLEMS

In these problems all cam displacements are given in degrees, measured from the reference line. All follower displacements are given in inches, measured from the starting position. Follower displacements marked plus ($+$) are to be measured in a direction outward from the cam axis (so as to increase the cam radius), and those marked minus ($-$) are to be measured in a general direction toward the cam axis (so as to decrease the cam radius).

*Experience has shown that this can be done with very high accuracy. The two tangent curves serve as an excellent guide to the eye.

9.1. Design a radial cam to turn clockwise about a fixed axis A and drive a pointed follower, with point at P, according to the following displacement schedule:

Cam displacement, θ	Follower displacement, S	Motion of follower, F
0 to 90	0 to $+1$	Uniform velocity
90 to 180	$+1$ to $+0.5$	Uniform velocity
180 to 270	$+0.5$ to $+1$	Uniform velocity
270 to 360	$+1$ to 0	Uniform velocity

Make the base circle radius AP equal to 2 in.

9.2. Design a radial cam to turn counterclockwise about axis A and drive a roller follower F as specified below. The roller is 1 in. in diameter with center at B. Distance AB is to be 2 in.

θ	S	Motion of F
0 to 120	0 to $+1$	Harmonic motion
120 to 180	$+1$	Dwell
180 to 240	$+1$ to 0	Harmonic motion
240 to 360	0	Dwell

9.3. Design a cam to turn about axis A and drive a flat translating follower F as specified below. The center line of the fixed guides for the follower passes through A. The contacting surface of F is perpendicular to this center line. The radius of the base circle is 2 in.

θ	S	Motion of F
0 to 180	0 to $+0.9$	Uniform acceleration and deceleration
180 to 360	$+0.9$ to 0	Uniform acceleration and deceleration

9.4. In the linkage shown in Figure P9.4, BC is 2 in. long and CF is 3 in. long. A cam turning clockwise about axis A drives the linkage through contact with roller C, of $\frac{3}{4}$ in. diameter, causing F to move up and down the vertical guides. Starting in the position shown, F rises 1.2 in. in 4 sec at uniform velocity, remains stationary for 2 sec, then returns to the starting position with uniform velocity in 5 sec and remains in this position for 1 sec. This motion is repeated every 12 sec. Design the cam and specify its angular velocity.

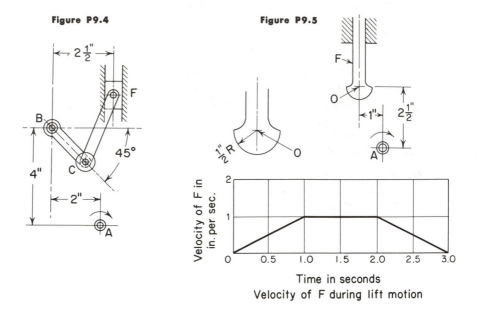

Figure P9.4

Figure P9.5

Velocity of F during lift motion

9.5. The round follower F shown in Figure P9.5 is driven by a plate cam turning clockwise at constant speed about fixed axis A. The lift motion of F takes place during the first $180°$ of cam rotation. The velocity of F during the lift motion is shown in the velocity-time plot. The follower dwells for the next $60°$ and returns to the starting position with simple harmonic motion in the final $120°$ of cam rotation.

(a) Determine the displacements and design the cam.

(b) Report the angular speed of the cam.

(c) Determine the maximum acceleration of the follower during the lift motion.

(d) Determine the maximum pressure angle during the lift motion.

9.6. Design a positive motion cam with a follower of the type shown in Figure P9.6. The follower is to make 20 complete strokes (up and down) per min with simple harmonic motion. The distance from the upper to the lower position is to be 1.4 in. The cam is to turn about fixed axis A at constant speed, maintaining continuous contact with surfaces B and C.

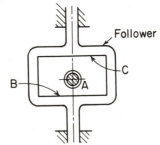

Figure P9.6

9.7. In Figure P9.7 the arms of the bell-crank lever BC and BE are attached to the same hub at B, so that they maintain the $105°$ angle between them. The lever turns a follower shaft in fixed bearings at B. $BC = BE = 3$ in. Rollers Q and R at C and E are $1\frac{1}{2}$ in. in diameter. Two cams, M and K, are fastened together. M drives roller Q, and cam K drives R. The cams turn clockwise at constant speed about fixed axis A. The shaft at A is $\frac{7}{8}$ in. diameter and the shaft at B is $\frac{3}{4}$ in. diameter.

Starting in the position shown, the bell-crank lever is to turn counterclockwise through an angle of $45°$ with uniformly accelerated and retarded motion in 1 sec, then dwell for $\frac{1}{2}$ sec, then turn clockwise for $45°$ to the original position with uniformly accelerated and retarded motion in 1 sec and dwell in this position for $\frac{1}{2}$ sec. This motion is to be repeated every 3 sec.

Design the cam K to drive R and cam M to contact Q during this motion, thus making a positive drive. Specify speed of cams.

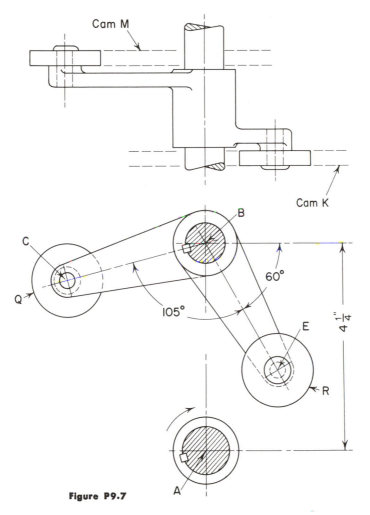

Figure P9.7

9.8. The mechanism shown in Figure P9.8 consists of a horizontal bar C which slides in fixed guides G and carries a roller cam follower T on pin E at its left end. The vertical guides S are also mounted on C. A second roller follower R turns about pin F on block J, which in turn slides freely in the guides S. A grooved-disk positive-motion cam, turning about fixed axis A, drives roller T, and a plate cam, turning about fixed axis B, drives R. Both rollers are 1 in. in diameter, and both cams turn counterclockwise at the same constant speed.

The cams are to be designed so that point F follows the D-shaped path during each revolution, starting in the position shown and traversing the straight portion first. The cam at A provides the horizontal motion of C, and the cam at B supports F at the correct height. Use 150° of cam rotation for the straight vertical path and 210° for the remaining portion. (Use 10 equally spaced design stations along the straight vertical path and 14 stations equally spaced along the rest of the path.)

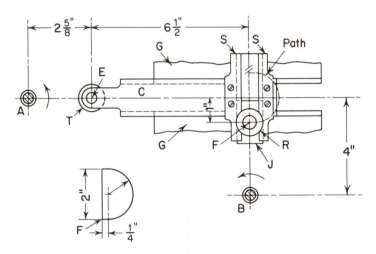

Figure P9.8

9.9. Bar M in Figure P9.9 slides in the horizontal guides G. A link OP is pinned to M at O and carries a $\frac{3}{4}$ in. diameter roller R on a pin at P. There is a pointed follower on the left end of M. Point P is to follow the outline of the circular segment shown, from A to B to C to A. Two cams turning counterclockwise at the same speed about axes D and E and contacting the pointed follower F and the roller R are to drive P along this path. The cams are to turn 120° while P traverses line AB and 120° while P traverses arc BC, completing the revolution as P returns to A. (Use eight equally spaced design stations on each side of the segment.) The cam at D provides the horizontal displacements of P, while the cam at E maintains P at the proper level. Springs (not shown) ensure contact between cams and followers.

9.10. Design a high-speed cam to open and close a valve with each revolution. The lift is 2 in. and occurs during 90° of cam rotation. The follower dwells for 90°, then drops 2 in. during 90° of cam rotation, and dwells for the last 90°. Use a

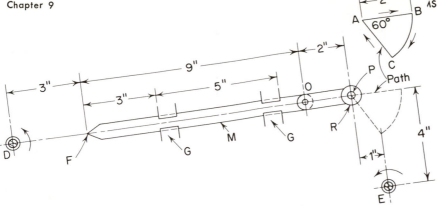

Figure P9.9

radial cam with a 1-in. roller follower. The pressure angle is not to exceed 30°. Use the minimum permissible base circle. The lift motion is to be cycloidal.

9.11. The design of a machine requires that a point P traverse the path shown in Figure P9.11. Starting at A, it is to travel the straight line path to B in 3 sec and then return by the curved path from B back to A in 6 sec. This motion is to be repeated every 9 sec. From A to B (straight path), point P is to move in simple harmonic motion. From B to C, point P is to be uniformly accelerated, and from C to A, it is to be uniformly decelerated along the curved path. This problem calls for the design of a mechanism to guide and propel point P along the prescribed path at the specified speeds. A cam-driven mechanism which is adapted to this task is shown in Figure P9.11. The pin P is capable of following any path in a plane when driven by two radial plate cams. The return spring will insure contact of the roller followers with the cams. The design of the mechanism as to the sizes of members, location of the path relative to the mechanism, and the location of fixed axes and spring anchor is to be worked out by the student. It is important that the device be as compact

Figure P9.11

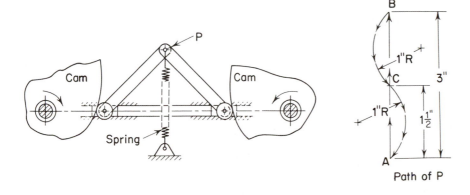

Path of P

as is reasonable, so the base circles of the cams should be minimized. The length and extension of the spring should be realistic. A full-size assembly drawing of the mechanism is required (no detailed dimensions will be necessary), including a precise layout of the cam surfaces.

9.12. As is shown in Figure P9.12, cam C turns about fixed axis O. Follower F swings about P, always maintaining contact with the cam. Derive an equation for the angular velocity of F in terms of the angular velocity of C and certain accessible measured distances.

If cam C turns 10 rpm, determine the angular velocity of F in the position shown, using the equation above.

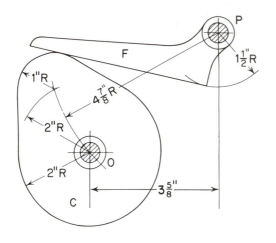

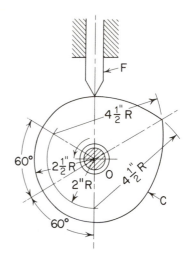

Figure P9.12   **Figure P9.13**

9.13. The radial cam C, shown in Figure P9.13, turns counterclockwise about fixed axis O at a constant speed of $8\frac{1}{3}$ rpm, driving pointed follower F.

(a) Devise an equation for the velocity of F in terms of ω_C and accessible measurements.

(b) Using this equation, determine velocities of F during the lift motion and plot a velocity-time curve. (Use velocity scale of 1 in. = 0.5 in./sec and time scale of 1 in. = 0.2 sec.)

(c) Determine the maximum acceleration of F during the lift.

9.14. A mechanism is needed to guide and propel a pin P along a *3-dimensional path* composed of 4 half-circles which are positioned in space as if they were lying on 4 faces of a $1\frac{1}{4}$-in. cube. See the path 1, 2, 3, 4 in Figure P9.14. A design of such a mechanism is shown in the illustration. Pin P joins 2 links L (4″ long from pin to pin), which are attached to sliders S. These sliders are splined to shaft E so that they must turn with the shaft, yet are free to

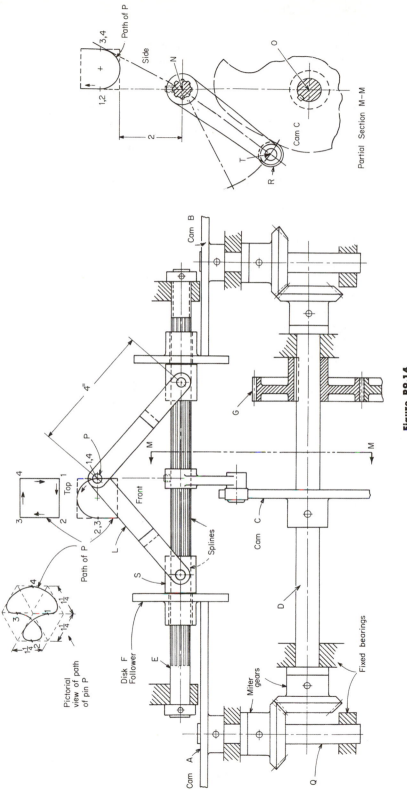

Figure P9.14

slide along it axially. These sliders are positioned by 2 circular plates or disks, *F*. These are carried by the shaft *E* but ride on top of the splines so that they may turn freely about the shaft. These disks *F* are in reality flat followers for cams *A* and *B*, which turn on shafts *Q*, positioned at right angles to the splined shaft *E*. Shafts *Q* are driven from the main drive shaft *D* by miter gears as shown. Shaft *D* is driven by the spur gears *G*. The splined shaft *E* is positioned radially by the arm *NT*, which is turned by cam *C*. This cam is mounted upon and keyed to the drive shaft *D*. A roller follower is mounted on pin *T* in contact with cam *C*.

This problem requires the design of cams *A*, *B*, and *C* so that the center of the pin *P* will traverse the path described above from point 1 to point 2 to point 3 to point 4 and back to point 1. *P* is to travel at constant speed throughout the motion. Point *P* is required to complete one motion cycle in 12 sec, traversing each arc in 3 sec. (It is suggested that stations along this path be chosen at $\frac{1}{2}$-second intervals when designing the cams.) The position of the cam shafts relative to the follower which is shown on the layout is intended to be reasonable but should be verified by the student as part of the design process (a 3-in. radius base circle for cams *A* and *B* and a $2\frac{1}{2}$-in. radius base circle for cam *C* is suggested). The speed of the drive shaft is to be reported in rpm, and the direction of rotation of all cams (shafts) is to be indicated.

No provision has been shown on the drawing to insure contact between the cams and their followers at all times. A solution to this requirement is to be submitted in the form of a sketched diagram with explanatory notes. It is suggested that separate drawings of the cam profiles be made, including the position of followers and their paths relative to the cam axes. It is not necessary to redraw the entire mechanism showing cams in proper assembly.

Mechanisms

10

The design of every machine or device starts with the design of a mechanism which will produce the specified motion. Over the years, many mechanisms have been designed to do innumerable tasks and a number of mechanisms have been found to be adaptable to many different purposes. It is not necessary that an entirely unique mechanism be developed for each new task to be performed. Neither is it reasonable to go back and "redesign the wheel" for each assignment, when there is known to be an existing mechanism that is well-suited to the new application.

It has been pointed out that the acquisition of a "vocabulary" of mechanisms is an important part of a designer's education. This is not ordinarily acquired by a systematic study of a collection of classic mechanisms but is usually absorbed gradually, through exposure to typical examples when motivated by specific design assignments.

The author has sought to familiarize the reader with a number of well-known mechanisms by using these as examples for the application of techniques of exploration, analysis, and design. Many of the problems involve the kinematic analysis of classic mechanisms and design assignments demanding the selection and adaptation of these devices to fill specified needs.

It is, of course, not possible to expose the student to very many different mechanisms in this way. Courses are becoming shorter and shorter and assignments cannot possibly offer experience with enough different mechanisms to provide more than a nucleus upon which to build.

While this chapter will by no means provide a comprehensive review of known useful mechanisms, it is intended to display examples of a number of basic mechanisms to further acquaint the reader with what has gone before.

It is expected that this part of the book will be used largely for reference—especially as a source of ideas and a stimulus to the ingenuity of one who is involved in a design problem.

The presentation will be substantially pictorial and the text will be brief, with no attempt to offer explicit or detailed kinematic analyses. This chapter might well serve as a source of challenging problems in exploration, analysis, and design to supplement those classified assignments included in the text. It is expected that the theories and methods expounded in the earlier chapters will provide concepts and techniques which the student may effectively apply to evaluate the additional mechanisms which follow.

10.1 Crank and Rocker Linkage   (Figure 10.1)

A *crank-and-rocker linkage* is a four-bar linkage in which:

$$a + b + c > d$$
$$a + d + c > b$$
$$a + b - c < d$$
$$b - a + c < d$$

Input crank a makes complete revolutions. Output crank c oscillates through angle θ.

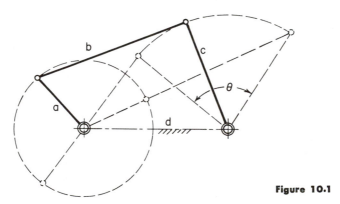

Figure 10.1

10.2 Drag Link   (Figure 10.2)

A *drag link* is a four-bar linkage in which both cranks make complete revolutions, and in which the members are proportioned as follows:

$$b > c + d - a$$
$$b > a + c - d$$

Figure 10.2

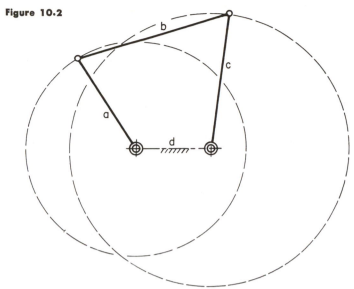

10.3 Double Rocking Levers

With certain proportions of bar lengths, both cranks of four-bar linkages are limited to oscillating or rocking motion. Depending on the position in which the links are assembled, the angle of oscillation of the cranks will differ for the same linkage. Since it is possible for this type of linkage to assume "dead-center" positions, it is necessary that stops be provided to limit crank strokes. In this way the bars can be prevented from reaching those critical positions which lead to indeterminate motion, reversal of crank stroke, or jamming.

In Figure 10.3(a)

Figure 10.3(a)   **Figure 10.3(b)**

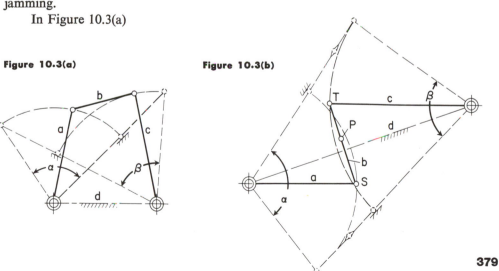

$$b + c < a + d \quad \text{and} \quad a + b < c + d$$

The limiting angle of oscillation of a is angle α, and a stop must be provided as shown to prevent links b and c from reaching a straight line position. It is also necessary to prevent a and b from reaching a similar dead-center position.

Figure 10.3(b) shows another form of the rocking-lever linkage. The limiting positions of the cranks are shown with the stops to prevent cranks from swinging too far. The maximum angle of oscillation of a is α and that of c is β. These linkages are rarely used as driving or propelling mechanisms but are used rather as guides to the motion of significant points or members. This linkage is a form of Watts' straight-line mechanism, in which a portion of the path of point P on link b is an approximate straight line. This is assured by the ratio of a to c being made equal to the ratio of TP to PS (see also Figure 2.7).

10.4 Parallel-Crank Linkage

The *parallel-crank linkage* is a four-bar linkage in which both cranks make complete revolutions at the same speed and the coupler is always parallel to the frame.

$$a = c \quad \text{and} \quad a \text{ is parallel to } c$$
$$b = d \quad \text{and} \quad b \text{ is parallel to } d$$

Cranks a and c always turn at the same speed. See Figure 10.4.

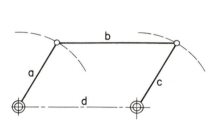

Figure 10.4

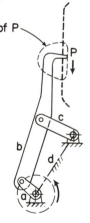

Path of P

Figure 10.5

10.5 Transport Linkage

The *transport linkage* is a four-bar linkage in which the coupler has been extended to point P, so located that a portion of the path of P is an approximate straight line and the rest is curved, as shown. The hook at P may be used to intermittently transport film or perforated tape as shown in Figure 10.5.

10.6 Pin-Joint Toggle Linkage

This linkage is used where a high mechanical advantage (ratio of output force to input force) is desired. Toggle linkages may have many different forms, but a simple example is the four-bar linkage in Figure 10.6(a). A small force F_{IN} applied at pin B will produce a very large force F_{OUT} at pin C, when the angle β approaches 180°. This would result in a high torque on the crank CD applied through a very small angle. This linkage is adapted to clamping devices by providing a stop at surface S where pin B has just passed to the other side of a straight line from A to C. A clamp employing another toggle linkage is shown in Figure 10.6(b).

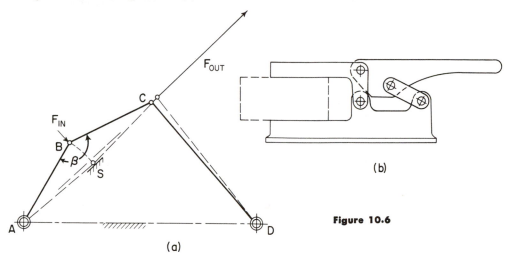

(a)

(b)

Figure 10.6

10.7 Straight-Line Mechanism

This is an inverted Peaucellier linkage (see Figure 2.10) in which the members labeled a are equal in length. Members labeled b are equal to one another but are longer than the a links. The input crank is AB, and the output path traced by pin P is an exact straight line, perpendicular to the fixed frame AC. (Figure 10.7)

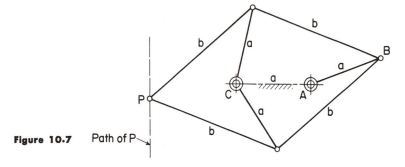

Figure 10.7   Path of P

10.8 Pantograph

These linkages are used to magnify or reduce displacements proportionately. Two of many examples are shown. In Figure 10.8(a), RM and LS are continuous rigid members turning about fixed axis O.

$$TR = RO = OS = ST \quad \text{and} \quad PM = MO = OL = LP$$

In Figure 10.8(b), OQ and OW are continuous members turning about O.

$$OQ = OW, ON = OY, PN = PY, \quad \text{and} \quad TQ = TW$$

In each case, any figure traced by point P will be reproduced at large scale by point T. The enlargement is in the ratio of distances $OP : OT$ (see also Figure 2.16).

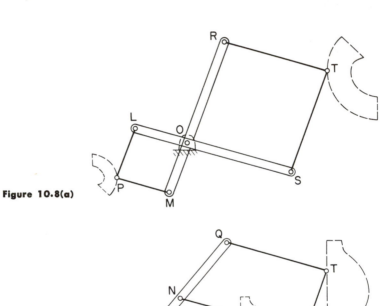

Figure 10.8(a)

Figure 10.8(b)

10.9 Slider Crank Linkage

This is a basic variation of the four-bar linkage, in which one crank translates and is replaced by a sliding block. This mechanism has countless applications, many of which may be recognized in the examples to follow.

Crank *AB* can make complete revolutions, as in Figure 10.9(a), or it can oscillate, as in Figure 10.9(b), about fixed axis *A*. Slider pin *C* reciprocates along straight guides, fixed in the desired position. The application shown in Figure 10.9(b) is interesting in that it provides a double stroke motion of the block *C*. In this case, the angle of oscillation of crank *AB* must be limited so that the block will not jam due to friction with the guides or attain a position where its motion becomes indeterminate. The swing to the left must be stopped when block *C* is at C_L, before angle α becomes much less than 120°, to prevent excessive friction. (If α diminishes to 90°, the motion of *C* will be indeterminate.) On the swing to the right it is possible for crank *AB* to pass beyond the position in which *AB* and *BC* are in line, unless a stop limits the size of angle β. The limit of *C*'s motion to the right is C_R, where *AB* and *BC* are in line, so as this in-line position is passed, block *C* will reverse direction and move a short distance to the left. When *AB* reaches this end of its oscillation, block *C* will return to position C_R before starting the long stroke back to C_L. Thus, block *C* travels one long and one short stroke during one typical cycle of motion. An auxiliary four-bar linkage (*ABDE*) is shown in Figure 10.9(b),

Figure 10.9(a)

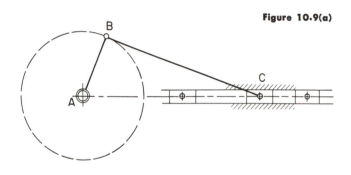

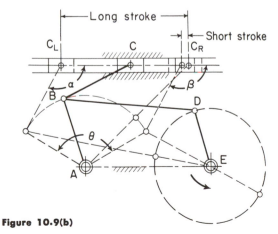

Figure 10.9(b)

which drives *AB* through the desired angle of oscillation with continuous rotation of the input crank *ED*. (See also Figure 2.18.) This two-stroke motion is sometimes very useful.

10.10 Sliding-Coupler Linkage

In Figure 10.10(a) crank *AB* turns about fixed axis *A*. The block swings freely about a fixed pin at *D*. Coupler *BC* slides through the block, causing it to oscillate. This linkage has also been used on oscillating cylinder devices with the coupler driven by a piston and the crank *AB* as the output member. Figure 10.10(b) shows that, if the crank *AB* is longer than the distance *AD*, then *BC* will cause the block to make complete revolutions about *D*.

Figure 10.10(a)

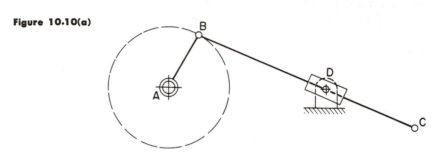

Figure 10.10(b)

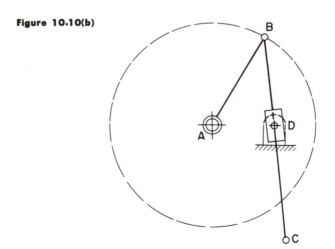

10.11 Oscillating-Beam Mechanism

This is another inversion of the slider-crank linkage, one in which a block B is pinned to the input crank AC so as to turn freely about pin C. The beam EF slides through block B and rotates about fixed axis E. If the crank AC is shorter than the distance AE between the fixed axes, the beam will oscillate as in Figure 10.11(a). If AC is longer than AE, as in Figure 10.11(b), the beam will make complete revolutions about E. This linkage has been extensively used on quick-return mechanisms. (See Figures 8.25 and 8.29.) Figure 10.11(c) shows the use of this linkage as an angular motion amplifier. Here the input crank EF oscillates through a small angle θ while AB turns through $180°$.

Figure 10.11(a)

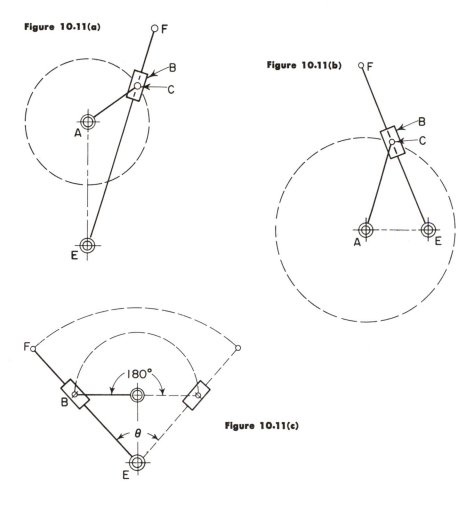

Figure 10.11(b)

Figure 10.11(c)

10.12 Scotch Yoke

This is a classic mechanism for the conversion of rotation to translation in which the translating member moves in simple harmonic motion. (See Figure 8.23 and 9.28.) The driving crank *AC* turns at constant speed about fixed axis *A*. Block *B* is pinned to turn on *C* and also slides in the straight slot in the yoke *Y*, causing the yoke to translate up and down. See Figure 10.12. The slot in the yoke is at 90° with the fixed guides, and the stroke of *Y* will be equal to twice the crank length *AC*.

Figure 10.12

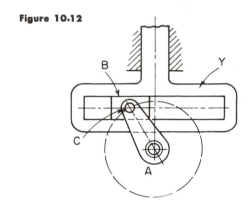

10.13 Modified Scotch Yokes

If the slot in the yoke is not at 90° with the fixed guides, as shown in Figure 10.13(a), the motion of *Y* will not be harmonic and the stroke of *Y* will be equal to $2 \times AC \div \sin \alpha$. Figure 10.13(b) shows another modification of

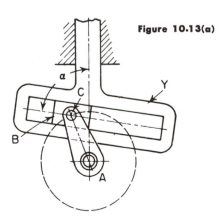

Figure 10.13(a)

Figure 10.13(b)

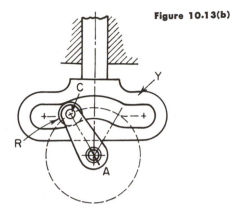

this mechanism which introduces a dwell into the reciprocating motion of the yoke. The slot is partly circular with the central portion at the same radius as the crank-pin circle. A roller R replaces the rectangular block B in order to adapt to this shape of slot. If the extent of the curved slot is 90° of the crank circle, the yoke will remain stationary at the top of the stroke for one quarter of a revolution of the crank. When moving, the motion of the yoke is not harmonic, but has similar characteristics. The dwell period may be varied, but it should not exceed 90° of crank rotation.

10.14 Sliding Toggle Linkages

The slider-crank linkage may be used for toggle mechanisms. Figure 10.14(a) shows one form of open toggle, while Figure 10.14(b) displays an inverted toggle which is more compact. In each case, a small force F_{in}, applied as shown at pin B, exerts a very large output force F_{out} at the block C when links AB and BC are nearly in line. The locking feature may be achieved by the fixed stop S.

Figure 10.14(a)

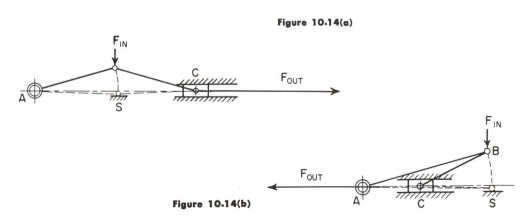

Figure 10.14(b)

10.15 Isosceles Linkage

This is another application of the slider-crank mechanism, one which can perform two useful functions. Figure 10.15(a) shows an isosceles linkage (so called because *AB* and *BC* are equal) in which the coupler *BC* is extended as a rigid member from *C* to *F* so that *BC = BF*. As the crank *AB* turns, *F* will describe an exact straight line perpendicular to *AC*.*

In Figure 10.15(b), the same isosceles linkage *ABC* has an extended rigid coupler *BCE* with *E* located anywhere except at *B* or *C* on the coupler. As the crank *AB* turns, *E* will describe a true ellipse with a major axis of 2 times *AB + BE* and a minor axis of 2 times *CE*.

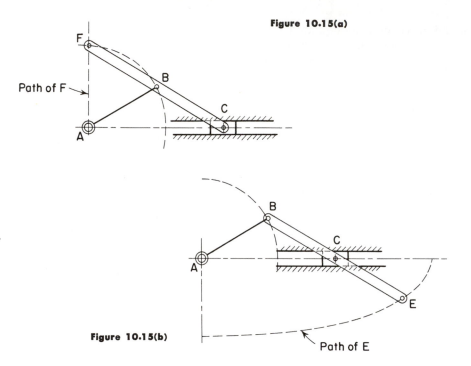

Figure 10.15(a)

Path of F

Figure 10.15(b)

Path of E

10.16 Revolving Crank with Dwells

The input crank *AB* makes complete revolutions (see Figure 10.16).† The output crank *GF* also makes complete revolutions but dwells for a short interval when *F* is in the position shown and again when *F* is at *F'*, this time only

*This is commonly known as the Scott-Russell linkage.
†A similar device is manufactured by the Beswick Engineering Co., Beverly, Mass.

momentarily. Point E, on the bent coupler bar BCE, traces the path indicated. One portion of this path, from points 1 to 2, is a circular arc about F, and another portion, from points 3 to 4, is circular about F'. In these positions, pin F and the output crank GF remain stationary while E traverses the circular paths. This type of linkage is not easy to design. One must first find a four-bar linkage in which a point on the coupler traces a path that has a circular segment. The link corresponding to EF in the mechanism illustrated here must then be made equal in length to the radius of the circular segment of the path of E. Next, the output crank, like GF, must be so located as to permit pin F to reach the center of curvature of the arc segment. Finally, the fixed center G must be placed so that crank GF makes continuous revolutions when driven by link EF. The relative dimensions of the links shown in Figure 10.16 are reported here so as to give some idea of proportions.* The dimension AG is 15.9 and, in the position shown, BE is 14.9.

Figure 10.16

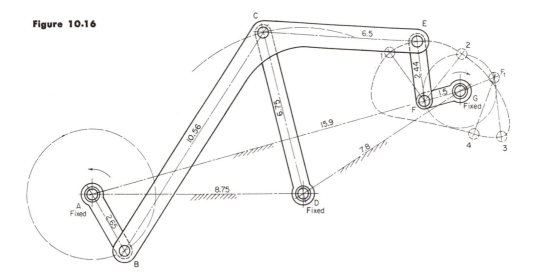

10.17 Oscillating Crank with Dwell

In the 5-bar linkage in Figure 10.17, the input crank AB makes complete revolutions. The output crank GF oscillates through angle θ but dwells momentarily when in the position shown. Point E on the coupler BEC traces the path indicated, a portion of which (from points E to E') is a circular arc about F. This pin F and crank GF will remain stationary while E traces the circular path from E to E'. In this example the angle of oscillation θ is 60°, and the time ratio of motion to dwell of the crank GF is 3 to 1.

*Hrones and Nelson's "Analysis of the Four-Bar Linkage" offers many examples of coupler curves having circular segments.

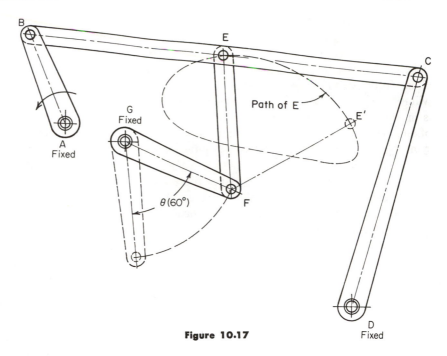

Figure 10.17

10.18 Oscillating Beam with Dwell

This is similar to the oscillating-beam mechanism but with a roller R in a curved slot in the beam M, instead of the straight rod with sliding block employed in 10.11. The curved portion of the slot in M is a circular arc having the same radius as the input crank AC. Thus, while R is traversing this arc-shaped slot, the beam M remains stationary.

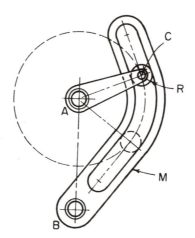

Figure 10.18

10.19 Symmetrical Coupler-Curve Linkage

The four-bar linkage *OLRM* in Figure 10.19 has fixed axes at *O* and *M*. Follower crank *MR* = *LR* and *RP* on the coupler. *LP* is the diameter of circle *C* on the semicircular coupler extension shown. A typical point *T* on circle *C* will trace a path which is symmetrical about axis *MT* drawn when the linkage is in the position shown, in which *OL* and *OM* are in line. This will be true of any tracer point on circle *C* on the coupler. This offers a wide variety of symmetrical curves which are often desired in design work.*

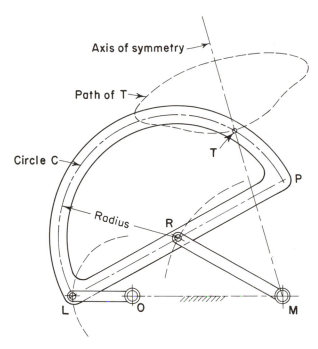

Figure 10.19

10.20 Linkage for Generating a Parabola

Figure 10.20 shows a linkage which will draw a parabolic curve of any desired focal distance. (See Figure 2.14 for another example.) The inclined *T*-shaped member has two slots in line and a third at right angles. Two equal links *LP* and *LM* are guided, as shown, by pins in these slots. The focus pin *F* is fixed and one slot in the *T*-member slides freely upon it. *F* is located at the required focal distance *OF* from point *O* which is the intersection of a

*Accredited to Stanley B. Tuttle in *Mechanisms for Engineering Design* (Wiley).

horizontal fixed slot S and a vertical fixed slot Q. Pin R in the T-bar engages slot Q, and pin M passes through a slot in the T-bar to engage in slot S. To draw a parabola, pin F is located at the desired focal distance (OF) from O and the tracing scriber is applied at point P. Sliding the T-bar in its various constraints causes P to describe a parabola with focus at F and vertex at O.*

Figure 10.20

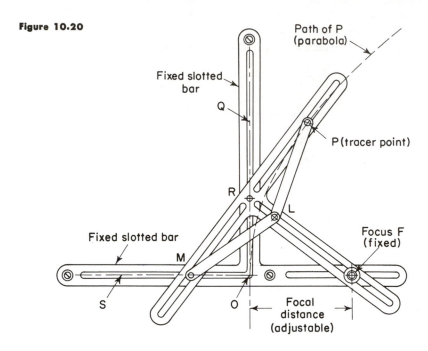

10.21 Geneva Mechanism

This is a classic mechanism for producing intermittent rotation which is always in the same direction. (See also Problem 9.10).

Figure 10.21 shows a 5-slot example in which the modified disk D is the driver and the Geneva wheel W is the follower. Disk D turns at constant speed carrying pin P, which is shown just about to enter a slot (angle BPA equals $90°$). As pin P swings, it slides in the slot, causing the Geneva wheel W to turn through an angle of $72°$, in this case. In position P', shown dotted, the pin leaves the slot and simultaneously the circular surface of the disk D engages the matching surface of W between the slotted lobes. This locks W in a stationary position until pin P is again in position to enter the next slot. The disk is cut away to clear the lobes while W is in motion. The number of slots can be varied to attain different motion to dwell ratios.

*Accredited to H. E. Schrank, Westinghouse Electric Corp.

Figure 10.21

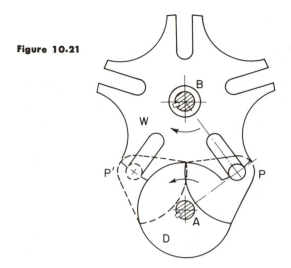

10.22 Linkage-Driven Geneva

Figure 10.22 shows another type of intermittent motion mechanism in which the slotted wheel is alternately driven and locked by a slider-crank linkage. In this example, the input crank *AB* turns clockwise at constant speed about fixed axis *A*, independently of the slotted wheel *W*. In the position shown, the driving pin *P* is just about to enter a slot in *W* while the locking pin *L* at the other end of the coupler is about to leave the horizontal slot and enter the aligned fixed guides *G*. (Pin *L* is attached to a block which is always constrained to slide in the guides.) The path of pin *P*, which is carried by the coupler, is shown dotted. With pin *L* disengaged from the other slot in *W*, pin *P* will rotate this wheel until it is in the position where it is just leaving the slot. As *P* emerges, the coupler will draw pin *L* into another slot, thereby locking *W* in position while *P* swings around to the next slot, ready to repeat the motion. *W* turns counterclockwise.

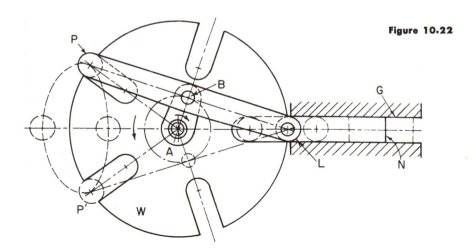

Figure 10.22

10.23 Pawl and Ratchet Mechanism

Figure 10.23 shows a simple mechanism for intermittent rotary motion. The toothed ratchet wheel R is driven by a link P, called a pawl. The pawl is carried on pin C of an oscillating driving crank AC, which swings about axis A, independently of the ratchet R. (If continuous rotational input is wanted, a supplementary linkage, shown dotted, may be used to drive crank AC.) When AC moves counterclockwise, pawl P is caused to swing down (by a spring not shown) and engage a tooth driving the ratchet through the desired angle. As this stroke of P ends, a retaining pawl S (actuated by another spring not shown) swings about fixed pin T to engage a convenient tooth. This prevents the ratchet from turning back when AC swings clockwise, disengaging P, leaving it ready for the next stroke. The pawl P will slide over the teeth in this return stroke as AC turns clockwise. Both pawls are so shaped as to engage the teeth only when moving in one direction. Smaller teeth and multiple offset pawls will provide a finer adjustment of indexing angle of the ratchet and will reduce backlash.

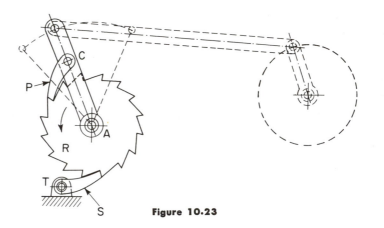

Figure 10.23

10.24 Friction Ratchet

It is possible to achieve intermittent motion with a ratchet which has no teeth. This has the advantage of limiting backlash and avoiding machining teeth. Figure 10.24 shows a friction ratchet, in which a smooth circular disk D is caused to rotate counterclockwise by an oscillating arm OP through the engagement of a shoe S, hinged at L to a link LM. Disk D and lever OP turn independently about fixed axis O. A similar holding link (TQ) is anchored to a fixed pin at T to prevent clockwise rotation during the return stroke of OP. Contact of the shoes with the disk is assured by springs not shown.

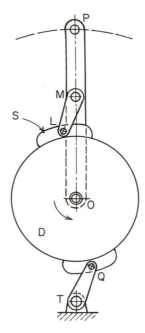

Figure 10.24

10.25 Cyclic-Motion Mechanism

In the mechanism shown in Figure 10.25, a constant-speed rotational input is applied to the flywheel *A*. The output member is the pinion *P*, which turns always in the same direction with an instantaneous stop at the end of each revolution of *A*. *A* and *P* turn about the same fixed axis. A rack *B* slides in a slot cut in the flywheel *A*, which keeps it in mesh with the pinion *P*. The rack is pinned at *E* to a circular strap which slides around a stationary disk *F*, which is positioned eccentric with flywheel *A* and the pinion *P*. This strap drives the rack back and forth in its slot, thereby rolling the pinion. A variable

Figure 10.25

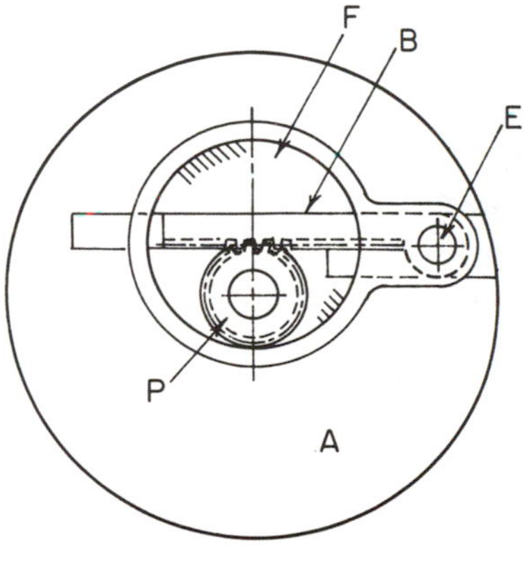

speed rotation is thus given to the output pinion with zero velocity and zero acceleration at the end of each cycle. This motion pattern is repeated with each revolution of the flywheel *A*.*

10.26 Eccentric-Driven Quick-Return Mechanism

Figure 10.26 shows an eccentric *W* which turns at constant speed around fixed axis *O*. A bar *RS* contains a strap which fits freely around *W*. *R* is connected to fixed pin *P* by link *PR*. A connecting rod *SK* is pinned to *S* and to a block at *K* which slides in fixed guides. If *W* turns clockwise, the slow stroke of *K* is left to right and the fast-return stroke is from right to left. In this example the time ratio is 11 : 7.

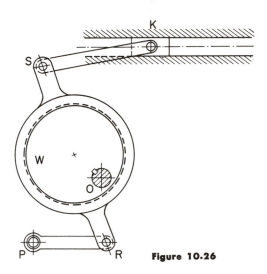

Figure 10.26

10.27 Differential Chain Hoist

In Figure 10.27 sprockets *A* and *B* are keyed to turn together about a fixed pin at *H*. (Sprockets are shown toothless for simplicity.) Pin *L* in sprocket *C* carries a load to be lifted. An endless chain *N* is wound around the 3 sprockets as shown with a downward input force applied at *P*. With sprockets in the proportions shown, the mechanical advantage (ratio of load lifted to force applied at *P*) is 8 : 1.

*Accredited to G. J. Talbourdet, United Shoe Machinery Corp., Beverly, Mass.

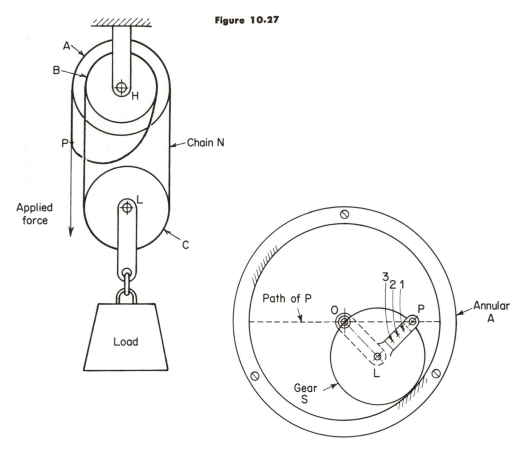

Figure 10.27

Figure 10.28

10.28 Epicyclic Straight-Line Mechanism

A spur gear S is constrained to roll around the inside of a fixed annular gear A, since it is mounted on the link OL shown in Figure 10.28. (Gears are shown here without teeth for simplicity.) The gear S has a pitch diameter exactly one half of that of annular A. A pin P is mounted on the pitch circle of S. As link OL is turned, P will traverse a straight line, which is a diameter of the annular A, and return. Pin P travels with simple harmonic motion.* This same mechanism is used on computing machines as a sine-cosine generator. The paths of points on the radius of S (see 1, 2, 3) are ellipses. The nearer the point to the pitch circle, the flatter the ellipse.

*This is commonly known as the Cardano circle mechanism.

10.29 Planetary Intermittent-Motion Mechanism

In Figure 10.29 a spur gear F is kept in mesh with a stationary spur gear D of identical size by the input crank AB. The planetary gear F rolls around D as crank AB turns about fixed axis A. A roller R is mounted on the pitch circle of F and moves in a slot in arm M describing an epicycloidal path as AB revolves. Slotted arm M turns about A independently of AB, its angular velocity going from zero to maximum and back to zero in one revolution of AB. This provides a very smooth motion of the output arm M.

Figure 10.29

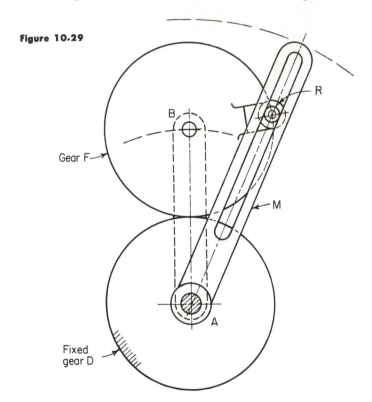

10.30 Cycloidal-Gear Speed Reducer

The mechanism shown in Figure 10.30 employs two rotating pinion gears that are driven by an eccentric cam in mesh with a stationary ring gear. The cam is on the high-speed input shaft. This gives a stepped-down rotation with uniform velocity to the concentric output shaft, which is turned by the drive pins rolling around holes in the pinion gears. It is possible with this mechanism to get a speed reduction of 174 : 1. Torques range from 90 in.-lbs to 1500 in.-lbs.*

*Manufactured by the Ferguson Machine Corp., St. Louis, Mo.

Figure 10.30 (Courtesy of Ferguson
Machine Corp., St. Louis, Mo.)

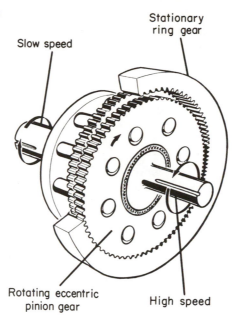

Stationary
ring gear

Slow speed

Rotating eccentric
pinion gear

High speed

10.31 Three-Gear Intermittent Drive

Gear A in Figure 10.31 turns at constant speed about the fixed eccentric
shaft O. Gear B is held in mesh with A by the link NP joining their centers.
B also meshes with output gear C, these being linked together by bar PR.
Gear C and bar PR turn independently about fixed axis R. Line ON on gear
A and link PR are cranks of a four-bar linkage in which NP is the coupler and
line OR the frame. The output motion of C depends on the distance OR. At
one critical length OR, gear C will rotate, accelerating and decelerating, and
will then dwell for a short period after which it will continue to rotate in the
same direction. If OR is less than this critical value, gear C will rotate, stop,

Figure 10.31

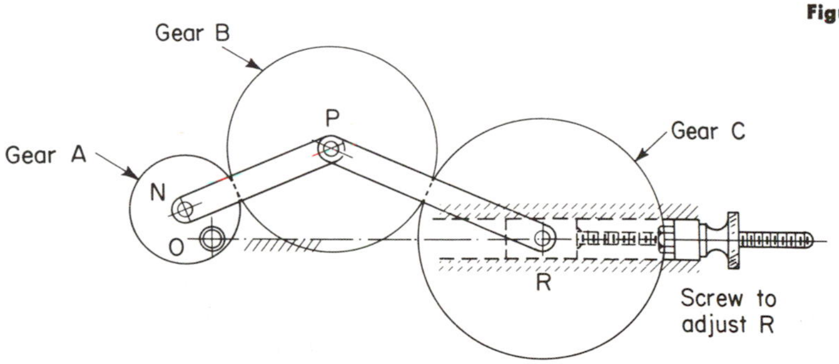

Gear B

P

Gear A

Gear C

N

O

R

Screw to
adjust R

rotate in the opposite direction, stop again, and then continue to turn in the original direction. If OR is greater than the critical value, gear C turns, then slows down, but does not stop. All of these motion cycles are repeated if the mechanism is run continuously. Provision for adjusting the distance OR may be provided as shown.

10.32 Random Reciprocating-Motion Mechanism

Gears E, F, and G are unequal in size. They turn in mesh as shown on fixed axes. A connecting rod is pinned to each gear. One rod connects F to a "walking beam" (JK) at J, and another rod connects G to K. The third connecting rod connects gear E to a second walking beam (LM) at L. A link PM connects beam JK to beam LM at pin M. The input may be applied at any one of the three gears, and the output is a reciprocating motion of block S which is connected to the upper walking beam by link OT. Block S slides in fixed guides, as shown, with an almost endless variety of reciprocating motions. This device could be useful in games of chance or in athletic training machines where the player must adjust to unpredictable motions.

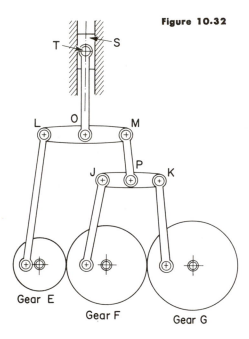

Figure 10.32

10.33 Translation Stroke Amplifier

This mechanism is designed to lengthen the stroke of a reciprocating slide or table in translation. Figure 10.33 shows two equal pinion gears, P_1 and P_2, connected by bar CE joining their centers. These pinions are rolled back and forth on a stationary rack S by the crank-and-slider linkage consisting of input crank AB (turning about fixed axis A) and coupler BD. A second rack T rides on the top of P_1 and P_2 in mesh with these pinions. The output member, a horizontal table M attached to rack T, is carried back and forth in translation. The table will move with a stroke equal to four times the length of crank AB.

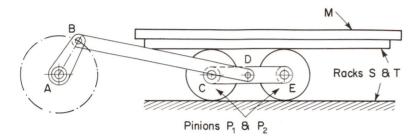

Figure 10.33

10.34 Rotational Motion Amplifier

In Figure 10.34 input crank CD revolves about fixed axis C. A coupler DM is pinned to CD at D and has a rack R attached to its underside. Rack R is held in mesh with pinion P, which turns about fixed axis E, by a link EF carrying roller G. The output pinion P turns several revolutions in each direction alternately as crank CD turns continuously in one direction. The amount of amplification of motion depends on the length of crank CD and the size of the pinion P.

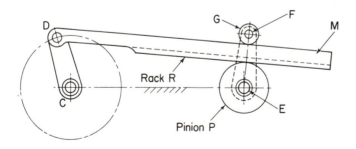

Figure 10.34

10.35 Mechanism to Generate a Square

The two-gear epicyclic train in Figure 10.35 may be used to generate a square with rounded corners. The input is applied at arm A, which turns about fixed axis O. The planet gear R is carried on pin B on A and meshes with stationary annular gear N. Point P, located as shown on R, describes the square path indicated. In this example the dimensions are in the following critical proportions: pitch diameter of annular $N = 4$; pitch diameter of planet $R = 1$; Radius $PB = 0.197$. Distance between sides of square $= 2.608$. Radius of rounded corners on square $= 0.376$. By varying the gear tooth ratios, it is possible to generate a number of different regular polygons with this mechanism.*

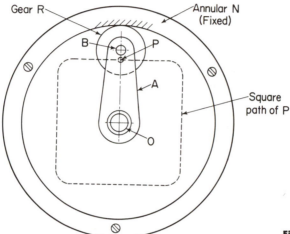

Figure 10.35

10.36 Disk-and-Wheel Variable Speed Drive

This is a friction drive in which the speeds are infinitely variable. Figure 10.36 shows two parallel disks turning shafts in fixed bearings. The disks are connected by an idler wheel as shown. The idler may be positioned as desired along its fixed shaft, thus changing the effective radii of the input and output disks, so as to obtain a variety of output speeds.

Figure 10.36

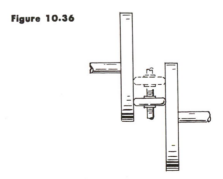

*Accredited to Stanley B. Tuttle in *Mechanisms for Engineering Design* (Wiley).

10.37 Cone-and-Ring Variable Speed Drive

Two identical truncated cones are mounted on fixed axes with clearance between them as shown in Figure 10.37. The cones are connected by a ring (ball bearing) which encircles them and is positioned by guides adjusted by screws. The screws are constrained by a chain and sprockets, not shown, so that they turn together. The ring may be positioned as desired along the outer surfaces of the cones, thus changing their effective radii and their input-output speed ratio.

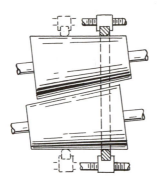

Figure 10.37

10.38 Spherical-Conoid Variable-Speed Drive

In Figure 10.38 two identical conoids are mounted on fixed shafts at right angles. The profiles of the conoids are circular arcs. An idler-wheel bearing pivots about a fixed axis, as shown crosshatched. Thus it can alter the effective radii of the conoids and provide a variety of input-output speed ratios.

Figure 10.38

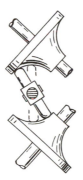

10.39 Disk-and-Sphere Speed Changer

Two disks are mounted with shafts in line, turning in fixed bearings. These are shown connected by two spheres in Figure 10.39. The bearings for the shafts on which the spheres are mounted may be adjusted to change the inclination of the axes of rotation of the balls by a cam mechanism not shown. This changes the effective radii of the spheres and makes possible a number of different speed ratios of the input and output disks.*

Figure 10.39

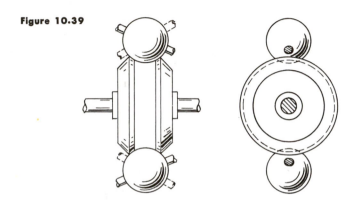

10.40 Ring-and-Roller Speed Changer

Each roller in Figure 10.40 is formed by two truncated cones in opposed position and keyed to turn the parallel input and output shafts. An adjusting device not shown changes the distance between the halves of each roller—

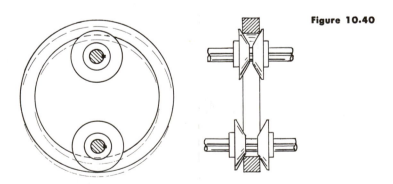

Figure 10.40

*This is the mechanism of the "Speed Variator," patented and manufactured by the Cleveland Worm & Gear Co., Cleveland, Ohio.

bringing them closer together in one case and separating them in the other. The rollers are connected by a ring which contacts the rollers on its inner edges only. The effective radius of each roller is altered by the adjustable distances between the roller halves. This alters the input-output speed ratio.*

10.41 *V*-Belt Variable-Speed Drive

Figure 10.41 shows a pair of parallel shafts driven by a flexible *V*-belt connecting two split pulleys that are keyed to the shafts. The pulley faces are conical so as to provide a *V*-shaped groove conforming to the belt. An adjusting mechanism not shown separates the halves of one pulley while closing up the parts of the other. This changes the radius of the belt as it passes around each pulley, thereby effecting a range of output speeds from a constant input. This is one of the most popular variable-speed friction drives.†

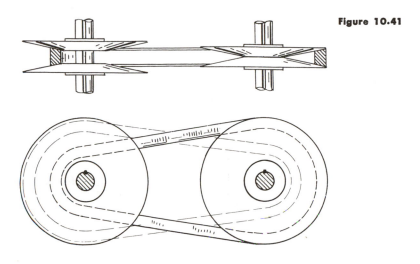

Figure 10.41

10.42 Swash-Plate Reciprocator

If a circular plate is mounted at an angle with a shaft in fixed bearings, as shown in Figure 10.42, constant-speed rotation of this input shaft will provide a reciprocating motion of an eccentrically mounted roller follower. The follower carriage must maintain the roller in a tangential position and must be

---

*This is the mechanism of the "Speedranger," patented and manufactured by the Master Electric Co.

†This is the basic mechanism used by a number of manufacturers of variable speed drives: Reeves Pulley Co., U. S. Electric Motor, Inc., Sterling Electric Motors, Inc., and others.

spring-loaded to maintain constant contact with the swash plate. The stroke of the follower may be adjusted by changing the distance between the input and follower shafts.

Figure 10.42

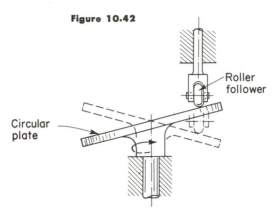

Roller follower

Circular plate

10.43 Yoke Cam Reciprocator

Figure 10.43 shows two examples of reciprocating drives employing plate cams in captive yokes. The input is a constant speed of the cam shaft in fixed bearings. These cams are in constant contact with the yoke surfaces on opposite sides (for this reason, they are often called constant-breadth cams), thus ensuring positive motion of the follower back and forth. Control of the follower motion is limited to a maximum of 180° of cam rotation since the remaining cam contour must be shaped to maintain contact with the opposite face of the follower. A special case is shown in Figure 10.43(b), in which there are two 60°-dwell periods in the motion of the follower with a very acceptable acceleration cycle, thus providing an intermittent reciprocating motion.

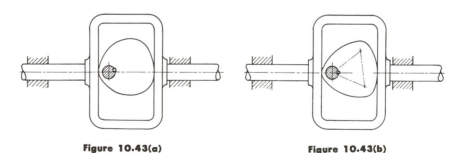

Figure 10.43(a)   **Figure 10.43(b)**

10.44 Barrel-Cam Indexing Mechanism

The mechanism pictured in Figure 10.44 provides for a variety of cyclic output motions from a constant speed input. The input shaft turns a form of cylindrical cam having a partly helical tapered rib (instead of a groove) which turns between two roller followers, thus providing a positive motion drive. The cams are interchangeable, offering a variety of rib configurations each imparting a different motion cycle to the output shaft carrying the followers. The velocity and acceleration of the follower shaft can thus be carefully controlled. In the example shown here, the system of equally spaced rollers provides for continuous rotation of the output shaft in one direction with a specific motion cycle repeated as each pair of rollers assumes the load. If desired, a short dwell may be obtained at specified intervals, making this an indexing mechanism.*

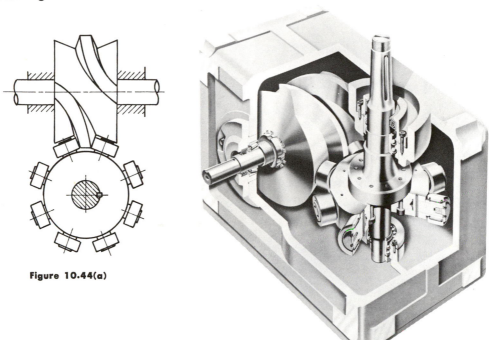

Figure 10.44(a)

Figure 10.44(b) (Courtesy of Ferguson Machine Corp., St. Louis, Mo.)

*This mechanism is used in the Ferguson Drives, manufactured and patented by Ferguson Machine Corp., St. Louis, Mo.

10.45 Cam-Driven Oscillating Drive

This mechanism is a variation of the indexing drive shown in Figure 10.44. The input shaft carries a cylindrical cam with a rib which passes between two roller followers on an output shaft at right angles (See Figure 10.45). In this case, the output shaft oscillates back and forth through a given angle repeatedly, with a motion which can be closely controlled by the design of the rib.*

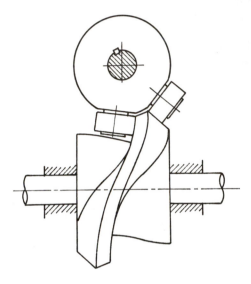

Figure 10.45(a)

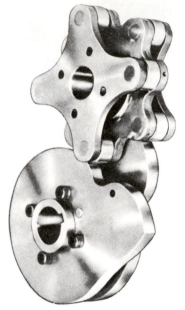

Figure 10.45(b) (Courtesy of Ferguson Machine Corp., St. Louis, Mo.)

*This mechanism is used on the Ferguson Oscillating Drives, patented and manufactured by the Ferguson Machine Corp., St. Louis, Mo.

10.46 Parallel-Shaft Indexing Mechanism

In this example, the input shaft carries two matched plate cams attached to one another in a fixed angular position. These turn at constant speed, each cam in contact with a roller follower which turns the parallel output shaft. Figure 10.46 shows the assembly with six roller followers alternately spaced on a spider on the output shaft. The design of the cam contours allows close control of the motion cycles, which may be exactly repeated.*

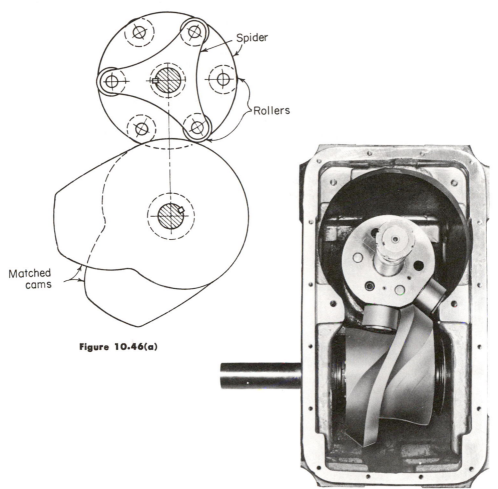

Figure 10.46(a)

Figure 10.46(b) (Courtesy of Commercial Cam & Machine Co, Chicago, Ill.)

*This mechanism is used on the Parallel Index Drives, patented and manufactured by the Commercial Cam and Machine Co., Chicago, Ill.

ACKNOWLEDGEMENTS

The collection of mechanisms appearing in Chapter 10 consists largely of classic examples that have been known to exist for many years. The less-familiar examples have come to the author's attention from various magazine articles, advertisements, and textbooks. The descriptions and the illustrations are original to this text, however similar they may, of necessity, appear to the texts or drawings of others. Many illustrations reflect the author's choice of configuration, but no claim is made or implied that they are of his invention.

In instances where the mechanism displayed is thought to be of recent issue and the source is known, acknowledgement has been made in the text to the inventor or author of the report. The author wishes especially to express his appreciation to Stanley B. Tuttle, Joseph Stiles Beggs, Roland T. Hinkle, G. J. Talbourdet, Preben W. Jensen, and H. E. Schrank.

In many cases the examples shown have been brought to mind by articles, design data sheets, and reviews of special purpose mechanisms that have been prepared by the editors of magazines without mention of individual authors. Grateful acknowledgements and appreciation are expressed to *Machine Design* and *Product Engineering* for these sources.

The generous cooperation of manufacturing concerns in offering the photographs used throughout the book is greatly appreciated. The computer linkage page was furnished by General Precision Systems Inc. and the gear train page by the Link-Belt Co.

Appendix

a	Linear acceleration
D	Driver
D^P	Pitch diameter
D^B	Base circle diameter
F	Follower
I	Instant center of rotation
L	Lift
P^C	Circular pitch
P^D	Diametral pitch
P^N	Normal pitch
S	Linear displacement
T	Time and also number of teeth
V	Linear velocity
α	Angular acceleration (alpha)
β	Label for angle (beta)
Δ	A small amount of and also a small change in (delta)
θ	Pressure angle of a gear (theta)
θ_L	Lift angle of a cam
ϕ	Pressure angle of a cam (phi)
π	A constant equal to 3.14 (pi)
ω	Angular velocity (omega)

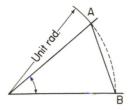

Table of Chords

Tabulated values are chord lengths of arcs of unit radius subtending the specified angles. Use multiples for accuracy.

Degrees	0′	10′	20′	30′	40′	50′
0	.0000	.0029	.0058	.0087	.0116	.0145
1	.0174	.0204	.0233	.0202	.0291	.0320
2	.0349	.0378	.0407	.0436	.0465	.0494
3	.0523	.0553	.0582	.0611	.0640	.0669
4	.0698	.0727	.0756	.0785	.0814	.0843
5	.0872	.0901	.0930	.0959	.0988	.1017
6	.1047	.1076	.1105	.1134	.1163	.1192
7	.1221	.1250	.1279	.1308	.1337	.1366
8	.1395	.1424	.1453	.1482	.1511	.1540
9	.1569	.1598	.1627	.1656	.1685	.1714
10	.1743	.1772	.1801	.1830	.1859	.1888
11	.1917	.1946	.1975	.2004	.2033	.2062
12	.2090	.2119	.2148	.2177	.2206	.2235
13	.2264	.2293	.2322	.2351	.2380	.2409
14	.2437	.2466	.2495	.2524	.2553	.2582
15	.2610	.2639	.2668	.2697	.2726	.2755
16	.2783	.2812	.2841	.2870	.2899	.2927
17	.2956	.2985	.3014	.3042	.3071	.3100
18	.3129	.3157	.3186	.3215	.3243	.3272
19	.3301	.3330	.3358	.3387	.3416	.3444
20	.3473	.3502	.3530	.3559	.3587	.3616
21	.3645	.3673	.3702	.3730	.3759	.3788
22	.3816	.3845	.3873	.3902	.3930	.3959
23	.3987	.4016	.4044	.4073	.4101	.4130
24	.4158	.4187	.4215	.4243	.4272	.4300
25	.4329	.4357	.4385	.4414	.4442	.4471
26	.4499	.4527	.4556	.4584	.4612	.4641
27	.4669	.4697	.4725	.4754	.4782	.4810
28	.4838	.4867	.4895	.4923	.4951	.4979
29	.5008	.5036	.5064	.5092	.5120	.5148
30	.5176	.5204	.5232	.5261	.5289	.5317
31	.5345	.5373	.5401	.5429	.5457	.5485
32	.5513	.5541	.5569	.5596	.5624	.5652
33	.5680	.5708	.5736	.5764	.5792	.5820
34	.5847	.5875	.5903	.5931	.5959	.5986
35	.6014	.6042	.6069	.6097	.6125	.6153
36	.6180	.6208	.6236	.6263	.6291	.6318
37	.6346	.6374	.6401	.6429	.6456	.6484
38	.6511	.6539	.6566	.6594	.6621	.6649
39	.6676	.6703	.6731	.6758	.6786	.6813
40	.6840	.6868	.6895	.6922	.6950	.6977
41	.7004	.7031	.7059	.7086	.7113	.7140
42	.7167	.7194	.7222	.7249	.7276	.7303
43	.7330	.7357	.7384	.7411	.7438	.7465
44	.7492	.7519	.7546	.7573	.7600	.7627
45	.7654	.7680	.7680	.7734	.7761	.7788

NOTES ON TRIGONOMETRY

Functions of an angle of a right triangle

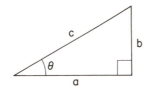

$$\sin \theta = \frac{b}{c} \qquad\qquad \text{cosecant } \theta = \frac{c}{b}$$

$$\text{cosine } \theta = \frac{a}{c} \qquad\qquad \text{secant } \theta = \frac{c}{a}$$

$$\text{tangent } \theta = \frac{b}{a} \qquad\qquad \text{cotangent } \theta = \frac{a}{b}$$

$$\text{tangent } \theta = \frac{\sin \theta}{\cos \theta}$$

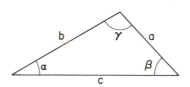

Laws for Oblique Triangles

Law of Sines:

$$\frac{a}{\sin \alpha} = \frac{b}{\sin \beta} = \frac{c}{\sin \gamma}$$

Law of Cosines:

$$a^2 = b^2 + c^2 - 2bc \cos \alpha$$

$$\text{Angle } \alpha + \text{angle } \beta + \text{angle } \gamma = 180°$$

Natural Trigonometric Functions

SINES, TANGENTS, COTANGENTS, COSINES
(*Read down for functions from 0° to 45°.*)
(*Read from bottom upward for functions from 45° to 90°.*)

	Sin	Tan	Cot	Cos	
0° 0′	.0000	.0000	Infinite	1.0000	90° 0′
10	.0029	.0029	343.7737	1.0000	50
20	.0058	.0058	171.8854	1.0000	40
30	.0087	.0087	114.5887	1.0000	30
40	.0116	.0116	85.9398	.9999	20
50	.0145	.0145	68.7501	.9999	10
1° 0′	.0175	.0175	57.2900	.9998	89° 0′
10	.0204	.0204	49.1039	.9998	50
20	.0233	.0233	42.9641	.9997	40
30	.0262	.0262	38.1885	.9997	30
40	.0291	.0291	34.3678	.9996	20
50	.0320	.0320	31.2416	.9995	10
2° 0′	.0349	.0349	28.6363	.9994	88° 0′
10	.0378	.0378	24.4316	.9993	50
20	.0407	.0407	24.5418	.9992	40
30	.0436	.0437	22.9038	.9990	30
40	.0465	.0466	21.4704	.9989	20
50	.0494	.0495	20.2056	.9988	10
3° 0′	.0523	.0524	19.0811	.9986	87° 0′
10	.0552	.0553	18.0750	.9985	50
20	.0581	.0582	17.1693	.9983	40
30	.0610	.0612	16.3499	.9981	30
40	.0640	.0641	15.6048	.9980	20
50	.0669	.0670	14.9244	.9978	10
4° 0′	.0698	.0699	14.3007	.9976	86° 0′
10	.0727	.0729	13.7267	.9974	50
20	.0756	.0758	13.1969	.9971	40
30	.0785	.0787	12.7062	.9969	30
40	.0814	.0816	12.2505	.9967	20
50	.0843	.0846	11.8262	.9964	10
5° 0′	.0872	.0875	11.4301	.9962	85° 0′
10	.0901	.0904	11.0594	.9959	50
20	.0929	.0934	10.7119	.9957	40
30	.0958	.0963	10.3854	.9954	30
40	.0987	.0992	10.0780	.9951	20
50	.1016	.1022	9.7882	.9948	10
6° 0′	.1045	.1051	9.5144	.9945	84° 0′
10	.1074	.1080	9.2553	.9942	50
20	.1103	.1110	9.0098	.9939	40
30	.1132	.1139	8.7769	.9936	30
40	.1161	.1169	8.5555	.9932	20
50	.1190	.1198	8.3450	.9929	10
7° 0′	.1219	.1228	8.1443	.9925	83° 0′
10	.1248	.1257	7.9530	.9922	50
20	.1276	.1287	7.7704	.9918	40
30	.1305	.1317	7.5958	.9914	30
40	.1334	.1346	7.4287	.9911	20
50	.1363	.1376	7.2687	.9907	10
8° 0′	.1392	.1405	7.1154	.9903	82° 0′
10	.1421	.1435	6.9682	.9899	50
20	.1449	.1465	6.8269	.9894	40
30	.1478	.1495	6.6912	.9890	30
40	.1507	.1524	6.5606	.9886	20
50	.1536	.1554	6.4348	.9881	10
9° 0′	.1564	.1584	6.3138	.9877	81° 0′
10	.1593	.1614	6.1970	.9872	50
20	.1622	.1644	6.0844	.9868	40
30	.1650	.1673	5.9758	.9863	30
40	.1679	.1703	5.8708	.9858	20
50	.1708	.1733	5.7694	.9853	10
10° 0′	.1736	.1763	5.6713	.9848	80° 0′
10	.1765	.1793	5.5764	.9843	50
20	.1794	.1823	5.4845	.9838	40
30	.1822	.1853	5.3955	.9833	30
40	.1851	.1883	5.3093	.9827	20
50	.1880	.1914	5.2257	.9822	79°10′
	Cos	Cot	Tan	Sin	

	Sin	Tan	Cot	Cos	
11° 0′	.1908	.1944	5.1446	.9816	79° 0′
10	.1937	.1974	5.0658	.9811	50
20	.1965	.2004	4.9894	.9805	40
30	.1994	.2035	4.9152	.9799	30
40	.2022	.2065	4.8430	.9793	20
50	.2051	.2095	4.7729	.9787	10
12° 0′	.2079	.2126	4.7046	.9781	78° 0′
10	.2108	.2156	4.6382	.9775	50
20	.2136	.2186	4.5736	.9769	40
30	.2164	.2217	4.5107	.9763	30
40	.2193	.2247	4.4494	.9757	20
50	.2221	.2278	4.3897	.9750	10
13° 0′	.2250	.2309	4.3315	.9744	77° 0′
10	.2278	.2339	4.2747	.9737	50
20	.2306	.2370	4.2193	.9730	40
30	.2334	.2401	4.1653	.9724	30
40	.2363	.2432	4.1126	.9717	20
50	.2391	.2462	4.0611	.9710	10
14° 0′	.2419	.2493	4.0108	.9703	76° 0′
10	.2447	.2524	3.9617	.9696	50
20	.2476	.2555	3.9136	.9689	40
30	.2504	.2586	3.8667	.9681	30
40	.2532	.2617	3.8208	.9674	20
50	.2560	.2648	3.7760	.9667	10
15° 0′	.2588	.2679	3.7321	.9659	75° 0′
10	.2616	.2711	3.6891	.9652	50
20	.2644	.2742	3.6470	.9644	40
30	.2672	.2773	3.6059	.9636	30
40	.2700	.2805	3.5656	.9628	20
50	.2728	.2836	3.5261	.9621	10
16° 0′	.2756	.2867	3.4874	.9613	74° 0′
10	.2784	.2899	3.4495	.9605	50
20	.2812	.2931	3.4124	.9596	40
30	.2840	.2962	3.3759	.9588	30
40	.2868	.2994	3.3402	.9580	20
50	.2896	.3026	3.3052	.9572	10
17° 0′	.2924	.3057	3.2709	.9563	73° 0′
10	.2952	.3089	3.2371	.9555	50
20	.2979	.3121	3.2041	.9546	40
30	.3007	.3153	3.1716	.9537	30
40	.3035	.3185	3.1397	.9528	20
50	.3062	.3217	3.1084	.9520	10
18° 0′	.3090	.3249	3.0777	.9511	72° 0′
10	.3118	.3281	3.0475	.9502	50
20	.3145	.3314	3.0178	.9492	40
30	.3173	.3346	2.9887	.9483	30
40	.3201	.3378	2.9600	.9474	20
50	.3228	.3411	2.9319	.9465	10
19° 0′	.3256	.3443	2.9042	.9455	71° 0′
10	.3283	.3476	2.8770	.9446	50
20	.3311	.3508	2.8502	.9436	40
30	.3338	.3541	2.8239	.9426	30
40	.3365	.3574	2.7980	.9417	20
50	.3393	.3607	2.7725	.9407	10
20° 0′	.3420	.3640	2.7475	.9397	70° 0′
10	.3448	.3673	2.7228	.9387	50
20	.3475	.3706	2.6985	.9377	40
30	.3502	.3739	2.6746	.9367	30
40	.3529	.3772	2.6511	.9356	20
50	.3557	.3805	2.6279	.9346	10
21° 0′	.3584	.3839	2.6051	.9336	69° 0′
10	.3611	.3872	2.5826	.9325	50
20	.3638	.3906	2.5605	.9315	40
30	.3665	.3939	2.5386	.9304	30
40	.3692	.3973	2.5172	.9293	20
50	.3719	.4006	2.4960	.9283	68°10′
	Cos	Cot	Tan	Sin	

	Sin	Tan	Cot	Cos	
22° 0′	.3746	.4040	2.4751	.9272	68° 0′
10	.3773	.4074	2.4545	.9261	50
20	.3800	.4108	2.4342	.9250	40
30	.3827	.4142	2.4142	.9239	30
40	.3854	.4176	2.3945	.9228	20
50	.3881	.4210	2.3750	.9216	10
23° 0′	.3907	.4245	2.3559	.9205	67° 0′
10	.3934	.4279	2.3369	.9194	50
20	.3961	.4314	2.3183	.9182	40
30	.3987	.4348	2.2998	.9171	30
40	.4014	.4383	2.2817	.9159	20
50	.4041	.4417	2.2637	.9147	10
24° 0′	.4067	.4452	2.2460	.9135	66° 0′
10	.4094	.4487	2.2286	.9124	50
20	.4120	.4522	2.2113	.9112	40
30	.4147	.4557	2.1943	.9100	30
40	.4173	.4592	2.1775	.9088	20
50	.4200	.4628	2.1609	.9075	10
25° 0′	.4226	.4663	2.1445	.9063	65° 0′
10	.4253	.4699	2.1283	.9051	50
20	.4279	.4734	2.1123	.9038	40
30	.4305	.4770	2.0965	.9026	30
40	.4331	.4806	2.0809	.9013	20
50	.4358	.4841	2.0655	.9001	10
26° 0′	.4384	.4877	2.0503	.8988	64° 0′
10	.4410	.4913	2.0353	.8975	50
20	.4436	.4950	2.0204	.8962	40
30	.4462	.4986	2.0057	.8949	30
40	.4488	.5022	1.9912	.8936	20
50	.4514	.5059	1.9768	.8923	10
27° 0′	.4540	.5095	1.9626	.8910	63° 0′
10	.4566	.5132	1.9486	.8897	50
20	.4592	.5169	1.9347	.8884	40
30	.4617	.5206	1.9210	.8870	30
40	.4643	.5243	1.9074	.8857	20
50	.4669	.5280	1.8940	.8843	10
28° 0′	.4695	.5317	1.8807	.8829	62° 0′
10	.4720	.5354	1.8676	.8816	50
20	.4746	.5392	1.8546	.8802	40
30	.4772	.5430	1.8418	.8788	30
40	.4797	.5467	1.8291	.8774	20
50	.4823	.5505	1.8165	.8760	10
29° 0′	.4848	.5543	1·8040	.8746	61° 0′
10	.4874	.5581	1.7917	.8732	50
20	.4899	.5619	1.7796	.8718	40
30	.4924	.5658	1.7675	.8704	30
40	.4950	.5696	1·7556	.8689	20
50	.4975	.5735	1.7437	.8675	10
30° 0′	.5000	.5774	1.7321	.8660	60° 0′
10	.5025	.5812	1.7205	.8646	50
20	.5050	.5851	1.7090	.8631	40
30	.5075	.5890	1.6977	.8616	30
40	.5100	.5930	1.6864	.8601	20
50	.5125	.5969	1.6753	.8587	10
31° 0′	.5150	.6009	1.6643	.8572	59° 0′
10	.5175	.6048	1.6534	.8557	50
20	.5200	.6088	1.6426	.8542	40
30	.5225	.6128	1.6319	.8526	30
40	.5250	.6168	1.6212	.8511	20
50	.5275	.6208	1.6107	.8496	10
32° 0′	.5299	.6249	1.6003	.8480	58° 0′
10	.5324	.6289	1.5900	.8465	50
20	.5348	.6330	1.5798	.8450	40
30	.5373	.6371	1.5697	.8434	30
40	.5398	.6412	1.5597	.8418	20
50	.5422	.6453	1.5497	.8403	10
33° 0′	.5446	.6494	1.5399	.8387	57° 0′
10	.5471	.6536	1.5301	.8371	50
20	.5495	.6577	1.5204	.8355	40
30	.5519	.6619	1.5108	.8339	30
40	.5544	.6661	1.5013	.8323	20
50	.5568	.6703	1.4919	.8307	57° 10′
	Cos	Cot	Tan	Sin	

	Sin	Tan	Cot	Cos	
34° 0′	.5592	.6745	1.4826	.8290	56° 0′
10	.5616	.6787	1.4733	.8274	50
20	.5640	.6830	1.4641	.8258	40
30	.5664	.6873	1.4550	.8241	30
40	.5688	.6916	1.4460	.8225	20
50	.5712	.6959	1.4370	.8208	10
35° 0′	.5736	.7002	1.4281	.8192	55° 0′
10	.5760	.7046	1.4193	.8175	50
20	.5783	.7089	1.4106	.8158	40
30	.5807	.7133	1.4019	.8141	30
40	.5831	.7177	1.3934	.8124	20
50	.5854	.7221	1.3848	.8107	10
36° 0′	.5878	.7265	1.3764	.8090	54° 0′
10	.5901	.7310	1.3680	.8073	50
20	.5925	.7355	1.3597	.8056	40
30	.5948	.7400	1.3514	.8039	30
40	·5972	.7445	1.3432	.8021	20
50	.5995	.7490	1.3351	.8004	10
37° 0′	.6018	.7536	1.3270	.7986	53° 0′
10	.6041	.7581	1.3190	.7969	50
20	.6065	.7627	1.3111	.7951	40
30	.6088	.7673	1.3032	.7934	30
40	.6111	.7720	1.2954	.7916	20
50	.6134	.7766	1.2876	.7898	10
38° 0′	.6157	.7813	1.2799	.7880	52° 0′
10	.6180	.7860	1.2723	.7862	50
20	.6202	.7907	1.2647	.7844	40
30	.6225	.7954	1.2572	.7826	30
40	.6248	.8002	1.2497	.7808	20
50	.6271	.8050	1.2423	.7790	10
39° 0′	.6293	.8098	1.2349	.7771	51° 0′
10	.6316	.8146	1.2276	.7753	50
20	.6338	.8195	1.2203	.7735	40
30	.6361	.8243	1.2131	.7716	30
40	.6383	.8292	1.2059	.7698	20
50	.6406	.8342	1.1988	.7679	10
40° 0′	.6428	.8391	1.1918	.7660	50° 0′
10	.6450	.8441	1.1847	.7642	50
20	.6472	.8491	1.1778	.7623	40
30	.6494	.8541	1.1708	.7604	30
40	.6517	.8591	1.1640	.7585	20
50	.6539	.8642	1.1571	.7566	10
41° 0′	.6561	.8693	1.1504	.7547	49° 0′
10	.6583	.8744	1.1436	.7528	50
20	.6604	.8796	1.1369	.7509	40
30	.6626	.8847	1.1303	.7490	30
40	.6648	.8899	1.1237	.7470	20
50	.6670	.8952	1.1171	.7451	10
42° 0′	.6691	.9004	1.1106	.7431	48° 0′
10	.6713	.9057	1.1041	.7412	50
20	.6734	.9110	1.0977	.7392	40
30	.6756	.9163	1.0913	.7373	30
40	.6777	.9217	1.0850	.7353	20
50	.6799	.9271	1.0786	.7333	10
43° 0′	.6820	.9325	1.0724	.7314	47° 0′
10	.6841	.9380	1.0661	.7294	50
20	.6862	.9435	1.0599	.7274	40
30	.6884	.9490	1.0538	.7254	30
40	.6905	.9545	1.0477	.7234	20
50	.6926	.9601	1.0416	.7214	10
44° 0′	.6947	.9657	1.0355	.7193	46° 0′
10	.6967	.9713	1.0295	.7173	50
20	.6988	.9770	1.0235	.7153	40
30	.7009	.9827	1.0176	.7133	30
40	.7030	.9884	1.0117	.7112	20
50	.7050	.9942	1.0058	.7092	10
45° 0′	.7071	1.0000	1.0000	.7071	45° 0′
	Cos	Cot	Tan	Sin	

Secant = 1 ÷ cosine
Cosecant = 1 ÷ sine
Versed sine = 1 − cosine
Coversed sine = 1 − sine

Excerpts from Catalog of Standard Spur Gears $14\frac{1}{2}°$ Pressure Angle

12 diametral pitch face: $\frac{3}{4}$ in.		10 diametral pitch face: 1 in.		8 diametral pitch face: $1\frac{1}{4}$ in.		6 diametral pitch face: $1\frac{1}{2}$ in.	
Price*	Tooth no.	Price	Tooth no.	Price	Tooth no.	Price	Tooth no.
2.90	11	3.60	11	5.50	11	7.10	11
2.90	12	3.60	12	5.50	12	7.10	12
3.00	13	4.10	14	6.50	14	8.10	14
3.20	14	4.50	15	7.00	15	9.10	15
3.50	15	4.90	16	7.20	16	10.20	16
3.60	16	5.20	18	7.70	18	11.20	18
3.90	18	5.70	20	8.60	20	11.70	20
4.10	20	6.40	24	9.60	22	12.20	21
4.50	21	7.00	25	13.80	24	17.10	24
4.90	22	9.80	28	14.70	28	18.50	27
5.20	24	10.60	30	15.60	30	21.10	30
5.70	30	10.90	32	16.40	32	23.00	32
7.70	32	11.40	35	17.80	36	23.90	33
7.80	36	11.70	36	19.30	40	26.50	36
8.40	40	12.10	40	20.10	42	28.50	40
8.70	42	12.40	42	20.70	44	29.50	42
9.50	48	13.00	45	23.30	48	32.70	48
10.30	54	13.50	48	25.40	54	35.70	54
11.70	60	13.80	50	26.20	56	39.00	60
12.10	64	14.30	54	28.10	60	40.10	64
13.20	72	14.50	55	29.10	64	40.70	66
15.80	84	15.80	60	31.90	72	45.20	72
18.50	96	17.10	64	34.80	80	52.70	84
21.10	108	19.00	70	37.10	84	62.90	96
22.50	112	19.80	72	37.70	88	73.80	108
23.90	120	22.20	80	40.70	96	84.20	120
30.30	144	23.90	84	47.60	112	113.20	132
36.80	168	25.40	90	51.30	120	126.60	144
43.40	192	26.90	96	55.00	128		
		28.50	100	63.80	144		
		31.60	110	72.40	160		
		34.80	120				

*Note: Prices listed above indicate relative values only.

Excerpts from Catalog of Standard Spur Gears 20° Press. Angle
(full depth)

12 diametral pitch face: 1 in.		10 diametral pitch face: $1\frac{1}{4}$ in.		8 diametral pitch face: $1\frac{1}{2}$ in.		6 diametral pitch face: 2 in.	
Price*	Tooth no.	Price	Tooth no.	Price	Tooth no.	Price	Tooth no.
3.00	12	4.60	12	5.80	12	9.30	12
3.20	13	5.10	14	7.00	14	10.60	14
3.50	14	5.40	15	7.20	15	12.20	15
3.60	15	5.70	16	7.70	16	13.20	16
4.10	16	6.40	18	8.10	18	14.60	18
4.50	18	6.70	20	9.10	20	15.80	21
4.90	20	7.20	24	10.20	22	22.50	24
5.40	21	7.50	25	11.20	24	24.10	27
5.70	24	8.80	28	12.20	28	27.60	30
6.40	28	12.40	30	17.10	32	30.90	33
7.00	30	13.50	35	18.50	36	34.30	36
10.60	36	14.70	40	21.10	40	38.50	42
11.30	42	16.10	45	23.90	44	40.70	48
12.10	48	16.90	48	26.50	48	46.60	54
13.00	54	17.40	50	29.10	56	50.60	60
13.80	60	19.30	55	30.35	60	54.80	66
14.60	66	21.10	60	31.60	64	59.00	72
15.80	72	23.90	70	34.30	72	68.50	84
18.50	84	26.50	80	36.80	80	82.30	96
21.10	96	29.10	90	39.50	88	95.90	108
23.90	108	31.60	100	43.40	96	109.40	120
26.50	120	34.30	110	50.10	112	124.00	132
29.10	132	38.20	120	54.10	120	137.10	144
31.60	144	43.40	140	58.10	128		
38.20	168	50.10	160	65.90	144		
44.80	192	58.10	180	76.50	160		
51.30	216	65.90	200	87.00	176		
58.10	240			97.50	192		

*Note: Prices listed above indicate relative values only.

Index